美国数学会经典影印系列

出版者的话

近年来，我国的科学技术取得了长足进步，特别是在数学等自然科学基础领域不断涌现出一流的研究成果。与此同时，国内的科研队伍与国外的交流合作也越来越密切，越来越多的科研工作者可以熟练地阅读英文文献，并在国际顶级期刊发表英文学术文章，在国外出版社出版英文学术著作。

然而，在国内阅读海外原版英文图书仍不是非常便捷。一方面，这些原版图书主要集中在科技、教育比较发达的大中城市的大型综合图书馆以及科研院所的资料室中，普通读者借阅不甚容易；另一方面，原版书价格昂贵，动辄上百美元，购买也很不方便。这极大地限制了科技工作者对于国外先进科学技术知识的获取，间接阻碍了我国科技的发展。

高等教育出版社本着植根教育、弘扬学术的宗旨服务我国广大科技和教育工作者，同美国数学会（American Mathematical Society）合作，在征求海内外众多专家学者意见的基础上，精选该学会近年出版的数百种专业著作，组织出版了“美国数学会经典影印系列”丛书。美国数学会创建于 1888 年，是国际上极具影响力的专业学术组织，目前拥有近 30000 会员和 580 余个机构成员，出版图书 3500 多种，冯 · 诺依曼、莱夫谢茨、陶哲轩等世界级数学大家都是其作者。本影印系列涵盖了代数、几何、分析、方程、拓扑、概率、动力系统等所有主要数学分支以及新近发展的数学主题。

我们希望这套书的出版，能够对国内的科研工作者、教育工作者以及青年学生起到重要的学术引领作用，也希望今后能有更多的海外优秀英文著作被介绍到中国。

高 等 教 育 出 版 社

2016 年 12 月

Introduction to Smooth Ergodic Theory

光滑遍历理论导论

Luis Barreira
Yakov Pesin

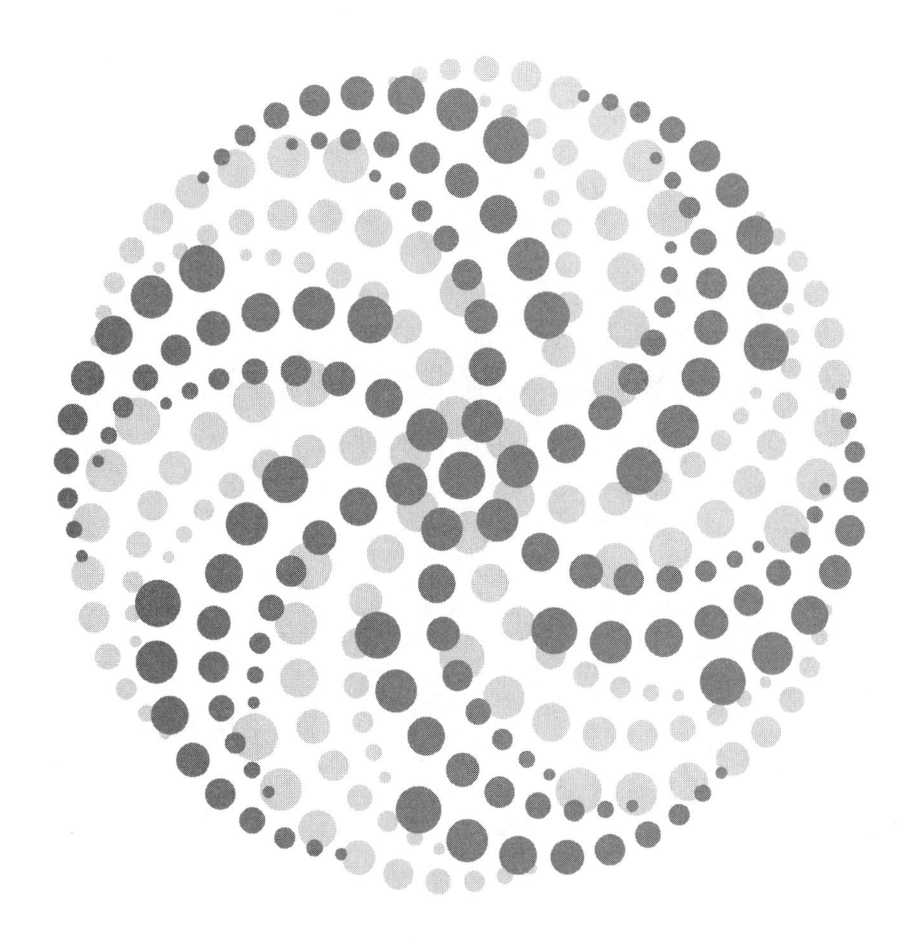

中国教育出版传媒集团
高等教育出版社 · 北京

Contents

Preface

This book is a revised and considerably expanded version of our book *Lyapunov Exponents and Smooth Ergodic Theory* [**7**]. When the latter was published, it became the only source of a systematic introduction to the core of smooth ergodic theory. It included the general theory of Lyapunov exponents and its applications to the stability theory of differential equations, nonuniform hyperbolicity theory, stable manifold theory (with emphasis on absolute continuity of invariant foliations), and the ergodic theory of dynamical systems with nonzero Lyapunov exponents, including geodesic flows. In the absence of other textbooks on the subject it was also used as a source or as supportive material for special topics courses on nonuniform hyperbolicity.

In 2007 we published the book *Nonuniform Hyperbolicity: Dynamics of Systems with Nonzero Lyapunov Exponents* [**9**], which contained an up-to-date exposition of smooth ergodic theory and was meant as a primary reference source in the field. However, despite an impressive amount of literature in the field, there has been until now no textbook containing a comprehensive introduction to the theory.

The present book is intended to cover this gap. It is aimed at graduate students specializing in dynamical systems and ergodic theory as well as anyone who wishes to acquire a working knowledge of smooth ergodic theory and to learn how to use its tools. While maintaining the essentials of most of the material in [**7**], we made the book more student-oriented by carefully selecting the topics, reorganizing the material, and substantially expanding the proofs of the core results. We also included a detailed description of essentially all known examples of conservative systems with nonzero Lyapunov exponents and throughout the book we added many exercises.

The book consists of two parts. While the first part introduces the reader to the basics of smooth ergodic theory, the second part discusses more advanced topics. This gives the reader a broader view of the theory and may help stimulate further study. This also provides nonexperts with a broader perspective of the field.

We emphasize that the new book is self-contained. Namely, we only assume that the reader has a basic knowledge of real analysis, measure theory, differential equations, and topology and we provide the reader with necessary background definitions and state related results.

On the other hand, in view of the considerable size of the theory we were forced to make a selection of the material. As a result, some interesting topics are barely mentioned or not covered at all. We recommend the books [**9, 15**] and the surveys [**8, 58**] for a description of many other developments and some recent work. In particular, we do not consider random dynamical systems (see the books [**5, 51, 56**] and the survey [**52**]), dynamical systems with singularities, including "chaotic" billiards (see the book [**50**]), the theory of nonuniformly expanding maps (see the survey [**57**]), and one-dimensional "chaotic" maps (such as the logistic family; see [**42**]).

Smooth ergodic theory studies the ergodic properties of smooth dynamical systems on Riemannian manifolds with respect to "natural" invariant measures. Among these measures most important are smooth measures, i.e., measures that are equivalent to the Riemannian volume. There are various classes of smooth dynamical systems whose study requires different techniques. In this book we concentrate on systems whose trajectories are hyperbolic in some sense. Roughly speaking, this means that the behavior of trajectories near a given orbit resembles the behavior of trajectories near a saddle point. In particular, to every hyperbolic trajectory one can associate two complementary subspaces such that the system acts as a contraction along one of them (called the stable subspace) and as an expansion along the other (called the unstable subspace).

A hyperbolic trajectory is unstable—almost every nearby trajectory moves away from it with time. If the set of hyperbolic trajectories is sufficiently large (for example, has positive or full measure), this instability forces trajectories to become separated. On the other hand, compactness of the phase space forces them back together; the consequent unending dispersal and return of nearby trajectories is one of the hallmarks of chaos.

Indeed, hyperbolic theory provides a mathematical foundation for the paradigm that is widely known as "deterministic chaos"—the appearance of irregular chaotic motions in purely deterministic dynamical systems. This paradigm asserts that conclusions about global properties of a nonlinear dynamical system with sufficiently strong hyperbolic behavior can be deduced from studying the linearized systems along its trajectories.

The study of hyperbolic phenomena originated in the seminal work of Artin, Morse, Hedlund, and Hopf on the instability and ergodic properties of geodesic flows on compact surfaces (see the survey [**37**] for a detailed description of results obtained at that time and for references). Later, hyperbolic behavior was observed in other situations (for example, Smale horseshoes and hyperbolic toral automorphisms).

The systematic study of hyperbolic dynamical systems was initiated by Smale (who mainly considered the problem of structural stability of hyperbolic systems; see [**83**]) and by Anosov and Sinai (who were mainly concerned with ergodic properties of hyperbolic systems with respect to smooth invariant measures; see [**3, 4**]). The hyperbolicity conditions describe the action of the linearized system along the stable and unstable subspaces and impose quite strong requirements on the system. The dynamical systems that satisfy these hyperbolicity conditions uniformly over all orbits are called Anosov systems.

In this book we consider the weakest (hence, most general) form of hyperbolicity, known as nonuniform hyperbolicity. It was introduced and studied by Pesin in a series of papers [**67, 68, 69, 70, 71**]. The nonuniform hyperbolicity theory (which is sometimes referred to as Pesin theory) is closely related to the theory of Lyapunov exponents. The latter originated in works of Lyapunov [**59**] and Perron [**66**] and was developed further in [**23**]. We provide an extended excursion into the theory of Lyapunov exponents and, in particular, introduce and study the crucial concept of Lyapunov–Perron regularity. The theory of Lyapunov exponents enables one to obtain many subtle results on the stability of differential equations.

Using the language of Lyapunov exponents, one can view nonuniformly hyperbolic dynamical systems as those systems where the set of points for which *all* Lyapunov exponents are nonzero is "large"—for example, has full measure with respect to an invariant Borel measure. In this case the Multiplicative Ergodic Theorem of Oseledets [**65**] implies that almost every point is Lyapunov–Perron regular. The powerful theory of Lyapunov exponents then yields a profound description of the local stability of trajectories, which, in turn, serves as grounds for studying the ergodic properties of these systems.

Luis Barreira, Lisboa, Portugal   Yakov Pesin, State College, PA USA

February 2013

Part 1

The Core of the Theory

Chapter 1

Examples of Hyperbolic Dynamical Systems

We begin our journey into smooth ergodic theory by constructing some principal examples of dynamical systems (both with discrete and continuous time), which illustrate various fundamental phenomena associated with uniform as well as nonuniform hyperbolicity. We shall only consider smooth systems which preserve a smooth measure (i.e., a measure which is equivalent to the Riemannian volume) leaving aside an important class of dissipative systems (e.g., the Hénon attractor).

While experts in the field believe that hyperbolic behavior is typical in some sense, it is usually difficult to rigorously establish that a given dynamical system is hyperbolic. In fact, known examples of uniformly hyperbolic diffeomorphisms include only those that act on tori and factors of some nilpotent Lie groups and it is believed – due to the strong requirement that every trajectory of such a system must be hyperbolic – that those are the only manifolds carrying uniformly hyperbolic diffeomorphisms. Despite such a shortage of examples (at least in the discrete-time case) the uniform hyperbolicity theory provides great insight into many principal phenomena associated with hyperbolic behavior and, in particular, to obtaining an essentially complete description of stochastic behavior of such systems.

In quest for more examples of dynamical systems with hyperbolic behavior one examines those that are not uniformly hyperbolic but possess "just enough" hyperbolicity to exhibit a high level of stochastic behavior.

A "typical" trajectory in these systems is hyperbolic although hyperbolicity is weaker than the one observed in Anosov systems and some trajectories are not hyperbolic at all. This is the case of nonuniform hyperbolicity. It is believed that nonuniformly hyperbolic systems are generic in some sense although this remains one of the most challenging problems in the area. In much contrast with Anosov diffeomorphisms, this belief is supported by the fact that every compact manifold of dimension ≥ 2 carries a nonuniformly hyperbolic diffeomorphism and that every compact manifold of dimension ≥ 3 carries a nonuniformly hyperbolic flow.

1.1. Anosov diffeomorphisms

We begin with dynamical systems exhibiting hyperbolic behavior in the strongest form. They were introduced by Anosov and are known as Anosov systems (see [**3**] and also [**4**]; for modern expositions of uniform hyperbolicity theory see [**19, 47**]).

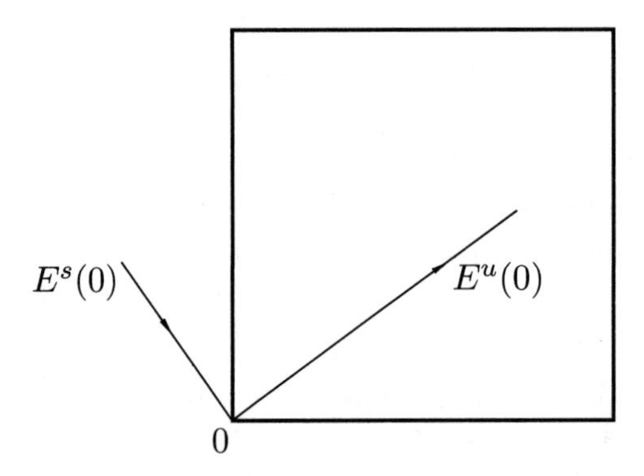

Figure 1.1. A hyperbolic toral automorphism

To describe the simplest example of such a system, consider the matrix $A = \left(\begin{smallmatrix} 2 & 1 \\ 1 & 1 \end{smallmatrix}\right)$. It induces a linear transformation T of the two-dimensional torus $\mathbb{T}^2 = \mathbb{R}^2/\mathbb{Z}^2$. The eigenvalues of A are λ^{-1} and λ where $\lambda = (3+\sqrt{5})/2$, and the corresponding eigendirections are given by the orthogonal vectors $(\frac{1+\sqrt{5}}{2}, 1)$ and $(\frac{1-\sqrt{5}}{2}, 1)$ (see Figure 1.1). For each $x \in \mathbb{T}^2$, we denote by $E^s(x)$ and $E^u(x)$ the one-dimensional subspaces of the tangent space $T_x\mathbb{T}^2$ (which can be identified with $\mathbb{R}^2$) obtained by translating the eigenlines of A. These subspaces are called, respectively, *stable* and *unstable subspaces* at x in view of the following estimates:

$$\|d_xT^n v\| = \lambda^{-n}\|v\| \quad \text{whenever} \quad v \in E^s(x),\ n \geq 0,$$

$$\|d_xT^{-n} v\| = \lambda^{-n}\|v\| \quad \text{whenever} \quad v \in E^u(x),\ n \geq 0.$$

Furthermore, since $E^s(x)$ and $E^u(x)$ are parallel to the eigendirections, they form dT-invariant bundles ($dT = A$), i.e., for every $x \in \mathbb{T}^2$,

$$d_xTE^s(x) = E^s(T(x)) \quad \text{and} \quad d_xTE^u(x) = E^u(T(x)).$$

Note that the origin is a fixed saddle (hyperbolic) point of T and that the trajectory of any point $x \in \mathbb{T}^2$ lies along a hyperbola; see Figure 1.2.

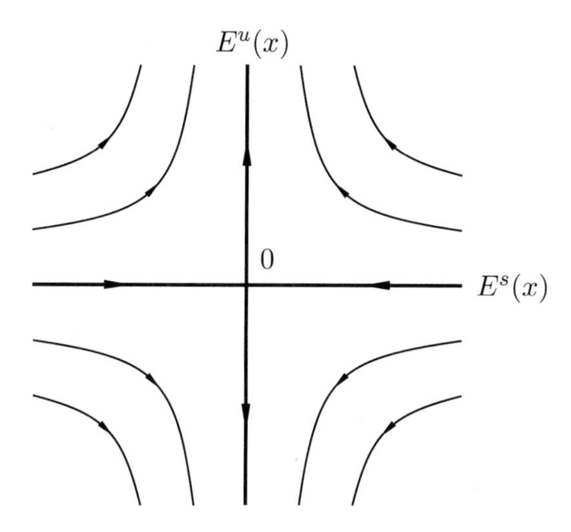

Figure 1.2. The structure of orbits near a hyperbolic fixed point

Let $\pi\colon \mathbb{R}^2 \to \mathbb{T}^2$ be the canonical projection. For each $x \in \mathbb{T}^2$ the sets $\pi(E^s(x))$ and $\pi(E^u(x))$ are C^∞ curves that are called *global stable* and *unstable curves (manifolds)* at x. They depend C^∞ smoothly on the base point x and form two partitions of $\mathbb{T}^2$.

The above example gives rise to the general notion of Anosov map. Let $f\colon M \to M$ be a diffeomorphism of a compact Riemannian manifold. We say that f is an *Anosov diffeomorphism* if for each $x \in M$ there exist a decomposition $T_xM = E^s(x) \oplus E^u(x)$ and constants $c > 0$ and $\mu \in (0, 1)$ such that for each $x \in M$:

(1) $d_xfE^s(x) = E^s(f(x))$ and $d_xfE^u(x) = E^u(f(x))$;

(2) $\|d_xf^nv\| \leq c\mu^n\|v\|$ whenever $v \in E^s(x)$ and $n \geq 0$;

(3) $\|d_xf^{-n}v\| \leq c\mu^n\|v\|$ whenever $v \in E^u(x)$ and $n \geq 0$.

In other words M is a uniformly hyperbolic set for f. In particular, the stable and unstable subspaces vary (Hölder) continuously with x.

If x is a fixed point for f, then the classical Grobman–Hartman and Hadamard–Perron theorems assert that the behavior of orbits in a sufficiently small neighborhood of x imitates the behavior of orbits near a saddle

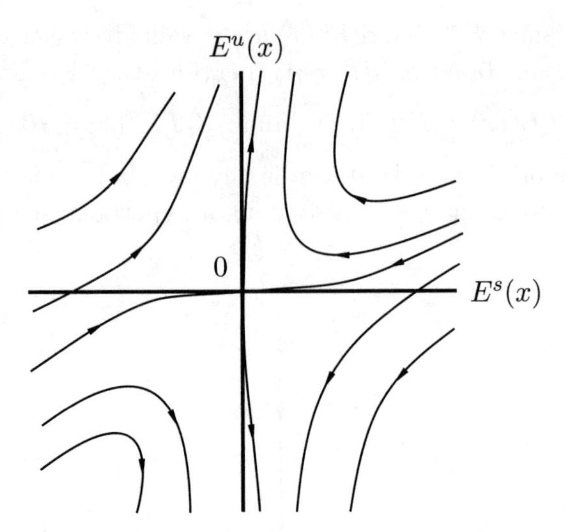

Figure 1.3. Invariant curves near a hyperbolic fixed point

point (see Figure 1.3). More precisely, there exists $\delta > 0$ such that for each $x \in M$ the sets

$$V^s(x) = \{y \in B(x,\delta) : d(f^n(y), f^n(x)) \to 0 \text{ as } n \to +\infty\},$$

$$V^u(x) = \{y \in B(x,\delta) : d(f^n(y), f^n(x)) \to 0 \text{ as } n \to -\infty\}$$

are immersed local smooth manifolds for which

$$d_x f V^s(x) = E^s(x) \quad \text{and} \quad d_x f V^u(x) = E^u(x).$$

The manifolds $V^s(x)$ and $V^u(x)$ are called, respectively, *local stable* and *unstable manifolds* at x of size δ. In general they depend only Hölder continuously on x (see Theorem 4.11). If two local stable manifolds intersect, then one of them is a continuation of the other and hence, they can be "glued" together. Continuing in this fashion we obtain the *global stable manifold* $W^s(x)$ at x. It can also be defined by the formula

$$W^s(x) = \bigcup_{n\geq 0} f^n(V^s(f^{-n}(x)))$$

and consists of all the points in the manifold whose trajectories converge to the trajectory of x, i.e.,

$$W^s(x) = \big\{y \in M : d(f^n(y), f^n(x)) \to 0 \text{ as } n \to +\infty\big\}.$$

The global stable leaves form a partition W^s of M called a foliation. More precisely, a partition W of M is called a *continuous foliation of M with smooth leaves* or simply a *foliation* if there exist $\delta > 0$ and $\ell > 0$ such that for each $x \in M$:

(1) the element $W(x)$ of the partition W containing x is a smooth ℓ-dimensional immersed submanifold; it is called the *(global) leaf* of the foliation at x; the connected component of the intersection $W(x) \cap B(x, \delta)$ that contains x is called the *local leaf* at x and is denoted by $V(x)$;

(2) there exists a continuous map $\varphi_x \colon B(x, \delta) \to C^1(D, M)$ (where $D \subset \mathbb{R}^\ell$ is the unit ball) such that for every $y \in B(x, \delta)$ the manifold $V(y)$ is the image of the map $\varphi_x(y) \colon D \to M$.

The function $\varphi_x(y, z) = \varphi_x(y)(z)$ is called the *foliation coordinate chart.* This function is continuous and has continuous derivative $\frac{\partial}{\partial z}\varphi_x$.

A continuous distribution E on TM is said to be *integrable* if there exists a foliation W of M such that $E(x) = T_x W(x)$ for every $x \in M$.

The stable foliation W^s is an integrable foliation for the stable distribution E^s. It is also invariant under f, i.e.,

$$f(W^s(x)) = W^s(f(x)).$$

We stress that this foliation in general is not smooth.

In a similar fashion one can glue unstable local leaves to obtain global unstable leaves. They form an invariant unstable foliation W^u of M that integrates the unstable distribution E^u. The stable and unstable foliations are transverse at every point on the manifold and each of them possesses the absolute continuity property (see Chapter 8). This property of invariant foliations is crucial in proving that an Anosov diffeomorphism of a compact connected smooth manifold is ergodic with respect to a smooth measure.

Recall that an f-invariant measure ν is *ergodic* if for each f-invariant measurable set $A \subset M$ either A or $M \setminus A$ has measure zero. Equivalently, ν is ergodic if and only if every f-invariant (mod 0) measurable function (i.e., a measurable function φ such that $\varphi \circ f = \varphi$ almost everywhere) is constant almost everywhere.

We also recall the notion of Bernoulli automorphism. Let (X, μ) be a Lebesgue space with a probability measure μ. Assume that μ has at most countably many atoms whose union $Y \subset X$ is such that $\mu|(X \setminus Y)$ is metrically isomorphic to the Lebesgue measure on the unit interval $[0, 1]$. One can associate to (X, μ) the two-sided Bernoulli shift $\sigma \colon X^{\mathbb{Z}} \to X^{\mathbb{Z}}$ defined by $(\sigma x)_n = x_{n+1}$, $n \in \mathbb{Z}$, which preserves the convolution $\bigoplus_{\mathbb{Z}} \mu$. A *Bernoulli automorphism* (S, ν) is an invertible (mod 0) measure-preserving transformation which is metrically isomorphic to the Bernoulli shift associated to some Lebesgue space (X, μ).

Theorem 1.1. *Any C^2 Anosov diffeomorphism*[1] *f of a compact smooth connected Riemannian manifold M preserving a smooth measure μ is ergodic. Furthermore, there is a number $n > 0$ and a subset $A \subset M$ such that:*

(1) *$f^k(A) \cap A = \varnothing$ for $k = 1, \ldots, n-1$ and $f^n(A) = A$;*

(2) *$\bigcup_{k=0}^{n-1} f^k(A) = M \pmod 0$;*

(3) *$f^n|A$ is a Bernoulli automorphism.*

Anosov diffeomorphisms are *structurally stable.* This means that any sufficiently small C^1 perturbation $g\colon M \to M$ of an Anosov diffeomorphism is still an Anosov diffeomorphism and there exists a Hölder homeomorphism $h\colon M \to M$ such that $g \circ h = h \circ f$. In particular, the family of Anosov diffeomorphisms of class C^1 is open in the C^1 topology.

1.2. Anosov flows

Now we consider uniformly hyperbolic dynamical systems with continuous time. It is well known that every smooth vector field $\mathfrak{X}$ on a compact smooth manifold M can be uniquely integrated; i.e., given $x \in M$, there exists a uniquely defined smooth curve $\gamma_x(t)$, $t \in \mathbb{R}$, such that $\gamma_x(0) = x$ and

$$\frac{d}{dt}\gamma_x(t) = \mathfrak{X}(\gamma_x(t))$$

for every t. One can now define a smooth flow $\varphi_t\colon M \to M$ such that $\mathfrak{X}(x) = \frac{d}{dt}(\varphi_t(x))|_{t=0}$ by $\varphi_t(x) = \gamma_x(t)$. We say that a flow φ_t is an *Anosov flow* if for each $x \in M$ there exist a decomposition

$$T_xM = E^s(x) \oplus E^0(x) \oplus E^u(x)$$

and constants $c > 0$ and $\mu \in (0,1)$ such that for each $x \in M$:

(1) $E^0(x)$ is the one-dimensional subspace generated by the vector field $\mathfrak{X}(x)$;

(2) $d_x\varphi_t E^s(x) = E^s(\varphi_t(x))$ and $d_x\varphi_t E^u(x) = E^u(\varphi_t(x))$ for $t \in \mathbb{R}$;

(3) for $t \geq 0$,

$$\|d_x\varphi_t v\| \leq c\mu^t\|v\| \text{ whenever } v \in E^s(x),$$

$$\|d_x\varphi_{-t} v\| \leq c\mu^t\|v\| \text{ whenever } v \in E^u(x).$$

---

[1]The requirement that f be of class of smoothness C^2 can be weakened to the requirement that f be of class of smoothness $C^{1+\alpha}$ for some $\alpha > 0$ (i.e., the differential of f is Hölder continuous). It is not known, however, whether this theorem holds for C^1 Anosov diffeomorphisms.

The stable and unstable distributions E^s and E^u integrate to continuous stable and unstable invariant foliations W^s and W^u for the flow. We also have the weakly stable and weakly unstable invariant foliations W^{st} and W^{ut} whose leaves at a point $x \in M$ are

$$W^{st}(x) = \bigcup_{t\in\mathbb{R}} W^s(\varphi_t(x)), \quad W^{ut}(x) = \bigcup_{t\in\mathbb{R}} W^u(\varphi_t(x)).$$

A simple example of an Anosov flow is a special flow over an Anosov diffeomorphism. We recall the definition of a special flow. Given a diffeomorphism f of a compact Riemannian manifold M and a smooth positive function h on M, consider the quotient space

$$M_h = \big\{(x,t) \in M \times \mathbb{R}_0^+ : 0 \le t \le h(x)\big\}/\equiv,$$

where $\equiv$ is the equivalence relation $(x, h(x)) \equiv (f(x), 0)$. The special flow $\varphi_t \colon M_h \to M_h$ with *roof function* h moves a point $(x,0)$ along $\{x\} \times \mathbb{R}_0^+$ to $(x, h(x))$, then jumps to $(f(x), 0)$ and continues along $\{f(x)\} \times \mathbb{R}_0^+$ (see Figure 1.4). More precisely,

$$\varphi_t(x,\tau) = (f^n(x), \tau'),$$

where the numbers n, τ, and τ' satisfy $0 \le \tau' \le h(f^n(x))$ and

$$\sum_{i=0}^{n-1} h(f^i(x)) + \tau' = t + \tau.$$

A special flow with constant roof function is called a suspension flow.

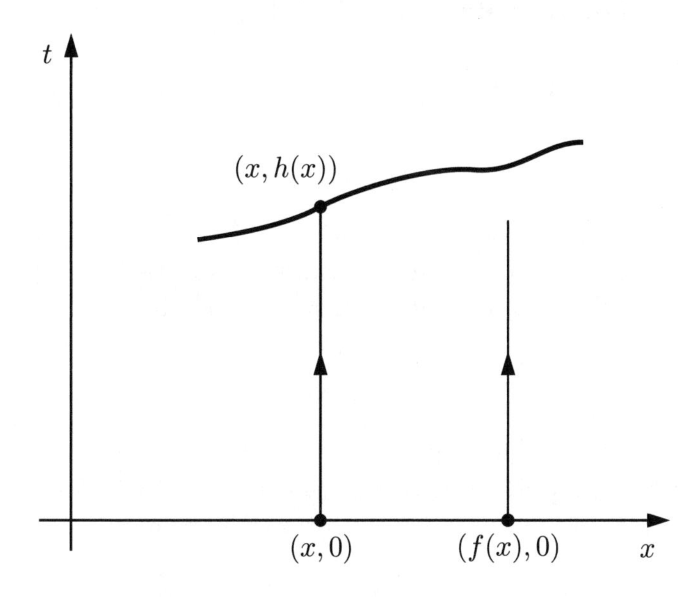

Figure 1.4. The special flow $\varphi_t \colon M_h \to M_h$ with roof function h.

Exercise 1.2. Let f be an Anosov diffeomorphism of a compact Riemannian manifold M and let h be a smooth positive function on M. Show that the special flow φ_t with the roof function h is an Anosov flow on M_h.

Recall that a φ_t-invariant measure ν is *ergodic* if for each φ_t-invariant measurable set $A \subset M$ either A or $M \setminus A$ has measure zero. Equivalently, ν is ergodic if and only if every φ_t-invariant (mod 0) measurable function is constant almost everywhere. Recall also that a flow φ_t is a Bernoulli flow if for each $t \neq 0$ the map φ_t is a Bernoulli automorphism.

An Anosov flow preserving a smooth measure may not be ergodic (just consider a special flow over an Anosov diffeomorphism with constant roof function). We say that a flow φ_t on a compact smooth manifold is *topologically transitive* if for any two nonempty open sets U and V there exists $n \in \mathbb{Z}$ such that $f^n(U) \cap V \neq \varnothing$, and we say that φ_t is *topologically mixing* if for any two nonempty open sets U and V there exists $N \geq 0$ such that for every $n \geq N$ we have that $f^n(U) \cap V \neq \varnothing$.

Theorem 1.3. *If an Anosov flow φ_t on a compact smooth connected Riemannian manifold preserves a smooth measure and is topologically transitive, then it is ergodic, and if φ_t is topologically mixing, then it is Bernoulli.*

Another example of an Anosov flow that stems from Riemannian geometry is the geodesic flow on a compact smooth Riemannian manifold of negative curvature. We will discuss here the simplest case of geodesic flows on surfaces of constant negative curvature, referring the reader to Chapter 10 for a more general case.

We shall use the Poincaré model of the Lobachevsky geometry on the upper-half plane

$$\mathbb{H} = \{z = x + iy \in \mathbb{C} : \operatorname{Im} z = y > 0\}.$$

The line $y = 0$ is called the *ideal boundary* of $\mathbb{H}$ and is denoted by $\mathbb{H}(\infty)$. The Riemannian metric on $\mathbb{H}$ with constant curvature $-k$ is given by the inner product in the tangent space $T_z\mathbb{H} = \mathbb{C}$ defined by

$$\langle v, w\rangle_z = \frac{k}{(\operatorname{Im} z)^2}\langle v, w\rangle,$$

where $\langle v, w\rangle$ is the standard inner product in $\mathbb{R}^2$.

Consider the group $SL(2,\mathbb{R})$ of matrices $A = \left(\begin{smallmatrix} a & b \\ c & d \end{smallmatrix}\right)$ with real entries and determinant 1 or -1, and define Möbius transformations T_A of $\mathbb{H}$ by

$$T_A(z) = \frac{az+b}{cz+d} \quad \text{and} \quad T_A(z) = \frac{a\bar{z}+b}{c\bar{z}+d}, \tag{1.1}$$

respectively, when $ad - bc = 1$ and $ad - bc = -1$. Clearly, $T_{-A} = T_A$.

Exercise 1.4. Show that the transformations T_A in (1.1) take $\mathbb{H}$ into itself and are isometries.

In fact, the group $G = SL(2,\mathbb{R})/\{\mathrm{Id}, -\mathrm{Id}\}$ is the group of all isometries of $\mathbb{H}$.

Exercise 1.5. Show that the geodesics between points in $\mathbb{H}$ are the vertical half-lines and the half-circles centered at points in the axis $\operatorname{Im} z = 0$. More precisely, given $z \in \mathbb{H}$ and $v \in \mathbb{C} \setminus \{0\}$:

(1) if v is parallel to the line $\operatorname{Re} z = 0$, then the geodesic passing through z in the direction of v is the half-line $\{z + vt : t \in \mathbb{R}\} \cap \mathbb{H}$;

(2) if v is not parallel to the line $\operatorname{Re} z = 0$, then the geodesic passing through z in the direction of v is the half-circle centered at the axis $\operatorname{Im} z = 0$ passing through z and with tangent v at that point.

Hint: First consider a path $\gamma\colon [0,1] \to \mathbb{H}$ between ic and id, with $d > c$, and note that its length ℓ_γ satisfies

$$\ell_\gamma = \int_0^T \frac{|\gamma'(t)|}{y(t)}\,dt \geq \int_0^T \frac{y'(t)}{y(t)}\,dt = \log\frac{d}{c},$$

where $y(t) = \operatorname{Im} \gamma(t)$. Since the vertical path between ic and id has precisely this length, it is in fact the geodesic joining the two points. For the general case, show that given $z, w \in \mathbb{H}$ with $z \neq w$, there exists a Möbius transformation taking the vertical line segment or the arc of the circle centered at the real axis joining z and w to a vertical line segment on the positive part of the imaginary axis and use the former result.

The *geodesic flow* is a flow acting on the unit tangent bundle

$$S\mathbb{H} = \big\{(z,v) \in \mathbb{H} \times \mathbb{C} : |v|_z = 1\big\}$$

and is defined as follows. Given $(z,v) \in S\mathbb{H}$, there exists a Möbius transformation T such that $T(z) = i$ and $T'(z)v = i$. It takes the geodesic passing through z in the direction of v into the geodesic ie^t traversing the positive part of the imaginary axis. The geodesic flow $\varphi_t\colon S\mathbb{H} \to S\mathbb{H}$ is given by

$$\varphi_t(z,v) = (\gamma(t), \gamma'(t)),$$

where $\gamma(t) = T^{-1}(ie^t)$.

Exercise 1.6. Show that:

(1) $S\mathbb{H}$ can be identified with G;

(2) $\varphi_t(z,v) = (z,v)g_t$ where $g_t = \begin{pmatrix} e^t & 0 \\ 0 & e^{-t} \end{pmatrix}$ is a one-parameter subgroup;

(3) the Haar measure μ on G is invariant under the geodesic flow and is the volume on SH.

Theorem 1.7. *The geodesic flow on a surface of constant negative curvature is an Anosov flow.*

To explain this result, we describe the global stable manifold through a point $(z, v) \in S\mathbb{H}$. Consider the oriented semicircle $\gamma_{z,v}(t)$ in $\mathbb{H}$ centered at a point on the ideal boundary $\mathbb{H}(\infty)$, which passes through z in the direction of v.[2] Let $\gamma_{z,v}(+\infty) \in \mathbb{H}(\infty)$ be the positive endpoint of the circle. Consider now the collection of all oriented semicircles in $\mathbb{H}$ centered at points in $\mathbb{H}(\infty)$ whose positive endpoints coincide with $\gamma_{z,v}(+\infty)$.[3]

Exercise 1.8. Show that for any two circles γ_1 and γ_2 in this collection,

$$\rho(\gamma_1(t), \gamma_2(t)) \le Ce^{\sqrt{-k}t} \tag{1.2}$$

for some $C > 0$ and all $t \ge 0$.

One can show that the circle $c(z, v)$ passing through the point z that is tangent to $\mathbb{H}(\infty)$ at the point $\gamma_{z,v}(+\infty)$ intersects each oriented semicircle in the collection orthogonally. It is called a *horocycle* through z (see Figure 1.5). It follows from (1.2) that the submanifold

$$W^s(z, v) = \big\{(z', v') \in S\mathbb{H}\colon z' \in c(z, v),\\ v' \text{ is orthogonal to } c(z, v) \text{ and points inward}\big\}$$

is the global stable leaf through (z, v) (see Figure 1.5).

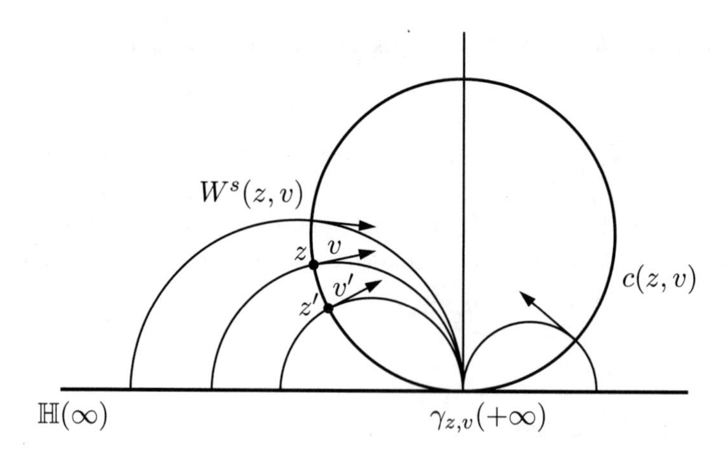

Figure 1.5. Horocycle $c(z, v)$ and global stable leaf $W^s(z, v)$

[2]That is, the geodesic through z in the direction of v.

[3]This collection also includes the line orthogonal to $\mathbb{H}(\infty)$ at $\gamma_{z,v}(+\infty)$.

A surface M of constant negative curvature is a factor space $M = \mathbb{H}/\Gamma$ where Γ is a discrete subgroup of G, and any compact surface of genus at least 2 can be obtained in this way. We stress that depending on the choice of the discrete subgroup Γ the factor space M can be a surface of finite area. The geodesic flow is now defined on the unit tangent bundle SM, and it preserves the Riemannian volume on SM.

The above construction of horocycles can be carried over to general surfaces of negative curvature and an estimate similar to (1.2) can be established. This allows one to extend Theorem 1.7 to surfaces of negative curvature and to study hyperbolic and ergodic properties of geodesic flows.

Theorem 1.9 ([**3**]). *The geodesic flow on a surface of negative curvature is an Anosov flow. Moreover, if the surface has finite area, then the geodesic flow is ergodic and indeed is a Bernoulli flow.*

We emphasize that the study of ergodic properties of geodesic flows on compact surfaces of nonpositive curvature—that lie on the boundary of the metrics of negative curvature—is substantially more complicated (see Chapter 10 where we describe the stable and unstable subspaces and foliations for the geodesic flow). To see this, let us observe that the geodesic flow on the flat torus has only zero Lyapunov exponents.

1.3. The Katok map of the 2-torus

In this section we describe a construction due to Katok [**45**] of an area-preserving nonuniformly hyperbolic diffeomorphism on the two-dimensional torus $\mathbb{T}^2$. The starting point of our constructions is a volume-preserving Anosov diffeomorphism, which is then perturbed. The perturbation is not small (otherwise the perturbation would still be an Anosov system) but is localized in a small neighborhood of a hyperbolic fixed or periodic point, say x, and it slows down trajectories, so that the time a given trajectory spends in this neighborhood of x increases. The Poincaré recurrence theorem ensures that for almost every trajectory (call them "good") the average increase in time is not significant to affect its hyperbolicity although the expansion and contraction rates get weaker. However, for some "bad" trajectories (which form a set of zero volume) the average increase in time is abnormally high and hyperbolicity gets destroyed. Along the way of our constructions we introduce one of our main concepts—the Lyapunov exponents—which measures the asymptotic contraction and expansion rates and distinguishes the "good" trajectories from the "bad" ones—at least some of the values of the Lyapunov exponents for the former are nonzero while all the values of the Lyapunov exponents for the latter are zero.

To effect our construction, consider again the matrix $A = \left(\begin{smallmatrix} 2 & 1 \\ 1 & 1 \end{smallmatrix}\right)$, which induces a linear transformation T of the two-dimensional torus $\mathbb{T}^2 = \mathbb{R}^2/\mathbb{Z}^2$. We shall obtain the desired map by slowing down T near the origin. This construction depends upon a real-valued function ψ, which is defined on the unit interval $[0, 1]$ and has the following properties:

(1) ψ is a C^∞ function except at the origin;

(2) $\psi(0) = 0$ and $\psi(u) = 1$ for $u \geq r_0$, for some $0 < r_0 < 1$;

(3) $\psi'(u) > 0$ for every $0 < u < r_0$;

(4) the following integral converges:

$$\int_0^1 \frac{du}{\psi(u)} < \infty.$$

Note that for the last requirement to hold the function $\psi(u)$ should be like u^α with $\alpha < 1$.

Let D_r be the disk centered at 0 of radius r. In the coordinate system (s_1, s_2) obtained from the eigendirections of A we have for small r that

$$D_r = \{(s_1, s_2) : {s_1}^2 + {s_2}^2 \leq r^2\}.$$

Let us choose numbers $r_2 > r_1 > r_0$ such that

$$T(D_{r_1}) \cup T^{-1}(D_{r_1}) \subset D_{r_2}, \quad D_{r_0} \subset \operatorname{int} T(D_{r_1}). \tag{1.3}$$

Consider the system of differential equations in D_{r_1}

$$\dot{s}_1 = s_1 \log \lambda, \quad \dot{s}_2 = -s_2 \log \lambda \tag{1.4}$$

and observe that the map T is the time-one map of the flow generated by this system of equations.

We now perturb the system (1.4) to obtain the system of differential equations in D_{r_2}

$$\dot{s}_1 = s_1 \psi({s_1}^2 + {s_2}^2) \log \lambda, \quad \dot{s}_2 = -s_2 \psi({s_1}^2 + {s_2}^2) \log \lambda. \tag{1.5}$$

Let V_ψ be the vector field generated by this system of equations and let g_t be the corresponding local flow. Denote by g the time-one map of the flow. Our choice of the function ψ and numbers r_0, r_1, and r_2 guarantees that the domain of definition of the map g contains the disk D_{r_1} and that $g(D_{r_1}) \subset D_{r_2}$. Furthermore, g is of class C^∞ in $D_{r_1} \setminus \{0\}$ and it coincides with T in some neighborhood of the boundary ∂D_{r_1}.[4] Therefore, the map

$$G(x) = \begin{cases} T(x) & \text{if } x \in \mathbb{T}^2 \setminus D_{r_1}, \\ g(x) & \text{if } x \in D_{r_1} \end{cases}$$

[4]To see this, consider a point $x \in \mathbb{T}^2 \setminus D_{r_1}$. Let $n > 0$ be the first moment the trajectory of x under T enters the disk D_{r_1} and let $m > n$ be the first moment when this trajectory exits D_{r_1}. It is easy to see that $G^n_{\mathbb{T}^2}(x) = T^n(x)$ and $G^k_{\mathbb{T}^2}(x) = T^m(x)$ where $k > m$ is the first moment when the trajectory of x under $G_{\mathbb{T}^2}$ exits D_{r_1}.

defines a homeomorphism of the torus $\mathbb{T}^2$, which is a C^∞ diffeomorphism everywhere except at the origin. It is the map $G(x)$ that is a slowdown of the automorphism T at 0.

The map G preserves the probability measure $d\nu = \kappa_0^{-1}\kappa\, dm$ where m is area and the density κ is a positive C^∞ function that is infinite at 0. It is defined by the formula

$$\kappa(s_1, s_2) = \begin{cases} (\psi({s_1}^2 + {s_2}^2))^{-1} & \text{if } (s_1, s_2) \in D_{r_1}, \\ 1 & \text{otherwise} \end{cases} \tag{1.6}$$

and

$$\kappa_0 = \int_{\mathbb{T}^2} \kappa\, dm.$$

Exercise 1.10. Show that the measure ν is G-invariant by verifying that $\operatorname{div}(\kappa v) = 0$ where v is the vector field generated in D_{r_2} by the system of differential equations (1.5).

Our next step is to further perturb the map G so that the new map $G_{\mathbb{T}^2}$ is an area-preserving C^∞ diffeomorphism. We will achieve this by changing the coordinate system in $\mathbb{T}^2$ using a map φ so that $G_{\mathbb{T}^2} = \varphi \circ G \circ \varphi^{-1}$. We define φ in D_{r_1} by the formula

$$\varphi(s_1, s_2) = \frac{1}{\sqrt{\kappa_0({s_1}^2 + {s_2}^2)}} \left(\int_0^{{s_1}^2 + {s_2}^2} \frac{du}{\psi(u)} \right)^{1/2} (s_1, s_2) \tag{1.7}$$

and we set $\varphi = \mathrm{Id}$ in $\mathbb{T}^2 \setminus D_{r_1}$. Clearly, φ is a homeomorphism and is a C^∞ diffeomorphism outside the origin.

To check that φ transfers the measure ν into the area, one must verify that the measure $\varphi_* \nu$ has a density $\rho(x)$ with respect to the measure ν that is equal to the Radon–Nikodym derivative $d\nu/dm$. In view of (1.6), for $x = (s_1, s_2) \in D_{r_1}$ this is equivalent to the identity

$$\rho(s_1, s_2) = \frac{1}{\kappa_0 \psi({s_1}^2 + {s_2}^2)}. \tag{1.8}$$

Exercise 1.11. Prove (1.8). Hint: Set $\varphi(s_1, s_2) = {\kappa_0}^{-1/2} f(r)(s_1/r, s_2/r)$, where $r = \sqrt{{s_1}^2 + {s_2}^2}$, and show that $(f(r)^2)' = 2r/\psi(r^2)$.

The behavior of the orbits of $G_{\mathbb{T}^2}$ is sketched in Figure 1.6.

We now show that $G_{\mathbb{T}^2}$ is infinitely many times differentiable at the origin and hence is a C^∞ diffeomorphism. Observe that the system (1.4) is *Hamiltonian with respect to area m* with the Hamiltonian function $H(s_1, s_2) = s_1 s_2 \log \lambda$. This means (we exploit the fact that the system is two-dimensional) that for every continuous vector field v on D_{r_1} we have

$$\Omega(v, V_H) = dHv,$$

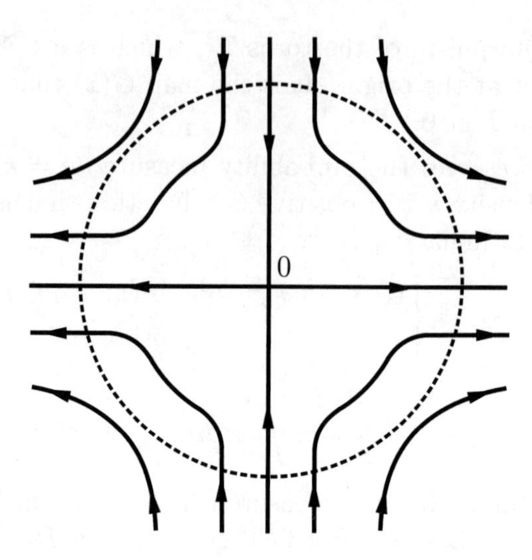

Figure 1.6. Invariant curves for $G_{\mathbb{T}^2}$ near 0

where V_H is the vector field generated by the system (1.4) and Ω is the volume two-form, i.e., $\Omega = ds_1 \wedge ds_2$. The vector field V_ψ, generated by the system (1.5), is obtained from the vector field V_H by a time change, so it is Hamiltonian with respect to the measure ν with the same Hamiltonian function H, i.e., $\Omega_\psi(v, V_\psi) = dHv$ where $\Omega_\psi = \kappa_0^{-1}\kappa\, ds_1 \wedge ds_2$. Near the origin the map $G_{\mathbb{T}^2}$ is the time-one map generated by the vector field $\varphi_* V_\psi$. Since $\varphi_*\nu = m$, this vector field is Hamiltonian with respect to the area m with the Hamiltonian function

$$H_1(s_1, s_2) = (H \circ \varphi^{-1})(s_1, s_2) = \frac{s_1 s_2 \beta(\sqrt{s_1{}^2 + s_2{}^2}) \log \lambda}{s_1{}^2 + s_2{}^2},$$

where $\beta(u)$ is the inverse function of

$$\gamma(u) = \kappa_0^{-1/2} \sqrt{\int_0^u \frac{d\tau}{\psi(\tau)}}. \tag{1.9}$$

Given a C^∞ function β, one can find the function ψ using the relation (1.9).

Exercise 1.12. Show that one can choose the function β such that the function ψ satisfies requirements (1)–(4) in the beginning of this section. Hint: Near zero the derivative of β must decrease sufficiently fast.

The next exercise describes the ergodic properties of the map $G_{\mathbb{T}^2}$.

Exercise 1.13. Show that the map $G_{\mathbb{T}^2}$ is ergodic with respect to area. Hint: Let A be a $G_{\mathbb{T}^2}$-invariant subset of positive area. Show that the set $B = A \cap (G_{\mathbb{T}^2} \setminus D_{r_1})$ has positive area. Consider then the set $C =$

$\bigcup_{n\geq 0} T^n(B)$, which is T-invariant and has positive area. Show that $B = C \cap (G_{\mathbb{T}^2} \setminus D_{r_1})$ and derive the desired result.

We examine the hyperbolic properties of the map $G_{\mathbb{T}^2}$. Since $G_{\mathbb{T}^2} = \varphi \circ G \circ \varphi^{-1}$, it is sufficient to do it for the map G. In particular, we shall show that G is not an Anosov diffeomorphism.

Let $x = (0, s_2) \in D_{r_1}$ be a point on the vertical segment through the origin. Note that

$$g_t(x) = (0, s_2(t)) \to 0 \quad \text{as} \quad t \to \infty,$$

where $s_2(t)$ is the solution of (1.5) with the initial condition $s_2(0) = s_2$. In view of the choice of the function ψ, we obtain that

$$\begin{aligned}\limsup_{t\to+\infty} \frac{\log d(g_t(x), 0)}{t} &= \limsup_{t\to+\infty} \frac{\log|s_2(t)|}{t} \\ &= \limsup_{t\to+\infty} (\log|s_2(t)|)' \\ &= \limsup_{t\to+\infty} (-\psi(s_2(t)^2)\log\lambda) = 0.\end{aligned}$$

This implies that

$$\limsup_{n\to+\infty} \frac{\log d(G^n(x), 0)}{n} = 0.$$

Similar arguments apply to points $x = (s_1, 0) \in D_{r_1}$ on the horizontal segment through the origin, showing that

$$\limsup_{n\to+\infty} \frac{\log d(G^{-n}(x), 0)}{n} = 0.$$

Thus G is not an Anosov diffeomorphism since otherwise the above limits would be negative.

Exercise 1.14. Show that $dG(0) = \mathrm{Id}$ where Id is the identity map.

For a typical trajectory, the situation is, however, quite different. To study it, we exploit the *cone techniques*—a collection of results aimed at establishing hyperbolic properties of the system. In Chapter 11 the reader can find a detailed exposition of these techniques.

Choose $x \in \mathbb{T}^2$ and define the *stable* and *unstable cones* in $T_x\mathbb{T}^2 = \mathbb{R}^2$ by

$$\begin{aligned} C^s(x) &= \{(v_1, v_2) \in \mathbb{R}^2 : |v_1| \leq |v_2|\}, \\ C^u(x) &= \{(v_1, v_2) \in \mathbb{R}^2 : |v_2| \leq |v_1|\}, \end{aligned}$$

where v_1 and v_2 are vectors in the eigenlines at x corresponding to λ and λ^{-1}.

Lemma 1.15. *For every $x \neq 0$ we have that*

$$dG^{-1}C^s(x) \subset C^s(G^{-1}(x)), \quad dGC^u(x) \subset C^u(G(x))$$

and the inclusions are strict.

Proof. We only consider the case of the unstable cones $C^u(x)$ as the other case is completely similar. Clearly, the desired inclusion is true for every x outside the disk D_{r_1}. To establish this property inside the disk, we consider the system of variational equations corresponding to the system (1.5) (i.e., the linear part of the vector field V_ψ)

$$\begin{aligned}\dot{\xi}_1 &= \log\lambda((\psi + 2s_1^2\psi')\xi_1 + 2s_1s_2\psi'\xi_2),\\ \dot{\xi}_2 &= -\log\lambda(2s_1s_2\psi'\xi_1 + (\psi + 2s_2^2\psi')\xi_2).\end{aligned}$$

This yields the following equation for the tangent $\eta = \xi_2/\xi_1$:

$$\frac{d\eta}{dt} = -2\log\lambda((\psi + (s_1^2 + s_2^2)\psi')\eta + s_1s_2\psi'(\eta^2 + 1)). \tag{1.10}$$

Substituting $\eta = 1$ and $\eta = -1$ in (1.10) gives

$$\begin{aligned}\frac{d\eta}{dt} &= -2\log\lambda(\psi + (s_1 + s_2)^2\psi') \le 0,\\ \frac{d\eta}{dt} &= 2\log\lambda(\psi + (s_1 - s_2)^2\psi') \ge 0.\end{aligned} \tag{1.11}$$

Moreover, these inequalities are strict everywhere except at the origin. The desired result follows. □

We define for every $x \neq 0$,

$$E^s(x) = \bigcap_{j=0}^{\infty} dG^{-j}C^s(G^j(x)), \quad E^u(x) = \bigcap_{j=0}^{\infty} dG^jC^u(G^{-j}(x)).$$

Lemma 1.16. *$E^s(x)$ and $E^u(x)$ are one-dimensional subspaces of $T_x\mathbb{T}^2$.*

Proof. We only prove the result for $E^u(x)$ as the argument for $E^s(x)$ is quite similar. To show that $E^u(x)$ is one-dimensional, we shall estimate from below the decrease of the angle between two arbitrary lines inside the cone $C^u(x)$ along the whole segment of trajectory of the local flow g_t for $x \in D_{r_2}$.

Let $\eta(t, \alpha_0) = \eta_{s_1^0, s_2^0}(t, \alpha_0)$ be the solution of the differential equation (1.11) with the initial condition $t = 0$ and $\alpha = \alpha_0$ along the trajectory of the flow through the point $(s_1^0, s_2^0) \in D_{r_2}$. We wish to estimate from above the ratio

$$\frac{|\eta(1, \alpha_1) - \eta(1, \alpha_2)|}{|\alpha_1 - \alpha_2|}$$

for all initial conditions α_1, α_2 such that $|\alpha_1| \le 1$, $|\alpha_2| \le 1$. To do so, we introduce the function

$$\hat{\eta}(t) = \hat{\eta}_{s_1^0, s_2^0}(t) = \exp\left(-2\log\lambda \int_0^t (\psi + (s_1^2 + s_2^2)\psi')\, du\right). \tag{1.12}$$

Exercise 1.17. Use equations (1.10) and (1.12) to show that

$$\eta(t,\alpha) = \alpha\hat{\eta}(t) - 2(\log\lambda)\hat{\eta}(t)\int_0^t \frac{s_1 s_2 \psi'(\eta^2(u,\alpha)+1)}{\hat{\eta}(u)}\, du. \tag{1.13}$$

It follows from (1.11) that the requirement $|\alpha| \le 1$ implies that $|\eta(t,\alpha)| \le 1$ for every t and hence by (1.13), we have that

$$\frac{|\eta(t,\alpha_1) - \eta(t,\alpha_2)|}{\hat{\eta}(t)} \le |\alpha_1 - \alpha_2| + 4\log\lambda \int_0^t \frac{|s_1 s_2|\psi'|\eta(u,\alpha_1) - \eta(u,\alpha_2)|}{\hat{\eta}(u)}\, du.$$

By the Gronwall inequality,

$$\frac{|\eta(t,\alpha_1) - \eta(t,\alpha_2)|}{\hat{\eta}(t)} \le |\alpha_1 - \alpha_2| \exp\left(4\log\lambda \int_0^t |s_1 s_2|\psi'\, du\right).$$

Substituting $t = 1$ and using (1.12), we obtain that

$$|\eta(1,\alpha_1) - \eta(1,\alpha_2)| \le |\alpha_1 - \alpha_2| \exp\left(-2\log\lambda \int_0^1 (\psi + \psi'(|s_1| - |s_2|)^2)\, du\right).$$

Let us fix positive numbers ε_0, β, δ and consider the region

$$Q_{\varepsilon_0}^{\beta} = \{(s_1, s_2) : |s_1 s_2| \le \varepsilon_0,\ |s_1| \le \beta,\ |s_2| \le \beta\} \subset D_{r_2}.$$

We wish to estimate the solution of the differential equation (1.10) along the segment of hyperbola

$$\{s_1 s_2 = \varepsilon,\ 0 \le s_1 \le \beta,\ 0 \le s_2 \le \beta\}$$

for all $0 < \varepsilon \le \varepsilon_0$. We consider s_1 as a parameter on the hyperbola. Equations (1.5) and (1.10) imply that

$$\frac{d\eta}{ds_1} = -\left(\frac{2}{s_1} + \frac{2(s_1^2 + s_2^2)}{s_1}\frac{\psi'}{\psi}\right)\eta + 2s_2\frac{\psi'}{\psi}(\eta^2 + 1). \tag{1.14}$$

For $i = 1, 2$ let $\eta(s_1, \varepsilon/\beta, \eta_i)$ be the solutions of this differential equation with the initial conditions $s_1 = \varepsilon/\beta$, $\eta = \eta_i$ for $|\eta_i| < 1$. Using equation (1.14), we obtain an equation for the difference $\eta(s_1, \varepsilon/\beta, \eta_1) - \eta(s_1, \varepsilon/\beta, \eta_2)$. Then the Gronwall inequality and the requirement that $|\eta(s_1, \varepsilon/\beta, \eta_i)| < 1$

imply that for every η_1, η_2, $|\eta_1| \le 1$, $|\eta_2| \le 1$ we have

$$\begin{aligned}
&\left|\eta\left(\beta, \frac{\varepsilon}{\beta}, \eta_1\right) - \eta\left(\beta, \frac{\varepsilon}{\beta}, \eta_2\right)\right| \\
&\le |\eta_1 - \eta_2| \exp\left(-2\int_{\varepsilon/\beta}^{\beta} \left(\frac{1}{s_1} + \frac{\psi'}{\psi}\frac{(s_1^2-\varepsilon)^2}{s_1^3}\right) ds_1\right) \\
&= |\eta_1 - \eta_2| \frac{\varepsilon^2}{\beta^4} \exp\left(-2\int_{\varepsilon/\beta}^{\beta} \frac{\psi'}{\psi}\frac{(s_1^2-\varepsilon)^2}{s_1^3} ds_1\right) \\
&\le |\eta_1 - \eta_2| \frac{\varepsilon^2}{\beta^4}.
\end{aligned}$$

The same inequalities hold in other quadrants, i.e., for $\eta(-\beta, -\varepsilon/\beta, \eta_0)$, etc.

Let $\{g^k(x), k = 0, \ldots, n-1\}$ be the segment of the trajectory of the map g through a point x lying inside $Q^{\beta}_{\varepsilon_0}$. This means that all these points belong to $Q^{\beta}_{\varepsilon_0}$ but the points $g^{-1}(x)$ and $g^n(x)$ do not. Suppose that $x = (s_1^0, s_2^0)$. Then for every η_1, η_2, $|\eta_1| < 1$, $|\eta_2| < 1$ we have that

$$\begin{aligned}
&|\eta_{s_1^0, s_2^0}(n, \eta_1) - \eta_{s_1^0, s_2^0}(n, \eta_2)| \\
&\qquad \le |\eta_1 - \eta_2| \frac{M(s_1^0 s_2^0)^2}{\beta^4} \\
&\qquad \le |\eta_1 - \eta_2| \frac{M\varepsilon^2}{\beta^4},
\end{aligned}$$

where M is a constant (independent of ε_0 and β provided that they are sufficiently small). This implies that every angle inside the cone $C^u(x)$ is contracted under the action of dg^n by at least $\mu = M\varepsilon^2/\beta^4$ times. If we choose ε_0 small enough to ensure that $\mu < 1$, we obtain the desired result for every point x for which the trajectory $g^n(x)$ does not go to the origin.

It remains to show that the conclusion of the lemma holds also for every trajectory converging to the origin. To this end we choose a number $\delta > 0$ and consider the point $x = (\delta, 0) \in D_{r_1}$ whose trajectory $g^n(x)$ goes to zero. On the line $s_2 = 0$, the differential equation (1.14) reduces to the linear differential equation

$$\frac{d\eta}{ds_1} = -\left(\frac{2}{s_1} + \frac{2s_1\psi'}{\psi}\right)\eta.$$

Consider the solution $\eta(\delta, \varepsilon, 1)$ of this equation with the initial condition $s_1 = \varepsilon < \delta$, $\eta = 1$ along the segment $[\varepsilon, \delta]$. For this solution we have

$$\eta(\delta, \varepsilon, 1) = \exp\left(-\int_{\varepsilon}^{1} \left(\frac{2}{s_1} + \frac{2s_1\psi'}{\psi}\right) ds_1\right) \le \frac{\varepsilon^2}{\delta^2}.$$

Therefore, $\eta(\delta, \varepsilon, 1) \to 0$ as $\varepsilon \to 0$ and by the linearity of the equation, we also have $\eta(\delta, \varepsilon, -1) \to 0$ as $\varepsilon \to 0$. The desired result follows. □

Clearly, the subspaces $E^s(x)$ and $E^u(x)$ are transverse to each other, i.e.,

$$T_x\mathbb{T}^2 = E^s(x) \oplus E^u(x),$$

and invariant under the action of the differential, i.e.,

$$dGE^s(x) = E^s(G(x)), \quad dGE^u(x) = E^u(G(x)).$$

We now introduce the Lyapunov exponent $\chi(x, v)$, $x \in \mathbb{T}^2$ and $v \in T_x\mathbb{T}^2$, which measures the asymptotic rate of growth (or decay) of the length of the vector $dG^n v$ with n:

$$\chi(x, v) = \limsup_{n\to\infty} \frac{1}{n} \log \|dG^n v\|.$$

In the next chapter we shall study properties of the Lyapunov exponents in great detail and we shall see later in the book the important role they play in studying stability of trajectories.

Exercise 1.18. Show that for every point $x \in D_{r_1}$ on the vertical (or horizontal) segment through the origin and every $v \in T_x\mathbb{T}^2$ the Lyapunov exponent $\chi(x, v) = 0$.

For a typical point $x \in \mathbb{T}^2$ the situation is quite different.

Lemma 1.19. *For almost every $x \in \mathbb{T}^2$ with respect to area we have that $\chi(x, v) < 0$ for $v \in E^s(x)$ and $\chi(x, v) > 0$ for $v \in E^u(x)$.*

Proof. By Lemma 1.15, the map G possesses two families of stable and unstable cones, $C^s(x)$ and $C^u(x)$, which are transverse and strictly invariant under dG^{-1} and dG, respectively, at every point $x \neq 0$. The desired statement now follows from a deep result by Wojtkowski (see Theorem 11.2): the mere fact that G preserves a measure that is absolutely continuous with respect to area and possesses such cone families implies that for almost every $x \in \mathbb{T}^2$ the Lyapunov exponent $\chi(x, v) < 0$ for $v \in C^s(x)$ and $\chi(x, v) > 0$ for $v \in C^u(x)$.

However, we outline a more direct proof of the lemma that does not use Wojtkowski's result. Fix a point $x \in \mathbb{T}^2 \setminus D_{r_1}$ and a vector $v \in C^u(x)$. Let $n_1 > 0$ be the first moment the trajectory of x under G enters the disk D_{r_1}. There is a number $\lambda > 1$ such that the vector $v_1 = dG^{n_1}v$ has norm $\|v_1\| \geq \lambda^{n_1}\|v\|$. Let now $m_1 > n_1$ be the first moment when this trajectory exits D_{r_1}. It is easy to see that the vector $w_1 = dG^{m_1}(x)$ has norm $\|w_1\| \geq \|v_1\|$. Continuing in the same fashion, we will construct a sequence n_k of entries and a sequence m_k of exits of the trajectory $\{G^l(x)\}$ to and from the disk D_{r_1} so that $n_k < m_k < n_{k+1}$ for every $k > 0$. Furthermore, setting

$$v_k = dG^{n_k}v, \quad w_k = dG^{m_k}v,$$

we have that

$$\|v_{k+1}\| \geq \lambda^{n_{k+1}-m_k}\|w_k\|, \quad \|w_{k+1}\| \geq \|v_{k+1}\|.$$

It follows that given $n > 0$, the length of the vector $\|dG^n v\| \geq \lambda^{N(n)}$ where $N(n) = \sum_{i=1}^{\ell}(n_i - m_i)$ is the largest integer such that $n_\ell \leq n$. This implies that

$$\chi(x, v) \geq \limsup_{n\to\infty} \frac{N(n)}{n} \log \lambda.$$

Since G is ergodic, by the Birkhoff Ergodic Theorem, we conclude that for almost every $x \in \mathbb{T}^2$,

$$\lim_{n\to\infty} \frac{N(n)}{n} = 1 - m(D_{r_1}).$$

This implies that $\chi(x, v) > 0$ for $v \in C^u(x)$. The proof that $\chi(x, v) < 0$ for $v \in C^s(x)$ is similar. □

Lemma 1.19 shows that the length of the vector $\|dG^n v\|$ with $v \in E^s(x)$ goes to zero as $n \to \infty$ and that the length of the vector $\|dG^n v\|$ with $v \in E^u(x)$ goes to ∞ as $n \to \infty$. In other words $E^s(x)$ is a stable subspace and $E^u(x)$ is an unstable subspace for df. Thus the map G, and hence also $G_{\mathbb{T}^2}$, admits an invariant splitting similar to the one for the hyperbolic automorphism T (except at the origin). However, the contraction and expansion rates along the stable and unstable directions are nonuniform in x.

We call the map $G_{\mathbb{T}^2}$ the *Katok map.*

We shall describe some other interesting properties of the map $G_{\mathbb{T}^2}$.

(1) The map $G_{\mathbb{T}^2}$ lies on the boundary of Anosov diffeomorphisms on $\mathbb{T}^2$; i.e., there is a sequence of Anosov diffeomorphisms G_n converging to $G_{\mathbb{T}^2}$ in the C^1 topology.

We outline the proof of this result (see [**45**] for details). Observe that the Katok map $G_{\mathbb{T}^2}$ depends on the choice of the function ψ and so we can write $G_{\mathbb{T}^2} = G_{\mathbb{T}^2}(\psi)$.

Let κ be a real-valued C^∞ function on $[0, 1]$ satisfying:

(a) $\kappa(0) > 0$ and $\kappa(u) = 1$ for $u \geq r_0$, for some $0 < r_0 < 1$;

(b) $\kappa'(u) > 0$ for every $0 < u < r_0$.

Starting with such a function κ, one can repeat the construction in this section and obtain an area-preserving C^∞ diffeomorphism $G_{\mathbb{T}^2}(\kappa)$ of the torus.

Exercise 1.20. Show that $G_{\mathbb{T}^2}(\kappa)$ is an Anosov diffeomorphism.

Let us now choose a function ψ that satisfies conditions (1)–(4) in the beginning of this section. There is a sequence of real-valued C^∞ functions κ_n on $[0,1]$ that converge uniformly to the function ψ. By Exercise 1.20, every map $G_n = G_{\mathbb{T}^2}(\kappa_n)$ is a volume-preserving Anosov diffeomorphism of the torus.

Exercise 1.21. Show that the sequence of maps G_n converges to $G_{\mathbb{T}^2}(\psi)$ in the C^1 topology.

Some other properties of the Katok map $G_{\mathbb{T}^2}$ are the following (for details see [**45**]):

(2) The map $G_{\mathbb{T}^2}$ is topologically conjugate to T; i.e., there exists a homeomorphism $h\colon \mathbb{T}^2 \to \mathbb{T}^2$ such that $G_{\mathbb{T}^2} \circ h = h \circ G_{\mathbb{T}^2}$.

(3) Let $W_T^u(x)$ and $W_T^s(x)$ be the projections of eigenlines through x corresponding to the eigenvalues λ and λ^{-1}. They form two smooth transverse invariant foliations that are unstable and stable foliations for T. The curves

$$W^u_{G_{\mathbb{T}^2}}(x) = h(W^u_T(x)), \quad W^s_{G_{\mathbb{T}^2}}(x) = h(W^s_T(x))$$

are smooth and form two continuous transverse invariant foliations for $G_{\mathbb{T}^2}$ that are tangent to $E^u(x)$ and $E^s(x)$ for $x \neq 0$, respectively. In particular, the subspaces $E^u(x)$ and $E^s(x)$ depend continuously on $x \neq 0$.

(4) $G_{\mathbb{T}^2}$ is a Bernoulli diffeomorphism.

1.4. Diffeomorphisms with nonzero Lyapunov exponents on surfaces

We now outline a construction of an area-preserving C^∞ diffeomorphism with nonzero Lyapunov exponents on the two-dimensional sphere. Observe that there is no Anosov diffeomorphisms on the sphere.

We begin with the automorphism T of the torus $\mathbb{T}^2$ induced by the matrix $A = \left(\begin{smallmatrix} 5 & 8 \\ 8 & 13 \end{smallmatrix}\right)$ that has four fixed points $x_1 = (0,0)$, $x_2 = (1/2, 0)$, $x_3 = (0, 1/2)$, and $x_4 = (1/2, 1/2)$.

For $i = 1,2,3,4$ consider the disk D_r^i centered at x_i of radius r. We choose numbers r_0, r_1, and r_2 such that the disks $D^i_{r_k}$, $k = 0,1,2$, satisfy (1.3) and $D^i_{r_2} \cap D^j_{r_2} = \varnothing$ for $i \neq j$. Repeating the arguments in the previous section, we construct a diffeomorphism g_i coinciding with T outside $D^i_{r_1}$. Therefore, the map

$$G_1(x) = \begin{cases} T(x) & \text{if } x \in \mathbb{T}^2 \setminus D, \\ g_i(x) & \text{if } x \in D^i_{r_1}, \end{cases}$$

where $D = \bigcup_{i=1}^4 D_{r_1}^i$, defines a homeomorphism of the torus $\mathbb{T}^2$ which is a C^∞ diffeomorphism everywhere except at the points x_i. The Lyapunov exponents of G_1 are nonzero almost everywhere with respect to the measure ν, which coincides with area m in $\mathbb{T}^2 \setminus D$ and is absolutely continuous with respect to m in D with the density function $\kappa(s_1, s_2)$ given in each $D_{r_1}^i$ by (1.6).

Using (1.7) in each disk $D_{r_2}^i$, $r_2 \geq r_1$, we introduce a coordinate change φ_i such that the map

$$\varphi(x) = \begin{cases} \varphi_i(x) & \text{if } x \in D_{r_1}^i, \\ x & \text{otherwise} \end{cases}$$

defines a homeomorphism of $\mathbb{T}^2$ which is a C^∞ diffeomorphism everywhere except at the points x_i. Repeating arguments of the previous section, it is easy to show that the map $G_2 = \varphi \circ G_1 \circ \varphi^{-1}$ is an area-preserving C^∞ diffeomorphism whose Lyapunov exponents are nonzero almost everywhere.

Using this map, we construct a diffeomorphism of the sphere S^2 with the desired properties. Consider the involution map $I\colon \mathbb{T}^2 \to \mathbb{T}^2$ given by $I(t_1, t_2) = (1 - t_1, 1 - t_2)$.

Exercise 1.22. Show that:

(1) the involution I has the points x_i, $i = 1, 2, 3, 4$, as its fixed points;

(2) I commutes with G_2; i.e., $G_2 \circ I = I \circ G_2$.

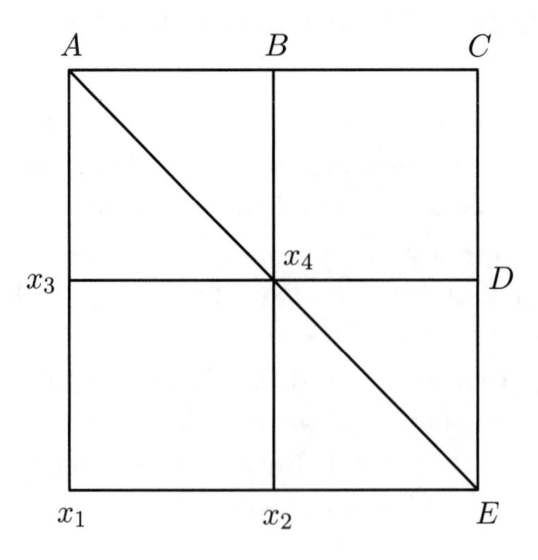

Figure 1.7. Mapping the factor space $\mathbb{T}^2/I$ to S^2 by making the following identifications: $AB \equiv Ex_2$, $BC \equiv x_2x_1$, $CD \equiv x_1x_3$, $DE \equiv x_3A$, $Ax_4 \equiv Ex_4$, $Bx_4 \equiv x_2x_4$, and $x_3x_4 \equiv Dx_4$.

The factor-space $\mathbb{T}^2/I$ is homeomorphic to the sphere S^2 (see Figure 1.7) and admits a natural smooth structure induced from the torus everywhere except for the points x_i, $i = 1, 2, 3, 4$. Moreover, the following statement holds.

Proposition 1.23 (Katok [**45**]). *There exists a map $\zeta\colon \mathbb{T}^2 \to S^2$ satisfying:*

(1) *ζ is a double branched covering, is regular (one-to-one on each branch), and is C^∞ everywhere except at the points x_i, $i = 1, 2, 3, 4$, where it branches;*

(2) *$\zeta \circ I = \zeta$;*

(3) *zeta preserves area; i.e., $\zeta_* m = \mu$ where μ is the area in S^2;*

(4) *there exists a local coordinate system in a neighborhood of each point $\zeta(x_i)$, $i = 1, 2, 3, 4$, in which*

$$\zeta(s_1, s_2) = \left(\frac{{s_1}^2 - {s_2}^2}{\sqrt{{s_1}^2 + {s_2}^2}}, \frac{2 s_1 s_2}{\sqrt{{s_1}^2 + {s_2}^2}} \right)$$

in each disk $D^i_{r_1}$.

Finally, one can show that the map $G_{S^2} = \zeta \circ G_2 \circ \zeta^{-1}$ is a C^∞ diffeomorphism which preserves area. It is easy to see that this map has nonzero Lyapunov exponents almost everywhere. It is also ergodic (and, indeed, is a Bernoulli diffeomorphism).

The diffeomorphism G_{S^2} can be used to build an area-preserving C^∞ diffeomorphism with nonzero Lyapunov exponents on any surface. This is a two step procedure. First, the sphere can be unfolded into the unit disk and the map G_{S^2} can be carried over to a C^∞ area-preserving diffeomorphism g of the disk. The second step is to cut the surface in a certain way so that the resulting surface with the boundary is diffeomorphic to the disk. This is a well-known topological construction but we need to complement it by showing that the map g can be carried over to produce the desired map on the surface. In doing so, a crucial fact is that g is the identity on the boundary of the disk and is "sufficiently flat" near the boundary. We shall briefly outline the procedure without going into details.

Set $p_i = \zeta(x_i)$, $i = 1, 2, 3, 4$. In a small neighborhood of the point p_4 we define a map η by

$$\eta(\tau_1, \tau_2) = \left(\frac{\tau_1 \sqrt{1 - \tau_1^2 - \tau_2^2}}{\sqrt{\tau_1^2 + \tau_2^2}}, \frac{\tau_2 \sqrt{1 - \tau_1^2 - \tau_2^2}}{\sqrt{\tau_1^2 + \tau_2^2}} \right).$$

One can extend it to an area-preserving C^∞ diffeomorphism η between $S^2 \setminus \{p_4\}$ and the unit disk $D^2 \subset \mathbb{R}^2$. The map

$$g = \eta \circ G_{S^2} \circ \eta^{-1} \tag{1.15}$$

is a diffeomorphism of the disk D^2 that preserves area and has nonzero Lyapunov exponents almost everywhere.

The disk D^2 can be embedded into any surface. This follows from a result by Katok [**45**] that we state without a proof.

Proposition 1.24. *Given a compact surface M, there exists a continuous map $h\colon D^2 \to M$ such that:*

(1) *the restriction $h|\operatorname{int} D^2$ is a diffeomorphic embedding;*

(2) $h(D^2) = M$;

(3) *the map h preserves area; more precisely, $h_*m = \mu$ where m is the area in D^2 and μ is the area in M. Moreover, $\mu(N \setminus h(D^2)) = 0$.*

Note that the map g in (1.15) is the identity on the boundary ∂D^2. Moreover, one can choose the function ψ in the construction of the map $G_{\mathbb{T}^2}$ such that g is "sufficiently flat" near the boundary of the disk. More precisely, let $\rho = \{\rho_n\}_n$ be a sequence of nonnegative real-valued continuous functions on D^2, which are strictly positive inside the disk. Let $C^\infty_\rho(D^2)$ be the set of all C^∞ functions $\varphi\colon D^2 \to \mathbb{R}$ satisfying the following condition: there exists a sequence of numbers $\varepsilon_n > 0$ such that for every $(x_1, x_2) \in D^2$ for which $x_1^2 + x_2^2 \geq (1-\varepsilon_n)^2$ we have

$$\left| \frac{\partial^n \varphi(x_1, x_2)}{\partial x_1^{i_1} \partial x_2^{i_2}} \right| < \rho_n(x_1, x_2),$$

where i_1, i_2 are nonnegative integers and $i_1 + i_2 = n$.

We write each diffeomorphism G of D^2 in the form

$$G(x_1, x_2) = (G_1(x_1, x_2), G_2(x_1, x_2))$$

and we denote by $\operatorname{Diff}^\infty_\rho(D^2)$ the set of diffeomorphisms $G \in \operatorname{Diff}^\infty(D^2)$ such that

$$G_i(x_1, x_2) - x_i \in C^\infty_\rho(D^2), \quad i = 1, 2.$$

Given $G \in \operatorname{Diff}^\infty_\rho(D^2)$ and a map h satisfying statements (1)–(3) of Proposition 1.24, consider the map k defined by $k(x) = h(G(h^{-1}(x)))$ for $x \in h(\operatorname{int} D^2)$ and $k(x) = x$ otherwise. In order to complete our construction, we need the following statement.

Proposition 1.25 (Katok [**45**]). *Given a compact surface M, there exists a sequence of functions ρ such that for any $G \in \operatorname{Diff}^\infty_\rho(D^2)$ the map k is a C^∞ diffeomorphism of M.*

The function ψ can be chosen so that $g \in \mathrm{Diff}^\infty_\rho(D^2)$ and hence the map f defined by

$$f(x) = \begin{cases} h(g(h^{-1}(x))), & x \in h(\operatorname{int} D^2), \\ x, & \text{otherwise} \end{cases}$$

has the desired properties: it preserves area and has nonzero Lyapunov exponents almost everywhere.

Thus we obtain the following result.

Theorem 1.26. *Given a compact surface M, there exists a C^∞ area-preserving ergodic diffeomorphism $f\colon M \to M$ with nonzero Lyapunov exponents almost everywhere. The map f is ergodic and indeed is a Bernoulli diffeomorphism.*

Dolgopyat and Pesin [**30**] have extended this result to any compact Riemannian manifold of dimension greater than 2.

Theorem 1.27. *Given a compact smooth Riemannian manifold M of dimension greater than 2, there exists a volume-preserving C^∞ diffeomorphism $f\colon M \to M$ with nonzero Lyapunov exponents almost everywhere which is a Bernoulli diffeomorphism.*

The mechanism of slowing down trajectories in a neighborhood of a hyperbolic fixed point is not robust in the C^1 topology as the following result by Bochi [**13**] demonstrates (see also [**14**]).

Theorem 1.28. *There exists an open set U in the space of area-preserving C^1 diffeomorphisms of $\mathbb{T}^2$ such that:*

1. *the map $G_{\mathbb{T}^2}$ lies on the boundary of U;*
2. *there is a G_δ subset $A \subset U$ such that every $f \in A$ is an area-preserving diffeomorphism whose Lyapunov exponents are all zero almost everywhere.*

1.5. A flow with nonzero Lyapunov exponents

We present an example of a dynamical system with continuous time which is nonuniformly hyperbolic. It was constructed in [**67**] by a "surgery" of an Anosov flow. Let φ_t be a volume-preserving ergodic Anosov flow on a compact three-dimensional manifold M and let X be the vector field of the flow.[5] Fix a point $p_0 \in M$ and introduce a coordinate system x, y, z in the ball $B(p_0, d)$ centered at p_0 of some radius $d > 0$ such that p_0 is the origin (i.e., $p_0 = 0$) and $X = \partial/\partial z$.

[5]For example, one can take φ_t to be the geodesic flow on a compact surface of negative curvature.

For each $\varepsilon > 0$, let $T_\varepsilon = S^1 \times D_\varepsilon \subset B(0,d)$ be the solid torus obtained by rotating the disk

$$D_\varepsilon = \{(x,y,z) \in B(0,d) : x = 0 \text{ and } (y - d/2)^2 + z^2 \le (\varepsilon d)^2\}$$

around the z-axis. Every point on the solid torus can be represented as (θ, y, z) with $\theta \in S^1$ and $(y,z) \in D_\varepsilon$.

For every $0 \le \alpha \le 2\pi$, we consider the cross-section of the solid torus $\Pi_\alpha = \{(\theta, y, z) : \theta = \alpha\}$ and construct a new vector field $\tilde{X}$ on $M \setminus T_\varepsilon$ (see Figure 1.8).

Lemma 1.29. *There exists a smooth vector field $\tilde{X}$ on $M \setminus T_\varepsilon$ such that the flow $\tilde{\varphi}_t$ generated by $\tilde{X}$ satisfies the following properties:*

(1) $\tilde{X}|(M \setminus T_{2\varepsilon}) = X|(M \setminus T_{2\varepsilon})$;

(2) *for any $0 \le \alpha, \beta \le 2\pi$, the vector field $\tilde{X}|\Pi_\beta$ is the image of the vector field $\tilde{X}|\Pi_\alpha$ under the rotation around the z-axis that moves Π_α onto Π_β;*

(3) *for every $0 \le \alpha \le 2\pi$, the unique two fixed points of the flow $\tilde{\varphi}_t|\Pi_\alpha$ are those in the intersection of Π_α with the planes $z = \pm\varepsilon d$;*

(4) *for every $0 \le \alpha \le 2\pi$ and $(y,z) \in D_{2\varepsilon} \setminus \operatorname{int} D_\varepsilon$, the trajectory of the flow $\tilde{\varphi}_t|\Pi_\alpha$ passing through the point (y,z) is invariant under the symmetry $(\alpha, y, z) \mapsto (\alpha, y, -z)$;*

(5) *the flow $\tilde{\varphi}_t|\Pi_\alpha$ preserves the conditional measure induced by volume on the set Π_α.*

Proof. We shall only describe the construction of $\tilde{X}$ on the cross-section Π_0. The vector field $\tilde{X}|\Pi_\alpha$ on an arbitrary cross-section is the image of $\tilde{X}|\Pi_0$ under the rotation around the z-axis that moves Π_0 onto Π_α.

Consider the Hamiltonian $H(y,z) = y(\varepsilon^2 - y^2 - z^2)$. It is straightforward to verify that in the annulus $\varepsilon^2 \le y^2 + z^2 \le 4\varepsilon^2$ the corresponding Hamiltonian flow is topologically conjugate to that shown in Figure 1.8. However, the Hamiltonian vector field $(-2yz, 3y^2 + z^2 - \varepsilon^2)$ is not everywhere vertical on the circle $y^2 + z^2 = 4\varepsilon^2$. To correct this, we consider a C^∞ function $p\colon [\varepsilon, \infty) \to [0,1]$ such that $p(t) = 1$ for $t \in [\varepsilon, 3\varepsilon/2]$, $p(t) = 0$ for $t \in [2\varepsilon, \infty)$, and p is strictly decreasing in $(3\varepsilon/2, 2\varepsilon)$. The flow generated by the system of differential equations

$$\begin{cases} y' = -2yzp(\sqrt{y^2+z^2}), \\ z' = (3y^2 + z^2 - \varepsilon^2)p(\sqrt{y^2+z^2}) + 1 - p(\sqrt{y^2+z^2}) \end{cases}$$

behaves as shown in Figure 1.8. We denote it by $\overline{\varphi}_t$ and we denote by $\overline{X}$ the corresponding vector field.

We now show that it is possible to effect a time change of the flow $\overline{\varphi}_t$ in the annulus $\varepsilon^2 \le y^2 + z^2 \le 4\varepsilon^2$ so that the new flow $\tilde{\varphi}_t$ preserves volume. To achieve this, we shall construct a C^1 function $\tau\colon (T_{2\varepsilon} \setminus \operatorname{int} T_\varepsilon) \times \mathbb{R} \to \mathbb{R}$ such that

$$\tilde{\varphi}_t(x) = \overline{\varphi}_{\tau(x,t)}(x).$$

It is easy to see that $\tilde{X}(x) = h(x)\overline{X}(x)$, where $h(x) = (\partial\tau/\partial t)(x)$. The flow $\tilde{\varphi}_t$ preserves volume if and only if

$$\operatorname{div}(h\overline{X}) = \nabla h \cdot \overline{X} + h \operatorname{div}(\overline{X}) = 0. \tag{1.16}$$

Since the two semicircles defined by $y^2 + z^2 = 4\varepsilon^2$, $y = \pm 2\varepsilon$ are not the characteristics, there exists a unique solution h of (1.16) such that $h = 1$ outside $\Pi_0 \cap T_{2\varepsilon}$. This completes the construction of the desired vector field $\tilde{X}$. □

One can see that the orbits of the flows φ_t and $\tilde{\varphi}_t$ coincide on $M \setminus T_{2\varepsilon}$, that the flow $\tilde{\varphi}_t$ preserves volume, and that the only fixed points of this flow are those on the circles $\{(\theta, y, z) : z = -\varepsilon d\}$ and $\{(\theta, y, z) : z = \varepsilon d\}$.

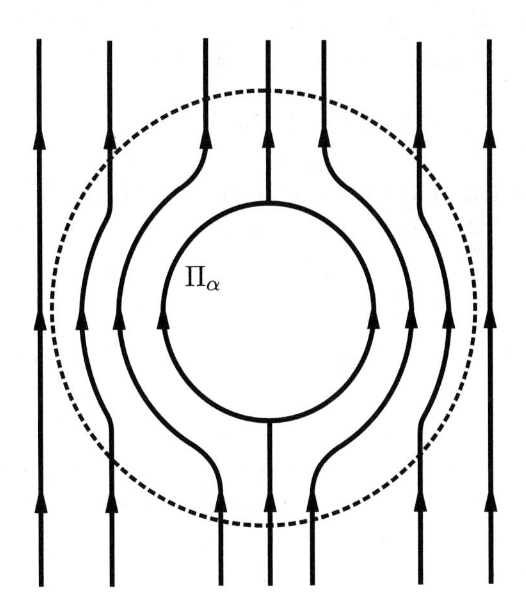

Figure 1.8. A cross-section Π_α and the flow $\tilde{\varphi}_t$

On the set $T_{2\varepsilon} \setminus \operatorname{int} T_\varepsilon$ we introduce coordinates θ_1, θ_2, r with $0 \le \theta_1, \theta_2 < 2\pi$ and $\varepsilon d \le r \le 2\varepsilon d$ such that the set of fixed points of $\tilde{\varphi}_t$ is composed of those for which $r = \varepsilon d$, and $\theta_1 = 0$ or $\theta_1 = \pi$.

Consider the flow on $T_{2\varepsilon} \setminus \operatorname{int} T_\varepsilon$ defined by

$$(\theta_1, \theta_2, r, t) \mapsto (\theta_1, \theta_2 + [2 - r/(\varepsilon d)]^4 t \cos\theta_1, r),$$

and let $\hat{X}$ be the corresponding vector field. Set

$$Y(x) = \begin{cases} X(x), & x \in M \setminus \operatorname{int} T_{2\varepsilon}, \\ \tilde{X}(x) + \hat{X}(x), & x \in \operatorname{int} T_{2\varepsilon} \setminus \operatorname{int} T_{\varepsilon}. \end{cases}$$

The vector field Y on $M \setminus \operatorname{int} T_\varepsilon$ generates the flow ψ_t on $M \setminus \operatorname{int} T_\varepsilon$.

Lemma 1.30. *The following properties hold:*

(1) *the flow ψ_t preserves volume and is ergodic;*

(2) *the flow ψ_t has no fixed points;*

(3) *for almost every $x \in M \setminus T_{2\varepsilon}$,*

$\chi(x,v) < 0$ *for* $v \in E^s(x)$, *and* $\chi(x,v) > 0$ *for* $v \in E^u(x)$,

where $E^u(x)$ and $E^s(x)$ are, respectively, the stable and unstable subspaces of the flow φ_t at the point x.

Proof. By construction, ψ_t preserves volume. Since the orbits of φ_t and ψ_t coincide in $M \setminus \operatorname{int} T_{2\varepsilon}$, the flow ψ_t is ergodic (see Exercise 1.13). The second statement follows from the construction of the flow ψ_t. In order to prove the third statement, consider the function

$$T(x,t) = \int_0^t I_{T_{2\varepsilon}}(\varphi_\tau x)\, d\tau,$$

where $I_{T_{2\varepsilon}}$ denotes the characteristic function of the set $T_{2\varepsilon}$. By the Birkhoff Ergodic Theorem, for almost every $x \in M$,

$$\lim_{t\to+\infty} \frac{T(x,t)}{t} = \mu(T_{2\varepsilon}).$$

Fix a point $x \in M \setminus T_{2\varepsilon}$. Consider the moment of time t_1 at which the trajectory $\psi_t(x)$ enters the set $T_{2\varepsilon}$ and the next moment of time t_2 at which this trajectory exits the set $T_{2\varepsilon}$. Given a vector $v \in E^u(x)$, denote by $\tilde{v}_i$ the orthogonal projection of the vector $d_x\psi_{t_i}v$ onto the (x,y)-plane for $i = 1, 2$. It follows from the construction of the flows $\tilde{\varphi}_t$ and ψ_t that $\|\tilde{v}_1\| \geq \|\tilde{v}_2\|$. Since the unstable subspaces $E^u(x)$ depend continuously on x, there exists $K \geq 1$ (independent of x, t_1, and t_2) such that

$$\|d_x\psi_t v\| \geq K\|d_x\varphi_{t-T(x,t)}v\|.$$

It follows that for almost every $x \in M \setminus T_{2\varepsilon}$ and $v \in E^u(x)$,

$$\chi(x,v) = \limsup_{t\to+\infty} \frac{1}{t} \log \|d_x\psi_t v\| \geq (1 - \mu(T_{2\varepsilon})) \limsup_{t\to+\infty} \frac{1}{t} \log \|d_x\varphi_t v\| > 0$$

provided that ε is sufficiently small. Repeating the above argument with respect to the inverse flow ψ_{-t}, one can show that $\chi(x,v) < 0$ for almost every $x \in M \setminus T_{2\varepsilon}$ and $v \in E^s(x)$. □

Set $M_1 = M \setminus T_\varepsilon$ and consider a copy $(\tilde{M}_1, \tilde{\psi}_t)$ of the flow (M_1, ψ_t). One can glue the manifolds M_1 and $\tilde{M}_1$ along their boundaries ∂T_ε and obtain a three-dimensional smooth Riemannian manifold D without boundary. We define a flow F_t on D by

$$F_t(x) = \begin{cases} \psi_t(x), & x \in M_1, \\ \tilde{\psi}_t(x), & x \in \tilde{M}_1. \end{cases}$$

It is clear that the flow F_t preserves volume and has nonzero Lyapunov exponents almost everywhere. Clearly, the flow $F_t|M_1$ is ergodic and so is the flow $F_t|M_2$.

Exercise 1.31. Modify the definition of the flow F_t to obtain a new flow that is ergodic on the whole manifold D.

One can show that any compact smooth Riemannian manifold M of dimension greater than 2 admits a volume-preserving C^∞ flow $\varphi_t \colon M \to M$ with nonzero Lyapunov exponents almost everywhere which is a Bernoulli flow[6]; see [**41**].

[6]A *Bernoulli flow* is a flow φ_t such that the transformation φ_t is a Bernoulli automorphism for every t.

Chapter 2

General Theory of Lyapunov Exponents

In this chapter we introduce the fundamental notion of Lyapunov exponent in a formal axiomatic setting and study its basic properties. We then discuss the crucial concept of Lyapunov–Perron regularity and describe various criteria that guarantee that a given trajectory of the system is Lyapunov–Perron regular. Finally we study Lyapunov exponents and Lyapunov–Perron regularity in some particular situations—linear differential equations and sequences of matrices.

2.1. Lyapunov exponents and their basic properties

We follow the approach developed in [**23**]. Let V be a p-dimensional real vector space. A function $\chi\colon V \to \mathbb{R} \cup \{-\infty\}$ is called a *Lyapunov characteristic exponent* or simply a *Lyapunov exponent* on V if:

(1) $\chi(\alpha v) = \chi(v)$ for each $v \in V$ and $\alpha \in \mathbb{R} \setminus \{0\}$;

(2) $\chi(v + w) \leq \max\{\chi(v), \chi(w)\}$ for each v, $w \in V$;

(3) $\chi(0) = -\infty$ (*normalization property*).

We describe some basic properties of Lyapunov exponents.

Theorem 2.1. *The following statements hold:*

(1) *if v, $w \in V$ are such that $\chi(v) \neq \chi(w)$, then*

$$\chi(v + w) = \max\{\chi(v), \chi(w)\};$$

(2) *if* $v_1, \dots, v_m \in V$ *and* $\alpha_1, \dots, \alpha_m \in \mathbb{R} \setminus \{0\}$, *then*

$$\chi(\alpha_1 v_1 + \cdots + \alpha_m v_m) \le \max\{\chi(v_i) : 1 \le i \le m\};$$

if, in addition, there exists i *such that* $\chi(v_i) > \chi(v_j)$ *for all* $j \ne i$, *then*

$$\chi(\alpha_1 v_1 + \cdots + \alpha_m v_m) = \chi(v_i);$$

(3) *if for some* $v_1, \dots, v_m \in V \setminus \{0\}$ *the numbers* $\chi(v_1), \dots, \chi(v_m)$ *are distinct, then the vectors* $v_1, \dots, v_m$ *are linearly independent;*

(4) *the function* χ *attains at most* p *distinct finite values.*

Proof. Suppose that $\chi(v) < \chi(w)$. We have

$$\chi(v + w) \le \chi(w) = \chi(v + w - v) \le \max\{\chi(v + w), \chi(v)\}.$$

It follows that if $\chi(v + w) < \chi(v)$, then $\chi(w) \le \chi(v)$, which contradicts our assumption. Hence, $\chi(v + w) \ge \chi(v)$, and thus, $\chi(v + w) = \chi(w)$. Statement (1) follows. Statement (2) is an immediate consequence of statement (1) and properties (1) and (2) in the definition of Lyapunov exponent.

In order to prove statement (3), assume on the contrary that the vectors $v_1, \dots, v_m$ are linearly dependent, i.e., $\alpha_1 v_1 + \cdots + \alpha_m v_m = 0$ with not all α_i equal to zero, while $\chi(v_1), \dots, \chi(v_m)$ are distinct. By statement (2) and property (3) in the definition of Lyapunov exponent, we obtain

$$\begin{aligned} -\infty &= \chi(\alpha_1 v_1 + \cdots + \alpha_m v_m) \\ &= \max\{\chi(v_i) : 1 \le i \le m \text{ and } \alpha_i \ne 0\} \ne -\infty. \end{aligned}$$

This contradiction implies statement (3). Statement (4) follows from statement (3). □

By Theorem 2.1, the Lyapunov exponent χ can take only finitely many distinct *values* on $V \setminus \{0\}$. We denote them by

$$\chi_1 < \cdots < \chi_s$$

for some $s \le p$. In general, χ_1 may be $-\infty$. For each $1 \le i \le s$, define

$$V_i = \{v \in V : \chi(v) \le \chi_i\}. \tag{2.1}$$

Put $V_0 = \{0\}$. It follows from Theorem 2.1 that V_i is a linear subspace of V and

$$\{0\} = V_0 \subsetneq V_1 \subsetneq \cdots \subsetneq V_s = V. \tag{2.2}$$

Any collection $\mathcal{V} = \{V_i : i = 0, \dots, s\}$ of linear subspaces of V satisfying (2.2) is called a *linear filtration* or simply a *filtration* of V.

The following result gives an equivalent characterization of Lyapunov exponents in terms of filtrations.

Theorem 2.2. *A function $\chi\colon V \to \mathbb{R} \cup \{-\infty\}$ is a Lyapunov exponent if and only if there exist numbers $\chi_1 < \cdots < \chi_s$ for some $1 \le s \le p$, and a filtration $\mathcal{V} = \{V_i : i = 0, \dots, s\}$ of V such that:*

(1) *$\chi(v) \le \chi_i$ for every $v \in V_i$;*

(2) *$\chi(v) = \chi_i$ for every $v \in V_i \setminus V_{i-1}$;*

(3) *$\chi(0) = -\infty$.*

Proof. If χ is a Lyapunov exponent, then the filtration

$$\mathcal{V} = \{V_i : i = 0, \dots, s\}$$

defined by (2.1) satisfies conditions (1) and (3) of the theorem. Moreover, for any $v \in V_i \setminus V_{i-1}$ we have $\chi_{i-1} < \chi(v) \le \chi_i$. Since χ attains no value strictly between χ_{i-1} and χ_i, we obtain $\chi(v) = \chi_i$ and condition (2) follows.

Now suppose that a function χ and filtration $\mathcal{V}$ satisfy the conditions of the theorem. Observe that $v \in V_i \setminus V_{i-1}$ if and only if $\alpha v \in V_i \setminus V_{i-1}$ for any $\alpha \in \mathbb{R} \setminus \{0\}$. Therefore, by condition (2), $\chi(\alpha v) = \chi(v)$. Now choose vectors $v_1, v_2 \in V$. Let $\chi(v_j) = \chi_{i_j}$ for $j = 1, 2$. It follows from conditions (1) and (2) that the subspace V_i can be characterized by (2.1). Therefore, $v_j \in V_{i_j}$ for $j = 1, 2$. Without loss of generality we may assume that $i_1 < i_2$. This implies that $v_1 + v_2 \in V_{i_1} \cup V_{i_2} = V_{i_2}$. Hence, by condition (1), we have

$$\chi(v_1 + v_2) \le \chi_{i_2} = \max\{\chi(v_1), \chi(v_2)\},$$

and thus, χ is a Lyapunov exponent. □

We refer to the filtration $\mathcal{V} = \{V_i : i = 0, \dots, s\}$ defined by (2.1), as the *filtration of V associated to χ* and denote it by $\mathcal{V}_\chi$. The number

$$k_i = \dim V_i - \dim V_{i-1}$$

is called the *multiplicity* of the value χ_i, and the collection of pairs

$$\operatorname{Sp} \chi = \{(\chi_i, k_i) : 1 \le i \le s\}$$

is called the *Lyapunov spectrum* of χ.

Given a filtration $\mathcal{V} = \{V_i : i = 0, \dots, s\}$ of V and numbers $\chi_1 < \cdots < \chi_s$, define a function $\chi\colon V \to \mathbb{R} \cup \{-\infty\}$ by $\chi(v) = \chi_i$ for every $v \in V_i \setminus V_{i-1}$, $1 \le i \le s$, and $\chi(0) = -\infty$. It is easy to see that the function χ and the filtration $\mathcal{V}$ satisfy the conditions of Theorem 2.2 and thus, determine a Lyapunov exponent on V.

To every filtration $\mathcal{V} = \{V_i : i = 0, \dots, s\}$ of V one can associate a special class of bases which are well adapted to the filtration. A basis $\mathbf{v} = (v_1, \dots, v_n)$ of V is said to be *subordinate* to $\mathcal{V}$ if for every $1 \le i \le s$ there exists a basis of V_i composed of n_i vectors from $\{v_1, \dots, v_n\}$.[1] A subordinate basis $\mathbf{v}$ is *ordered* if for every $1 \le i \le s$, the vectors $v_1, \dots, v_{n_i}$ form a basis of V_i.

Exercise 2.3. Show that for every filtration $\mathcal{V}$ there exists a basis that is subordinate to $\mathcal{V}$ and is ordered. Moreover, two filtrations coincide if and only if each basis that is subordinate to one of them is also subordinate to the other one.

Let $\mathcal{V}_\chi$ be the filtration associated to a Lyapunov exponent χ and let $\mathbf{v}$ be a basis that is subordinate to $\mathcal{V}_\chi$. We note that among the numbers $\chi(v_1), \dots, \chi(v_n)$ the value χ_i occurs exactly k_i times, for each $i = 1, \dots, s$. Hence,

$$\sum_{j=1}^{n} \chi(v_j) = \sum_{i=1}^{s} k_i \chi_i. \tag{2.3}$$

We use this observation to prove the following result.

Theorem 2.4. *A basis $\mathbf{v}$ is subordinate to a filtration $\mathcal{V}_\chi$ if and only if*

$$\inf \left\{ \sum_{j=1}^{n} \chi(w_j) : \mathbf{w} \text{ is a basis of } V \right\} = \sum_{j=1}^{n} \chi(v_j). \tag{2.4}$$

Proof. We begin with the following statement.

Lemma 2.5. *Let $\mathbf{v}$ be a basis that is subordinate to the filtration $\mathcal{V}_\chi$ that is ordered and let $\mathbf{w}$ be a basis of V for which $\chi(w_1) \le \dots \le \chi(w_p)$. Then:*

(1) *$\chi(w_j) \ge \chi(v_j)$ for every $1 \le j \le p$, and $\chi(w_p) = \chi(v_p)$;*

(2) *$\sum_{j=1}^{p} \chi(w_j) \ge \sum_{j=1}^{p} \chi(v_j)$;*

(3) *$\mathbf{w}$ is subordinate to the filtration $\mathcal{V}_\chi$ if and only if $\chi(w_j) = \chi(v_j)$ for every $1 \le j \le p$;*

(4) *$\mathbf{w}$ is subordinate to the filtration $\mathcal{V}_\chi$ if and only if*

$$\sum_{j=1}^{p} \chi(w_j) = \sum_{j=1}^{p} \chi(v_j).$$

Proof of the lemma. Since χ_1 is the minimal value of χ on $V \setminus \{0\}$, we have $\chi(w_j) \ge \chi(v_j) = \chi_1$ for every $j = 1, \dots, p_1$. Assume that $\chi(w_{p_1+1}) = \chi_1$. Then $\chi(w_1) = \dots = \chi(w_{p_1+1}) = \chi_1$ and

$$p_1 \ge \dim \operatorname{span}\{w_1, \dots, w_{p_1+1}\} = p_1 + 1,$$

[1]In the literature, subordinate bases are also called *normal* (see [**7**, **9**]). However, we use here the term "subordinate" since it reflects better the meaning of this notion.

where span Z denotes the linear space generated by the set of vectors in Z. This contradiction implies that $\chi(w_{p_1+1}) \geq \chi_2$ and hence, $\chi(w_j) \geq \chi(v_j) = \chi_2$ for every $j = p_1 + 1, \dots, n_2$.

Repeating the same argument finitely many times, we obtain $\chi(w_j) \geq \chi(v_j)$ for every $1 \leq j \leq p$. In particular, $\chi(w_p) = \chi(v_p)$ since $\chi(v_p)$ is the maximum value of χ. Statement (1) follows. Statement (2) is an immediate consequence of statement (1).

By (2.1), $\chi(w_j) = \chi(v_j)$ for every $1 \leq j \leq p$ if and only if $w_1, \dots, w_{p_i}$ is a basis of V_i for every $1 \leq i \leq s$ and hence, if and only if the basis $\mathbf{w}$ is subordinate to the filtration $\mathcal{V}_\chi$. This implies statement (3). The last statement is a consequence of statements (2) and (3). □

We turn to the proof of the theorem. By the lemma, the infimum in (2.4) is equal to

$$\inf \left\{ \sum_{j=1}^{p} \chi(w_j) : \mathbf{w} \text{ is subordinate to the filtration } \mathcal{V}_\chi \right\} = \sum_{i=1}^{s} k_i \chi_i.$$

In view of statement (4) of the lemma, the basis $\mathbf{v}$ is subordinate to the filtration $\mathcal{V}_\chi$ if and only if the relation (2.3) holds, and hence, if and only if the relation (2.4) holds. The theorem follows. □

Given a filtration $\mathcal{V}$, there is a useful construction of subordinate bases due to Lyapunov, which we now describe. Starting with a basis $\mathbf{v} = (v_1, \dots, v_p)$, we define a sequence of bases $\mathbf{w}_n = (w_{n1}, \dots, w_{np})$, $n \geq 0$, $\mathbf{w}_0 = \mathbf{v}$, which makes the sum $\sum_{j=1}^{p} \chi(w_{nj})$ decrease as n increases. Since χ takes on only finitely many values, this process ends up after a finite number of steps. More precisely, assume that there is a linear combination $x = \sum_{j=1}^{p} \alpha_j v_j$, with not all α_j being zero, such that

$$\chi(x) < \max \{ \chi(v_j) : 1 \leq j \leq p \text{ and } \alpha_j \neq 0 \}.$$

Choose a vector v_k for which

$$\chi(v_k) = \max \{ \chi(v_j) : 1 \leq j \leq p \text{ and } \alpha_j \neq 0 \}$$

and replace it with the vector x. Using the fact that $\alpha_k \neq 0$, we obtain that the vectors $v_1, \dots, v_{k-1}, x, v_{k+1}, \dots, v_p$ are linearly independent and hence, form a basis $\mathbf{w}_1$. Continuing in a similar fashion, we obtain a basis $\mathbf{w} = (w_1, \dots, w_p)$ with the property that for any linear combination $x = \sum_{j=1}^{p} \alpha_j w_j$, with not all α_j being zero,

$$\chi(x) = \max \{ \chi(w_j) : 1 \leq j \leq p \text{ and } \alpha_j \neq 0 \}. \tag{2.5}$$

We claim that the basis $\mathbf{w}$ is subordinate to the filtration $\mathcal{V}$. Otherwise, there exists i such that the number of vectors in the basis lying in the space

V_i is strictly less than $\dim V_i$. Hence, these vectors fail to form a basis of V_i and there is a vector $x \in V_i$ which is linearly independent of these vectors and $\chi(x) \le \chi_i$. On the other hand, we can write $x = \sum_{j=1}^{p} \alpha_j w_j$ with some $\alpha_j \ne 0$ such that the vector w_j lies outside of the space V_i. Since $\chi(w_j) > \chi_i$, this contradicts (2.5) and hence the basis $\mathbf{w}$ is subordinate to $\mathcal{V}$.

We now describe the behavior of subordinate bases under linear transformations. Let $A\colon V \to V$ be an invertible linear transformation of a vector space V. Moreover, let $\mathcal{V} = \{V_i : i = 0, \dots, s\}$ be a filtration of V and let $\mathbf{v}$ be a basis that is subordinate to $\mathcal{V}$ and is ordered.

Exercise 2.6. Show that the following properties are equivalent:

(1) the basis $(Av_1, \dots, Av_p)$ is subordinate to $\mathcal{V}$ and is ordered;

(2) the transformation A preserves the filtration $\mathcal{V}$; i.e., $AV_i = V_i$ for every $1 \le i \le s$;

(3) the transformation A, with respect to the basis $\mathbf{v}$, has the lower block-triangular form

$$\begin{pmatrix} A_1 & 0 & \cdots & 0 \\ & A_2 & \ddots & \vdots \\ & & \ddots & 0 \\ & & & A_s \end{pmatrix},$$

where each A_i is a $k_i \times k_i$ matrix with $\det A_i \ne 0$.

Let $\mathcal{V}$ and $\mathcal{W}$ be filtrations of V. As a consequence of the above statement one can show that *there exists a basis that is subordinate to both filtrations.* Indeed, let $\mathbf{v}$ be a basis that is subordinate to $\mathcal{V}$ and is ordered. There exists a lower triangular $p \times p$ matrix such that the basis $\mathbf{w} = (Av_1, \dots, Av_p)$ is subordinate to $\mathcal{W}$. It follows from Exercise 2.6 that this basis is also subordinate to $\mathcal{V}$.

2.2. The Lyapunov and Perron regularity coefficients

Let V be a vector space. Consider the dual vector space V^* to V which consists of the linear functionals on V. If $v \in V$ and $v^* \in V^*$, then $\langle v, v^* \rangle$ denotes the value of v^* on v. Let $\mathbf{v} = (v_1, \dots, v_p)$ be a basis in V and let $\mathbf{v}^* = (v_1^*, \dots, v_p^*)$ be a basis in V^*. We say that $\mathbf{v}$ is *dual* to $\mathbf{v}^*$ and write $\mathbf{v} \sim \mathbf{v}^*$ if $\langle v_i, v_j^* \rangle = v_j^*(v_i) = \delta_{ij}$ for each i and j.

Let χ be a Lyapunov exponent on V and let χ^* be a Lyapunov exponent on V^*. We say that the exponents χ and χ^* are *dual* and write $\chi \sim \chi^*$ if for any pair of dual bases $\mathbf{v}$ and $\mathbf{v}^*$ and every $1 \le i \le p$ we have

$$\chi(v_i) + \chi^*(v_i^*) \ge 0.$$

We denote by $\chi'_1 \leq \cdots \leq \chi'_p$ the values of χ counted with their multiplicities. Similarly, we denote by $\chi^{*\prime}_1 \geq \cdots \geq \chi^{*\prime}_p$ the values of χ^* counted with their multiplicities. We define:

(1) the *regularity coefficient*

$$\gamma(\chi, \chi^*) = \min \max \{\chi(v_i) + \chi^*(v_i^*) : 1 \leq i \leq p\}, \tag{2.6}$$

where the minimum is taken over all pairs of dual bases $\mathbf{v}$ and $\mathbf{v}^*$ of V and V^*;

(2) the *Perron coefficient* of χ and χ^*

$$\pi(\chi, \chi^*) = \max \{\chi'_i + \chi^{*\prime}_i : 1 \leq i \leq p\}.$$

Exercise 2.7. Show that there is always a pair of dual bases $\mathbf{v} = (v_i)$ and $\mathbf{v}^* = (v_i^*)$ for which the minimum is achieved in (2.6), i.e.,

$$\gamma(\chi, \chi^*) = \max \{\chi(v_i) + \chi^*(v_i^*) : 1 \leq i \leq p\}.$$

The following theorem establishes some relations between the two coefficients.

Theorem 2.8. *The following statements hold:*

(1) $\pi(\chi, \chi^*) \leq \gamma(\chi, \chi^*)$;

(2) *if* $\chi \sim \chi^*$, *then* $0 \leq \pi(\chi, \chi^*) \leq \gamma(\chi, \chi^*) \leq p\pi(\chi, \chi^*)$.

Proof. We begin with the following lemma.

Lemma 2.9. *Given numbers* $\lambda_1 \leq \cdots \leq \lambda_p$ *and* $\mu_1 \geq \cdots \geq \mu_p$ *and a permutation* σ *of* $\{1, \ldots, p\}$, *we have*

$$\min\{\lambda_j + \mu_{\sigma(j)} : 1 \leq j \leq p\} \leq \min\{\lambda_i + \mu_i : 1 \leq i \leq p\},$$

$$\max\{\lambda_i + \mu_{\sigma(i)} : 1 \leq i \leq p\} \geq \max\{\lambda_i + \mu_i : 1 \leq i \leq p\}.$$

Proof of the lemma. Notice that the second inequality follows from the first one in view of the following relations:

$$\begin{aligned}
\max\{\lambda_i + \mu_{\sigma(i)} : 1 \leq i \leq p\} &= -\min\{-\mu_{\sigma(i)} - \lambda_i : 1 \leq i \leq p\} \\
&= -\min\{-\mu_i - \lambda_{\sigma^{-1}(i)} : 1 \leq i \leq p\} \\
&\geq -\min\{-\mu_i - \lambda_i : 1 \leq i \leq p\}, \\
&= \max\{\lambda_i + \mu_i : 1 \leq i \leq p\}.
\end{aligned}$$

We now prove the first inequality. We may assume that σ is not the identity permutation (otherwise the result is trivial). Fix an integer i such that $1 \leq i \leq p$. If $i \leq \sigma(i)$, then $\mu_{\sigma(i)} \leq \mu_i$ and

$$\min\{\lambda_j + \mu_{\sigma(j)} : 1 \leq j \leq p\} \leq \lambda_i + \mu_{\sigma(i)} \leq \lambda_i + \mu_i.$$

If $i > \sigma(i)$, then there exists $k < i$ such that $i \le \sigma(k)$. Otherwise, we would have $\sigma(1), \dots, \sigma(i-1) \le i-1$ and hence, $\sigma(i) \ge i$. It follows that

$$\min\{\lambda_i + \mu_{\sigma(i)} : 1 \le i \le p\} \le \lambda_k + \mu_{\sigma(k)} \le \lambda_i + \mu_i.$$

The desired result now follows. □

We proceed with the proof of the theorem. Consider dual bases $\mathbf{v}$ and $\mathbf{v}^*$. Without loss of generality, we may assume that $\chi(v_1) \le \cdots \le \chi(v_p)$. Let σ be a permutation of $\{1, \dots, p\}$ such that the numbers $\mu_{\sigma(i)} = \chi^*(v_i^*)$ satisfy $\mu_1 \ge \cdots \ge \mu_p$. We have $\chi(v_i) \ge \chi_i'$ and $\mu_i \ge \chi_i^{*\prime}$. By Lemma 2.9, we obtain

$$\begin{aligned}\max\{\chi(v_i) + \chi^*(v_i^*) : 1 \le i \le p\} &\ge \max\{\chi(v_i) + \mu_i : 1 \le i \le p\} \\ &\ge \max\{\chi_i' + \chi_i^{*\prime} : 1 \le i \le p\} \\ &= \pi(\chi, \chi^*).\end{aligned}$$

Therefore, $\gamma(\chi, \chi^*) \ge \pi(\chi, \chi^*)$ and statement (1) follows.

We assume now that $\chi \sim \chi^*$. Let $\mathcal{V}_\chi = \{V_i, i = 1, \dots, s\}$ be the filtration associated to χ and let $\mathcal{V}_\chi^* = \{V_i^*, i = 1, \dots, s^*\}$ be the filtration associated to χ^*. One can choose dual bases $\mathbf{v}$ and $\mathbf{v}^*$ such that $\mathbf{v}$ is subordinate to the filtration $\mathcal{V}_\chi$ while $\mathbf{v}^*$ is subordinate to the filtration $\mathcal{V}_{\chi^*}$. Indeed, consider the filtration $\mathcal{V}_{\chi^*}^\perp$ that is comprised of the orthogonal complements $V_i^{*\perp}$ to the subspaces forming the filtration $\mathcal{V}_{\chi^*}$, i.e.,

$$V_1^{*\perp} \subset \cdots \subset V_{s^*}^{*\perp}.$$

There is a basis $\mathbf{v}$ of V which is subordinate to both filtrations $\mathcal{V}_\chi$ and $\mathcal{V}_{\chi^*}^\perp$. Then the basis $\mathbf{v}^*$ of V^* that is dual to $\mathbf{v}$ is subordinate to $\mathcal{V}_{\chi^*}$.

We assume that the basis $\mathbf{v}$ is ordered. It follows that $\chi(v_i) = \chi_i'$ and $\mu_i = \chi_i^{*\prime}$ for each i. Thus,

$$\begin{aligned}\gamma(\chi, \chi^*) &\le \max\{\chi(v_i) + \chi^*(v_i^*) : 1 \le i \le p\} \\ &\le \sum_{i=1}^p (\chi(v_i) + \chi^*(v_i^*)) = \sum_{i=1}^p (\chi_i' + \chi_i^{*\prime}) \\ &\le p \max\{\chi_i' + \chi_i^{*\prime} : 1 \le i \le p\} = p\,\pi(\chi, \chi^*).\end{aligned}$$

Finally, since $\chi \sim \chi^*$, we have $\gamma(\chi, \chi^*) \ge 0$. This implies that $\pi(\chi, \chi^*) \ge 0$, and statement (2) follows. □

We now introduce the crucial concept of regularity of a pair of Lyapunov exponents χ and χ^* in dual vector spaces V and V^*. Roughly speaking, regularity means that the filtrations $\mathcal{V}_\chi$ and $\mathcal{V}_{\chi^*}$ are well adapted to each other (in particular, they are orthogonal; see Theorem 2.10 below). This yields some special properties of Lyapunov exponents which determine their role in the stability theory. At first glance the regularity requirements seem quite strong and even a bit artificial. However, they hold in "typical" situations.

The pair of Lyapunov exponents (χ, χ^*) is said to be *regular* if $\chi \sim \chi^*$ and $\gamma(\chi, \chi^*) = 0$.[2] By Theorem 2.8, this holds if and only if $\pi(\chi, \chi^*) = 0$ and also if and only if $\chi_i^{*\prime} = -\chi_i'$.

Theorem 2.10. *If the pair (χ, χ^*) is regular, then the filtrations $\mathcal{V}_\chi$ and $\mathcal{V}_{\chi^*}$ are orthogonal, that is, $s = s^*$, $\dim V_i + \dim V_{s-i}^* = p$, and $\langle v, v^* \rangle = 0$ for every $v \in V_i$ and $v^* \in V_{s-i}^*$.*

Proof. Set $m_i = p - \dim V_{s^*-i}^* + 1$. Then

$$V_{s-i}^* = \{v^* \in V^* : \chi^*(v^*) \leq \chi_{m_i}^*{}'\}$$

and $\chi_{m_i}^*{}' = -\chi_{m_i}'$ in view of Theorem 2.8. Let $\mathbf{v}$ be a basis of V that is subordinate to $\mathcal{V}_\chi$ and is ordered, and let $\mathbf{v}^*$ be the basis of V^* that is dual to $\mathbf{v}$. Since $\chi \sim \chi^*$, we obtain

$$\begin{aligned} V_{s-i}^* &= \{v^* \in V^* : \chi_{m_i}' + \chi^*(v^*) \leq 0\} \\ &= \{v^* \in V^* : \chi(v_j) + \chi^*(v^*) < 0 \text{ if and only if } j < m_i\} \\ &= \operatorname{span}\{v_{m_i}^*, \ldots, v_p^*\}. \end{aligned}$$

This implies that the basis $\mathbf{v}^*$ is subordinate to $\mathcal{V}_{\chi^*}$ and is ordered. Fix i and consider any basis $\tilde{\mathbf{v}} = (\tilde{v}_1, \ldots, \tilde{v}_{p_i}, v_{p_i+1}, \ldots, v_p)$ of V that is subordinate to $\mathcal{V}_\chi$ and is ordered. Let $\tilde{\mathbf{v}}^*$ be the basis of V^* that is dual to $\tilde{\mathbf{v}}$. Then the last $p - p_i$ components of $\tilde{\mathbf{v}}^*$ coincide with those of $\mathbf{v}^*$. This implies that $s^* = s$ and

$$V_{s-i}^* = \operatorname{span}\{v_{n_i+1}^*, \ldots, v_p^*\} = V_i^\perp.$$

The desired result now follows. □

2.3. Lyapunov exponents for linear differential equations

Consider a linear differential equation

$$\dot{v} = A(t)v, \tag{2.7}$$

where $v(t) \in \mathbb{C}^p$ and $A(t)$ is a $p \times p$ matrix with complex entries depending continuously on $t \in \mathbb{R}$. For every $v_0 \in \mathbb{C}^p$ there exists a unique solution $v(t) = v(t, v_0)$ of equation (2.7) that is defined for every $t \in \mathbb{R}$ and satisfies the initial condition $v(0, v_0) = v_0$. We also assume that the matrix function $A(t)$ is bounded, i.e.,

$$\sup\{\|A(t)\| : t \in \mathbb{R}\} < \infty. \tag{2.8}$$

We consider the function $\chi^+ : \mathbb{C}^p \to \mathbb{R} \cup \{-\infty\}$ given by the formula

$$\chi^+(v_0) = \limsup_{t \to +\infty} \frac{1}{t} \log\|v(t)\|, \tag{2.9}$$

[2]While we adopt the traditional terminology in calling $\gamma(\chi, \chi^*)$ the regularity coefficient, in view of this definition it indeed measures the level of irregularity of the pair of Lyapunov exponents (χ, χ^*).

for each $v_0 \in \mathbb{C}^p$, where $v(t)$ is the unique solution of (2.7) satisfying the initial condition $v(0) = v_0$.

Exercise 2.11. Show that $\chi^+(v_0)$ is a Lyapunov exponent in $\mathbb{C}^p$.

By Theorem 2.1, the function χ^+ attains only finitely many distinct values $\chi_1^+ < \cdots < \chi_{s^+}^+$ on $\mathbb{C}^p \setminus \{0\}$ where $s^+ \leq p$. By (2.8), each number χ_i^+ is finite and occurs with some multiplicity k_i so that $\sum_{i=1}^{s^+} k_i = p$. We denote by $\mathcal{V}^+$ the filtration of $\mathbb{C}^p$ associated to χ^+:

$$\{0\} = V_0^+ \subsetneq V_1^+ \subsetneq \cdots \subsetneq V_{s^+}^+ = \mathbb{C}^p,$$

where

$$V_i^+ = \{v \in \mathbb{C}^p : \chi^+(v) \leq \chi_i^+\}.$$

Note that for every $\varepsilon > 0$ there exists $C_\varepsilon > 0$ such that for every solution $v(t)$ of (2.7) and any $t \geq 0$ we have

$$\|v(t)\| \leq C_\varepsilon e^{(\chi_{s^+}^+ + \varepsilon)t} \|v(0)\|.$$

In particular, if $\chi_{s^+}^+ < 0$, then the zero solution of equation (2.7) is exponentially stable.

We now discuss the regularity of the Lyapunov exponent χ^+. Consider the linear differential equation that is dual to (2.7),

$$\dot{w} = -A(t)^* w, \tag{2.10}$$

where $w(t) \in \mathbb{C}^p$ and $A(t)^*$ denotes the complex-conjugate transpose of $A(t)$. Let $w(t)$ be the unique solution of this equation with the initial condition $w(0) = w$. The function $\chi^{*+} \colon \mathbb{C}^p \to \mathbb{R} \cup \{-\infty\}$ given by

$$\chi^{*+}(w) = \limsup_{t \to +\infty} \frac{1}{t} \log\|w(t)\|$$

defines the Lyapunov exponent associated with equation (2.10). We note that the exponents χ^+ and χ^{*+} are dual. To see that, let $v(t)$ be a solution of the equation (2.7) and let $v^*(t)$ be a solution of the dual equation (2.10). Observe that for every $t \in \mathbb{R}$,

$$\begin{aligned} \frac{d}{dt}\langle v(t), v^*(t)\rangle &= \langle A(t)v(t), v^*(t)\rangle + \langle v(t), -A(t)^* v^*(t)\rangle \\ &= \langle A(t)v(t), v^*(t)\rangle - \langle A(t)v(t), v^*(t)\rangle = 0, \end{aligned}$$

where $\langle \cdot, \cdot \rangle$ denotes the standard inner product in $\mathbb{C}^p$. Hence,

$$\langle v(t), v^*(t)\rangle = \langle v(0), v^*(0)\rangle$$

for any $t \in \mathbb{R}$. Now choose dual bases $(v_1, \ldots, v_p)$ and $(v_1^*, \ldots, v_p^*)$ of $\mathbb{C}^p$. Let $v_i(t)$ be the unique solution of (2.7) such that $v_i(0) = v_i$, and let $v_i^*(t)$ be the unique solution of (2.10) such that $v_i^*(0) = v_i^*$, for each i. We obtain

$$\|v_i(t)\| \cdot \|v_i^*(t)\| \geq 1$$

for every $t \in \mathbb{R}$, and hence,

$$\chi^+(v_i) + \chi^{*+}(v_i^*) \geq 0$$

for every i. It follows that the exponents χ^+ and χ^{*+} are dual.

We discuss the regularity of the pair of Lyapunov exponents (χ^+, χ^{*+}). Let $\mathbf{v} = (v_1, \ldots, v_p)$ be a basis of $\mathbb{C}^p$. Denote by $\Gamma_m(t) = \Gamma_m^{\mathbf{v}}(t)$ the volume of the m-parallelepiped generated by the vectors $v_i(t)$, $i = 1, \ldots, m$, that are solutions of (2.7) satisfying the initial conditions $v_i(0) = v_i$. Let $V_m(t)$ be the $m \times m$ matrix whose entries are $\langle v_i(t), v_j(t) \rangle$. Then

$$\Gamma_m(t) = \Gamma_m^{\mathbf{v}}(t) = |\det V_m(t)|^{1/2}.$$

In particular, $\Gamma_1(t) = |v_1(t)|$ and $\Gamma_n(t) = \Gamma_n(0)\Delta(t)$ where

$$\Delta(t) = \exp\left(\int_0^t \operatorname{tr} A(\tau)\, d\tau\right). \tag{2.11}$$

The following theorem provides some crucial criteria for the pair (χ^+, χ^{*+}) to be regular.

Theorem 2.12. *The following statements are equivalent:*

(1) *the pair (χ^+, χ^{*+}) is regular;*

(2)

$$\lim_{t \to +\infty} \frac{1}{t} \log \Delta(t) = \sum_{i=1}^{s^+} k_i \chi_i^+; \tag{2.12}$$

(3) *for any basis $\mathbf{v}$ of $\mathbb{C}^p$ that is subordinate to $\mathcal{V}_{\chi^+}$ and is ordered and for any $1 \leq m \leq p$ the following limit exists:*

$$\lim_{t \to +\infty} \frac{1}{t} \log \Gamma_m^{\mathbf{v}}(t).$$

In addition, if the pair (χ^+, χ^{*+}) is regular, then for any basis $\mathbf{v}$ of $\mathbb{C}^p$ that is subordinate to $\mathcal{V}_{\chi^+}$ and is ordered and for any $1 \leq m \leq p$,

$$\lim_{t \to +\infty} \frac{1}{t} \log \Gamma_m^{\mathbf{v}}(t) = \sum_{i=1}^{m} \chi^+(v_i).$$

Proof. We adopt the following notation. Given $f\colon (0, \infty) \to \mathbb{R}$, we set

$$\overline{\chi}(f) = \limsup_{t \to +\infty} \frac{1}{t} \log |f(t)| \quad \text{and} \quad \underline{\chi}(f) = \liminf_{t \to +\infty} \frac{1}{t} \log |f(t)|.$$

If, in addition, f is integrable, we shall also write

$$\overline{f} = \limsup_{t \to +\infty} \frac{1}{t} \int_0^t f(\tau)\, d\tau \quad \text{and} \quad \underline{f} = \liminf_{t \to +\infty} \frac{1}{t} \int_0^t f(\tau)\, d\tau. \tag{2.13}$$

We first show that statement (1) implies statement (2). We start with an auxiliary result.

Lemma 2.13. *The following statements hold:*

(1) $\underline{\chi}(\Delta) = \operatorname{Re}\underline{\operatorname{tr} A}$ *and* $\overline{\chi}(\Delta) = \operatorname{Re}\overline{\operatorname{tr} A}$;

(2) *if* $(v_1, \dots, v_p)$ *is a basis of* $\mathbb{C}^p$, *then*

$$-\sum_{i=1}^{p} \chi^{*+}(v_i) \le \underline{\chi}(\Delta) \le \overline{\chi}(\Delta) \le \sum_{i=1}^{p} \chi^{+}(v_i).$$

Proof of the lemma. It follows from (2.11) that

$$\underline{\chi}(\Delta) = \liminf_{t\to+\infty} \frac{1}{t} \operatorname{Re} \int_0^t \operatorname{tr} A(\tau)\, d\tau = \operatorname{Re}\underline{\operatorname{tr} A}$$

and

$$\overline{\chi}(\Delta) = \limsup_{t\to+\infty} \frac{1}{t} \operatorname{Re} \int_0^t \operatorname{tr} A(\tau)\, d\tau = \operatorname{Re}\overline{\operatorname{tr} A}.$$

This proves the first statement. Since $\Gamma_p(0)\Delta(t)$ gives the volume of the parallelepiped determined by the vectors $v_1(t), \dots, v_p(t)$, we have

$$\Delta(t) \le \prod_{i=1}^{p} \|v_i(t)\|,$$

and hence,

$$\overline{\chi}(\Delta) \le \sum_{i=1}^{p} \chi^{+}(v_i).$$

In a similar way,

$$-\underline{\chi}(\Delta) = -\operatorname{Re}\underline{\operatorname{tr} A} = \operatorname{Re}\overline{\operatorname{tr}(-A^*)} = \overline{\chi}(\Delta^*) \le \sum_{i=1}^{p} \chi^{*+}(v_i),$$

where

$$\Delta^*(t) = \exp\left(-\int_0^t A(\tau)^*\, d\tau\right).$$

The lemma follows. □

We proceed with the proof of the theorem. Let χ_i' and $\chi_i^{*'}$ be the values of the Lyapunov exponents χ^+ and χ^{*+}, counted with their multiplicities. Choose a basis $(v_1, \dots, v_p)$ of $\mathbb{C}^p$ that is subordinate to $\mathcal{V}_{\chi^+}$. It follows from Lemma 2.13 that

$$-\sum_{i=1}^{p} \chi_i^{*'} \le \underline{\chi}(\Delta) \le \overline{\chi}(\Delta) \le \sum_{i=1}^{p} \chi_i'.$$

Therefore,

$$\overline{\chi}(\Delta) - \underline{\chi}(\Delta) \le \sum_{i=1}^{p} (\chi_i' + \chi_i^{*'}) \le p\,\pi(\chi^+, \chi^{*+}).$$

This shows that if the pair (χ^+, χ^{*+}) is regular, then

$$\underline{\chi}(\Delta) = \overline{\chi}(\Delta) = \sum_{i=1}^{p} \chi_i' = -\sum_{i=1}^{p} \chi_i^{*'},$$

and (2.12) holds.

We now show that statement (3) implies statement (1). We split the proof into two steps.

Step 1. For every $t \geq 0$ consider a linear coordinate change in $\mathbb{C}^p$ given by a differentiable matrix function $U(t)$. Setting $z(t) = U(t)^{-1}v(t)$, we obtain

$$\begin{aligned} \dot{v}(t) &= \dot{U}(t)z(t) + U(t)\dot{z}(t) \\ &= A(t)v(t) = A(t)U(t)z(t). \end{aligned}$$

It follows that $\dot{z} = B(t)z$ where the matrix function $B(t) = (b_{ij}(t))$ is defined by

$$B(t) = U(t)^{-1}A(t)U(t) - U(t)^{-1}\dot{U}(t). \tag{2.14}$$

We need the following lemma of Perron. Its main manifestation is to show how to reduce equation (2.7) with a general matrix function $A(t)$ to a linear differential equation with a triangular matrix function.

Lemma 2.14. *There exists a differentiable matrix function $U(t)$ such that:*

(1) *$U(t)$ is unitary;*

(2) *the matrix $B(t)$ is upper triangular;*

(3) $\sup\{|b_{ij}(t)| : t \geq 0, i \neq j\} < \infty$;

(4) *for $k = 1, \ldots, p$,*

$$\operatorname{Re} b_{kk}(t) = \frac{d}{dt} \log \frac{\Gamma_k^{\mathbf{v}}(t)}{\Gamma_{k-1}^{\mathbf{v}}(t)}.$$

Proof of the lemma. Given a basis $\mathbf{v} = (v_1, \ldots, v_p)$, we construct the desired matrix function $U(t)$ by applying the Gram–Schmidt orthogonalization procedure to the basis $\mathbf{v}(t) = (v_1(t), \ldots, v_p(t))$ where $v_i(t)$ is the solution of (2.7) satisfying the initial condition $v_i(0) = v_i$. Thus, we obtain a collection of functions $u_1(t), \ldots, u_p(t)$ such that $\langle u_i(t), u_j(t)\rangle = \delta_{ij}$ where δ_{ij} is the Kronecker symbol. Let $V(t)$ and $U(t)$ be the matrices with columns $v_1(t), \ldots, v_p(t)$ and $u_1(t), \ldots, u_p(t)$, respectively. The matrix $U(t)$ is unitary. Moreover, the Gram–Schmidt procedure can be effected in such a way that each function $u_k(t)$ is a linear combination of functions $v_1(t), \ldots, v_k(t)$. It follows that the matrix $Z(t) = U(t)^{-1}V(t)$ is upper triangular.

The columns $z_1(t) = U(t)^{-1}v_1(t), \ldots, z_p(t) = U(t)^{-1}v_p(t)$ of the matrix $Z(t)$ form a basis of the space of solutions of the linear differential equation

$\dot{z} = B(t)z$. Furthermore,

$$B(t) = \dot{Z}(t)Z(t)^{-1},$$

and as $Z(t)$ is upper triangular, so is the matrix $B(t)$. Since $U(t)$ is unitary, using (2.14) we obtain

$$\begin{aligned} B(t) + B(t)^* &= U(t)^*(A(t) + A(t)^*)U(t) - (U(t)^*\dot{U}(t) + \dot{U}(t)^*U(t)) \\ &= U(t)^*(A(t) + A(t)^*)U(t) - \frac{d}{dt}(U(t)^*U(t)) \\ &= U(t)^*(A(t) + A(t)^*)U(t). \end{aligned}$$

Since $B(t)$ is triangular, we conclude that $|b_{ij}(t)| \le 2\|A(t)\| < \infty$ uniformly over $t \ge 0$ and $i \ne j$, proving the third statement.

In order to prove the last statement of the lemma, assume first that all entries of the matrix $Z(t) = (z_{ij}(t))$ are real. Then the entries of the matrix $B(t)$ are also real and

$$b_{kk}(t) = \frac{\dot{z}_{kk}(t)}{z_{kk}(t)} = \frac{d}{dt}\log z_{kk}(t).$$

Observe that

$$v_i(t) = \sum_{1\le \ell \le i} u_\ell(t) z_{\ell i}(t).$$

Therefore,

$$\begin{aligned} \langle v_i(t), v_j(t)\rangle &= \sum_{1\le \ell\le i,\, 1\le m\le j} \delta_{\ell m} z_{\ell i}(t)\overline{z_{mj}(t)} \\ &= \sum_{1\le \ell\le \min\{i,j\}} z_{\ell i}(t)\overline{z_{\ell j}(t)} = \langle z_i(t), z_j(t)\rangle. \end{aligned}$$

Set $\mathbf{z} = (z_1(0), \dots, z_n(0))$. This implies that $\Gamma_k^{\mathbf{v}}(t) = \Gamma_k^{\mathbf{z}}(t)$ for each k and thus,

$$\frac{\Gamma_k^{\mathbf{v}}(t)}{\Gamma_{k-1}^{\mathbf{v}}(t)} = \frac{\Gamma_k^{\mathbf{z}}(t)}{\Gamma_{k-1}^{\mathbf{z}}(t)} = z_{kk}(t).$$

In the general case (when the entries of $Z(t)$ are not necessarily real) we can write

$$\frac{\Gamma_k^{\mathbf{v}}(t)}{\Gamma_{k-1}^{\mathbf{v}}(t)} = |z_{kk}(t)|$$

and

$$\begin{aligned} \frac{d}{dt}\log|z_{kk}(t)| &= \frac{1}{2}\frac{d}{dt}\log(\overline{z_{kk}(t)}z_{kk}(t)) \\ &= \frac{1}{2}\left(\frac{\overline{\dot{z}_{kk}(t)}}{\overline{z_{kk}(t)}} + \frac{\dot{z}_{kk}(t)}{z_{kk}(t)}\right) \\ &= \frac{1}{2}(\overline{b_{kk}(t)} + b_{kk}(t)) = \operatorname{Re} b_{kk}(t). \end{aligned}$$

This completes the proof of the lemma. □

We define a $p \times p$ matrix function $Z(t) = (z_{ij}(t))$ as follows: $z_{ij}(t) = 0$ if $j < i$,

$$z_{ij}(t) = \exp\left(\int_0^t b_{ii}(\tau)\, d\tau\right)$$

if $j = i$, and

$$z_{ij}(t) = \int_{a_{ij}}^t \sum_{k=i+1}^{j} b_{ik}(s) z_{kj}(s) e^{\int_s^t b_{ii}(\tau)\, d\tau} ds$$

if $j > i$.

Lemma 2.15. *For any constants a_{ij}, with $1 \le i < j \le p$, the columns of the matrix $Z(t)$ form a basis of solutions of the equation $\dot{z} = B(t)z$.*

Proof of the lemma. For each i we have $\dot{z}_{ii}(t) = b_{ii}(t) z_{ii}(t)$ and for each $j > i$,

$$\dot{z}_{ij}(t) = \sum_{k=i+1}^{j} b_{ik}(t) z_{kj}(t) + b_{ii}(t) z_{ij}(t) = \sum_{k=i}^{j} b_{ik}(t) z_{kj}(t).$$

This shows that

$$\dot{Z}(t) = B(t) Z(t)$$

and hence, the columns of $Z(t)$ (i.e., the vectors $\mathbf{z}_i(t) = (z_{1i}(t), \ldots, z_{pi}(t))$) are solutions of the equation $\dot{z} = B(t)z$. Since $Z(t)$ is upper triangular, we have

$$\det Z(t) = \exp\left(\sum_{i=1}^{n} \int_0^t b_{ii}(\tau)\, d\tau\right) \neq 0,$$

and hence, the vectors $\mathbf{z}_i(t)$ form a basis. □

Step 2. Assume that $\overline{\chi}(\Gamma_m^{\mathbf{v}}) = \underline{\chi}(\Gamma_m^{\mathbf{v}})$ for any basis $\mathbf{v}$ that is subordinate to $\mathcal{V}_{\chi^+}$ and is ordered and for any $1 \le m \le p$. We show that the pair (χ^+, χ^{*+}) is regular. By Lemma 2.14, it suffices to consider the equation $\dot{z} = B(t)z$ where $B(t)$ is a $p \times p$ upper triangular matrix for every t.

Lemma 2.16. *If $B_i := \overline{\operatorname{Re} b_{ii}} = \underline{\operatorname{Re} b_{ii}}$ for each $i = 1, \ldots, p$, then:*

(1) *the pair of Lyapunov exponents corresponding to the equations $\dot{z} = B(t)z$ and $\dot{w} = -B(t)^* w$ is regular;*

(2) *the numbers $B_1, \ldots, B_p$ are the values of the Lyapunov exponent χ^+;*

(3) *the numbers $-B_1, \ldots, -B_p$ are the values of the Lyapunov exponent χ^{*+}.*

Proof of the lemma. We consider the solutions of the equation $\dot{z} = B(t)z$ described in Lemma 2.15 and show that each column

$$z_i(t) = (z_{1i}(t), \dots, z_{ni}(t))$$

of $Z(t)$ satisfies

$$\chi^+(z_i) = \limsup_{t\to+\infty} \frac{1}{t} \log\|z_i(t)\| = B_i$$

for some choice of the constants a_{ij}. Since $\overline{\operatorname{Re} b_{ii}} = B_i$, we clearly have $\chi^+(z_{ii}) = B_i$. Assume now that $\chi^+(z_{kj}) \le B_j$ for each $i+1 \le k \le j$. We show that $\chi^+(z_{ij}) \le B_j$. Observe that for each $\varepsilon > 0$ we have

$$\begin{aligned}\chi^+(z_{ij}) &\le \limsup_{t\to+\infty} \frac{1}{t}\Bigg(\log\left|e^{\int_0^t b_{ii}(\tau)\,d\tau}\right| \\ &\quad + \log\left|\int_{a_{ij}}^t \sum_{k=i+1}^{j} b_{ik}(s) z_{kj}(s) e^{-\int_0^s b_{ii}(\tau)\,d\tau}\,ds\right|\Bigg) \\ &\le B_i + \limsup_{t\to+\infty} \frac{1}{t} \log\left|\int_{a_{ij}}^t Kpe^{(B_j - B_i + \varepsilon)s}\,ds\right|.\end{aligned}$$

We exploit here the fact that, by Lemma 2.14, $|b_{ij}(t)| \le K$ for some $K > 0$ independent of i, j, and t. For each $j > i$ set $a_{ij} = 0$ if $B_j - B_i \ge 0$ and $a_{ij} = +\infty$ if $B_j - B_i < 0$. Then for every sufficiently small $\varepsilon > 0$ we have

$$\chi^+(z_{ij}) \le B_i + \limsup_{t\to+\infty} \frac{1}{t} \log \frac{Kp(e^{(B_j - B_i + \varepsilon)t} - 1)}{B_j - B_i + \varepsilon}$$

if $B_j - B_i \ge 0$ and

$$\chi^+(z_{ij}) \le B_i + \limsup_{t\to+\infty} \frac{1}{t} \log \frac{Kpe^{(B_j - B_i + \varepsilon)t}}{B_j - B_i + \varepsilon}$$

if $B_j - B_i < 0$. Therefore,

$$\chi^+(z_{ij}) \le B_i + B_j - B_i + \varepsilon = B_j + \varepsilon.$$

Since ε is arbitrary, we obtain $\chi^+(z_{ij}) \le B_j$. This shows that $\chi^+(z_i) = B_i$ for each $1 \le i \le p$.

In a similar way, one can show that there exists a lower triangular matrix $W(t)$ such that

$$\dot{W}(t) = -B(t)^* W(t).$$

The entries of the matrix $W(t)$ are defined by $w_{ij}(t) = 0$ if $j > i$,

$$w_{ij}(t) = \exp\left(-\int_0^t \overline{b_{jj}(\tau)}\,d\tau\right)$$

if $j = i$, and

$$w_{ij}(t) = -\int_{a_{ji}}^{t} \sum_{k=j}^{i-1} \overline{b_{ki}(s)} w_{kj}(s) e^{-\int_s^t \overline{b_{ii}(\tau)}\, d\tau}\, ds$$

if $j < i$, where the constants a_{ji} are chosen as above. Since $\operatorname{Re} b_{ii} = B_i$, the columns $w_1(t), \ldots, w_p(t)$ of $W(t)$ satisfy

$$\chi^{*+}(w_i) = \limsup_{t \to +\infty} \frac{1}{t} \log\|w_i(t)\| = -B_i = -\chi^+(z_i).$$

Note that $\chi^+(z_i) + \chi^{*+}(w_i) = 0$ for each i. In order to prove that the pair (χ^+, χ^{*+}) is regular, it remains to show that the bases $\mathbf{z}$ and $\mathbf{w}$ are dual. Clearly, $\langle z_i(0), w_j(0)\rangle = 0$ for every $i < j$. Moreover, $\langle z_i(0), w_j(0)\rangle = 1$ for each $1 \le i \le p$. Fix $i > j$ and $t > 0$. We have that

$$\langle z_i(t), w_j(t)\rangle = \sum_{k=j}^{i} z_{ki}(t)\overline{w_{kj}(t)}. \tag{2.15}$$

Since $\chi^+(z_{ij}) \le B_j$ and $\chi^{*+}(w_{ij}) \le -B_j$ for every i, j and $\varepsilon > 0$, we obtain

$$\begin{aligned}
\overline{\chi}(\langle z_i(0), w_j(0)\rangle) &\le \max_{j+1 \le k \le i-1} \overline{\chi}(z_{ki}\overline{w_{kj}}) \\
&\le \max_{j+1 \le k \le i-1} \limsup_{t \to +\infty} \frac{1}{t}\left(\log\left|\int_{a_{ki}}^{t} Kpe^{(B_i - B_k + \varepsilon)s}\, ds\right|\right. \\
&\quad \left. + \log\left|\int_{a_{jk}}^{t} Kpe^{(-B_j + B_k + \varepsilon)s}\, ds\right|\right) \\
&\le \max_{j+1 \le k \le i-1} (B_i - B_k - B_j + B_k + 2\varepsilon) = B_i - B_j + 2\varepsilon.
\end{aligned}$$

Since ε is arbitrary, if $B_i - B_j < 0$, we obtain

$$\overline{\chi}(\langle z_i(0), w_j(0)\rangle) < 0$$

and

$$\langle z_i(0), w_j(0)\rangle = \lim_{t \to +\infty} \langle z_i(t), w_j(t)\rangle = 0.$$

If $B_i - B_j \ge 0$, then $a_{ji} = 0$. Moreover, for each k we have $B_i - B_k \ge 0$ or $B_k - B_j \ge 0$, and hence, $a_{ki} = 0$ or $a_{jk} = 0$. Letting $t \to 0$ in (2.15) yields

$$\langle z_i(0), w_j(0)\rangle = z_{ji}(0)\overline{w_{jj}(0)} + z_{ii}(0)\overline{w_{ij}(0)} + \sum_{k=j+1}^{i-1} z_{ki}(0)\overline{w_{kj}(0)}.$$

Since $i > j$ and $a_{ji} = 0$, we obtain $z_{ji}(0) = w_{ij}(0) = 0$. Moreover, for each k such that $j + 1 \le k \le i - 1$ we have $a_{ki} = 0$ or $a_{jk} = 0$, and hence, $z_{ki}(0) = 0$ or $w_{kj}(0) = 0$. Therefore, each term in the above sum is zero. Thus, $\langle z_i(0), w_j(0)\rangle = 0$, and the lemma follows. □

By Lemma 2.14 and statement (3) of the theorem, we have

$$\begin{aligned}\frac{1}{t}\int_0^t \operatorname{Re} b_{ii}(\tau)\,d\tau &= \frac{1}{t}\int_0^t \frac{d}{d\tau}\log\frac{\Gamma_i^{\mathbf{v}}(t)}{\Gamma_{i-1}^{\mathbf{v}}(t)}\,d\tau \\ &= \frac{1}{t}\log\frac{\Gamma_i^{\mathbf{v}}(t)/\Gamma_{i-1}^{\mathbf{v}}(t)}{\Gamma_i^{\mathbf{v}}(0)/\Gamma_{i-1}^{\mathbf{v}}(0)} \to \overline{\chi}(\Gamma_i^{\mathbf{v}}) - \overline{\chi}(\Gamma_{i-1}^{\mathbf{v}})\end{aligned}$$

as $t \to +\infty$. Since $\overline{\chi}(\Gamma_i^{\mathbf{v}}) = \underline{\chi}(\Gamma_i^{\mathbf{v}})$, this implies that

$$\begin{aligned}\overline{\operatorname{Re} b_{ii}} = \limsup_{t\to\infty}\frac{1}{t}\int_0^t \operatorname{Re} b_{ii}(\tau)\,d\tau &= \overline{\chi}(\Gamma_i^{\mathbf{v}}) - \overline{\chi}(\Gamma_{i-1}^{\mathbf{v}}) = \underline{\chi}(\Gamma_i^{\mathbf{v}}) - \underline{\chi}(\Gamma_{i-1}^{\mathbf{v}}) \\ &= \liminf_{t\to\infty}\frac{1}{t}\int_0^t \operatorname{Re} b_{ii}(\tau)\,d\tau = \underline{\operatorname{Re} b_{ii}}.\end{aligned}$$

We can therefore apply Lemma 2.16 and conclude that the Lyapunov exponent corresponding to the equation $\dot{z} = B(t)z$ is regular.

We now show that statement (2) implies statement (3). By Lemma 2.14, for every $1 \le m \le p$ we have

$$\Gamma_m^{\mathbf{v}}(t) = \Gamma_m^{\mathbf{z}}(t) = \prod_{k=1}^m z_{kk}(t). \tag{2.16}$$

By Lyapunov's construction of subordinate bases (see the description of the construction after Theorem 2.4), there exists a basis $(v_1(0), \dots, v_p(0))$ of $\mathbb{C}^p$ that is subordinate to $\mathcal{V}_{\chi^+}$ such that $z_k(0) = e_k + f_k$ for some $f_k \in \operatorname{span}\{e_1, \dots, e_{k-1}\}$ where $(e_1, \dots, e_p)$ is the canonical basis of $\mathbb{C}^p$. Since the matrix solution $Z(t)$ is upper triangular for each t, the vectors e_k and $Z(t)f_k$ are orthogonal. Therefore, $\|Z(t)f_k\| \le |z_{kk}(t)|$, and

$$\chi^+(z_k) \ge \limsup_{t\to\infty}\frac{1}{t}\log|z_{kk}(t)| =: \lambda_k. \tag{2.17}$$

Without loss of generality we may use the norm in $\mathbb{C}^p$ given by

$$\|(w_1, \dots, w_n)\| = |w_1| + \cdots + |w_p|.$$

By (2.17), we obtain

$$\lim_{t\to\infty}\frac{1}{t}\log\Gamma_p^{\mathbf{v}}(t) = \sum_{i=1}^s k_i\chi_i^+ \ge \sum_{k=1}^p \chi^+(z_k) \ge \sum_{k=1}^p \lambda_k. \tag{2.18}$$

Furthermore, by (2.11) and Lemma 2.15, we have

$$\Gamma_p^{\mathbf{z}}(t)/\Gamma_p^{\mathbf{z}}(0) = \exp\left(\int_0^t \operatorname{tr} B(\tau)\,d\tau\right) = \prod_{k=1}^p z_{kk}(t),$$

and hence, by (2.16),

$$\lim_{t\to\infty}\frac{1}{t}\log\Gamma_p^{\mathbf{v}}(t)\le\sum_{k=1}^{p}\lambda_k. \tag{2.19}$$

It follows from (2.17), (2.18), and (2.19) that

$$\chi^+(v_k)=\chi^+(z_k)=\lim_{t\to+\infty}\frac{1}{t}\log\|z_k(t)\|=\lambda_k$$

for each $k=1,\dots,p$. Thus, again by (2.16), for each $1\le m\le p$ we conclude that

$$\underline{\chi}(\Gamma_m^{\mathbf{v}})=\overline{\chi}(\Gamma_m^{\mathbf{v}})=\sum_{k=1}^{m}\chi^+(v_k).$$

This completes the proof of Theorem 2.12. □

2.4. Forward and backward regularity. The Lyapunov–Perron regularity

We stress that the relation (2.12) includes two requirements: (1) the limit $\lim_{t\to+\infty}\frac{1}{t}\log\Delta(t)$ exists and (2) it is equal to $\sum_{i=1}^{s^+}k_i\chi_i^+$.

The following example illustrates that the second requirement cannot be dropped. Consider the system of differential equations

$$\dot{v}_1=-p(t)v_1,\quad \dot{v}_2=p(t)v_2$$

for $t>0$ where

$$p(t)=\cos\log t-\sin\log t-1.$$

Exercise 2.17. Show that the general solution of the system can be written in the form

$$v_1(t)=C_1q(t)^{-1},\quad v_2(t)=C_2q(t),$$

for some constants C_1 and C_2, where

$$q(t)=\exp\left(\int_1^t p(\tau)\,d\tau\right)=\exp(t(\cos\log t-1)).$$

Observe that $\Delta(t)=1$ for every t, and hence, the first requirement holds.

Exercise 2.18. Show that

$$\liminf_{t\to+\infty}\frac{1}{t}\int_1^t p(\tau)\,d\tau=-2\quad\text{and}\quad\limsup_{t\to+\infty}\frac{1}{t}\int_1^t p(\tau)\,d\tau=0 \tag{2.20}$$

and hence, the limit of 1-volumes does not exist.

The last exercise shows that the pair of Lyapunov exponents (χ^+, χ^{*+}) is not regular. Note that for every $v = (v_1, v_2)$ with $v_1 \neq 0$ the Lyapunov exponent is $\chi^+(v) = 2$ and $\chi^+(v) = 0$ otherwise. Hence,

$$0 = \lim_{t \to +\infty} \frac{1}{t} \log \Delta(t) < \chi_1^+ + \chi_2^+ = 2.$$

We now illustrate that the pair (χ^+, χ^{*+}) may not be regular if the limit in (2.12) does not exist. Consider the system of differential equations

$$\dot{v}_1 = v_2, \quad \dot{v}_2 = p(t)v_2$$

for $t > 0$ with the same function $p(t)$ as above.

Exercise 2.19. Show that the general solution of the system can be written in the form

$$v_1(t) = C_1 + C_2 \int_1^t q(\tau)\, d\tau, \quad v_2(t) = C_2 q(t)$$

for some constants C_1 and C_2. Show that for every vector $v = (v_1, v_2) \neq 0$,

$$\chi^+(v) = \limsup_{t \to +\infty} \frac{1}{t} \log\|v(t)\| = 0.$$

By (2.11), we obtain that $\Delta(t) = \exp(\int_0^t p(\tau)\, d\tau)$, and hence,

$$\limsup_{t \to +\infty} \frac{1}{t} \log \Delta(t) = 0 = \chi_1^+ + \chi_2^+.$$

On the other hand, it follows from (2.20) that the limit in (2.12) does not exist, and hence, the pair of Lyapunov exponents (χ^+, χ^{*+}) is not regular.

An important manifestation of Theorem 2.12 is that one can verify the regularity property of the pair of Lyapunov exponents (χ^+, χ^{*+}) dealing with the Lyapunov exponent χ^+ only. This justifies calling the Lyapunov exponent χ^+ *forward regular* (to stress that we only allow positive time) if the pair of Lyapunov exponents (χ^+, χ^{*+}) is regular.

In an analogous manner, reversing the time, we introduce the Lyapunov exponent $\chi^- \colon \mathbb{C}^p \to \mathbb{R} \cup \{-\infty\}$,

$$\chi^-(v) = \limsup_{t \to -\infty} \frac{1}{|t|} \log \|v(t)\|,$$

where $v(t)$ is the solution of (2.7) satisfying the initial condition $v(0) = v$. The function χ^- takes on only finitely many values $\chi_1^- > \cdots > \chi_{s^-}^-$ where $s^- \leq p$. We denote by $\mathcal{V}^-$ the filtration of $\mathbb{C}^p$ associated to χ^-,

$$\mathbb{C}^p = V_0^- \supsetneq V_1^- \supsetneq \cdots \supsetneq V_{s^-}^- \supsetneq V_{s^-+1}^- = \{0\},$$

where

$$V_i^- = \{v \in \mathbb{C}^p : \chi^-(v) \leq \chi_i^-\},$$

and by $k_i^- = \dim V_i^- - \dim V_{i+1}^-$ the *multiplicity* of the value χ_i^- such that $\sum_{i=1}^{s^-} k_i^- = p$.

Consider the dual Lyapunov exponent $\chi^{*-}\colon \mathbb{C}^p \to \mathbb{R} \cup \{-\infty\}$ given by

$$\chi^{*-}(v^*) = \limsup_{t\to-\infty} \frac{1}{|t|} \log \|v^*(t)\|,$$

where $v^*(t)$ is the solution of the dual equation (2.10) satisfying the initial condition $v^*(0) = v^*$. We say that the Lyapunov exponent χ^- is *backward regular* if the pair of Lyapunov exponents (χ^-, χ^{*-}) is regular. Reversing the time in Theorem 2.12, one can verify the regularity of the pair $(\chi^-, \tilde{\chi}^-)$ dealing only with the Lyapunov exponent χ^-.

We now introduce the crucial concept of Lyapunov–Perron regularity that substantially strengthens the notions of forward and backward regularity. We say that the filtrations $\mathcal{V}^+$ and $\mathcal{V}^-$ are *coherent* if the following properties hold:

(1) $s^+ = s^- =: s$;

(2) there exists a decomposition

$$\mathbb{C}^p = \bigoplus_{i=1}^{s} E_i \tag{2.21}$$

into subspaces E_i such that

$$V_i^+ = \bigoplus_{j=1}^{i} E_j \quad \text{and} \quad V_i^- = \bigoplus_{j=i}^{s} E_j;$$

(3) $\chi_i^+ = -\chi_i^- =: \chi_i$;

(4) if $v \in E_i \setminus \{0\}$, then

$$\lim_{t\to\pm\infty} \frac{1}{t} \log \|v(t)\| = \chi_i$$

with uniform convergence on $\{v \in E_i : \|v\| = 1\}$ (recall that $v(t)$ is the solution of equation (2.7) with initial condition $v(0) = v$).

The decomposition (2.21) is called the *Oseledets decomposition* associated with the Lyapunov exponent χ^+ (or with the pair of Lyapunov exponents (χ^+, χ^-)).

We say that the Lyapunov exponent χ^+ is *Lyapunov–Perron regular* or simply *LP-regular* if the exponent χ^+ is forward regular, the exponent χ^- is backward regular, and the filtrations $\mathcal{V}^+$ and $\mathcal{V}^-$ are coherent.

Remark 2.20. We stress that simultaneous forward and backward regularity of the Lyapunov exponents does not imply the LP-regularity. Roughly speaking, whether the Lyapunov exponent is forward (respectively, backward) regular may not depend on the backward (respectively, forward) behavior of solutions of the system; i.e., the forward behavior of the system may "know nothing" about its backward behavior. To illustrate this, consider a flow φ_t of the sphere with the north pole a repelling fixed point with the rate of expansion $\lambda > 0$ and the south pole an attracting fixed point with the rate of contraction $\mu < 0$, so that every trajectory moves from the north pole to the south pole (see Figure 2.1). It is easy to see that every point of the flow is simultaneously forward and backward regular but if $\lambda \neq -\mu$ none of the points is LP-regular except for the north and south poles.

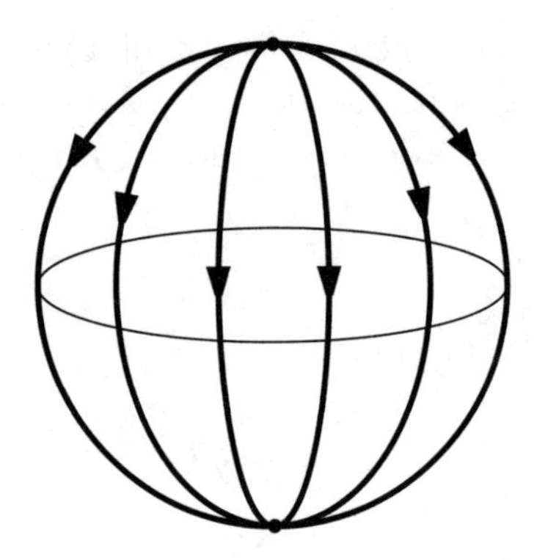

Figure 2.1. Flow φ_t

The LP-regularity requires some compatibility between the forward and backward behavior of solutions which is expressed in terms of the filtrations $\mathcal{V}^+$ and $\mathcal{V}^-$. Such compatibility can be expected if the trajectory of the flow is infinitely recurrent, i.e., it returns infinitely often to an arbitrarily small neighborhood of the initial point. This type of behavior occurs for trajectories that are typical with respect to an invariant measure for the flow due to the Poincaré Recurrence Theorem. While the recurrence property alone does not guarantee the LP-regularity, it turns out that given an invariant measure, almost every trajectory is LP-regular. This is due to the celebrated result by Oseledets known as the Multiplicative Ergodic Theorem (see Theorem 6.1).

Theorem 2.21. *The Lyapunov exponent χ^+ is LP-regular if and only if there exists a decomposition*

$$\mathbb{C}^p = \bigoplus_{i=1}^{s} E_i$$

and numbers $\chi_1 < \cdots < \chi_s$ *such that:*

(1) *if* $i = 1, \ldots, s$ *and* $v \in E_i \setminus \{0\}$, *then*

$$\lim_{t\to\pm\infty} \frac{1}{t} \log\|v(t)\| = \chi_i;$$

(2)

$$\lim_{t\to\pm\infty} \frac{1}{t} \log \Delta(t) = \sum_{i=1}^{s} \chi_i \dim E_i.$$

In addition, if the Lyapunov exponent χ^+ *is LP-regular, then the following statements hold:*

(1) *for any collection of vectors* $\mathbf{v} = (v_1, \ldots, v_k)$ *the limits*

$$\lim_{t\to\pm\infty} \frac{1}{t} \log \Gamma_k^{\mathbf{v}}(t)$$

exist and $\chi_i^+ = -\chi_i^- =: \chi_i$;

(2) $E_i = V_i^+ \cap V_i^-$ *and* $\dim E_i = k_i^+ = k_i^-$;

(3) *if* $\mathbf{v} = (v_1, \ldots, v_{k_i})$ *is a basis of* E_i, *then*

$$\lim_{t\to\pm\infty} \frac{1}{t} \log \Gamma_{k_i}^{\mathbf{v}}(t) = \chi_i k_i;$$

(4) *if* $v_i(t)$ *and* $v_j(t)$ *are solutions of equation* (2.7) *such that* $v_i(0) \in E_i \setminus \{0\}$ *and* $v_j(0) \in E_j \setminus \{0\}$ *with* $i \neq j$, *then*

$$\lim_{t\to\pm\infty} \frac{1}{t} \log \angle(v_i(t), v_j(t)) = 0.$$

Proof. All statement except the last one follow from Theorem 2.12. In order to prove the last statement, let $\Gamma(t)$ denote the area of the rectangle formed by the vectors $v_i(t)$ and $v_j(t)$. We have

$$\Gamma(t) = \|v_i(t)\| \cdot \|v_j(t)\| \sin \angle(v_i(t), v_j(t)).$$

Since $\sin \angle(v_i(t), v_j(t)) \leq 1$, we obtain

$$\begin{aligned}
\chi^+(v_i) + \chi^+(v_j) &= \lim_{t\to+\infty} \frac{1}{t} \log \Gamma(t) \\
&\leq \chi^+(v_i) + \chi^+(v_j) + \liminf_{t\to+\infty} \frac{1}{t} \log \sin \angle(v_i(t), v_j(t)) \\
&\leq \chi^+(v_i) + \chi^+(v_j) + \limsup_{t\to+\infty} \frac{1}{t} \log \sin \angle(v_i(t), v_j(t)) \\
&\leq \chi^+(v_i) + \chi^+(v_j).
\end{aligned}$$

A similar argument applies to the exponent χ^- and the desired result follows.

□

We describe an example of a nonautonomous linear differential equation whose Lyapunov exponent is both forward and backward regular but is *not* LP-regular. Consider the system of equations

$$\dot{v}_1 = a(t)v_1, \quad \dot{v}_2 = a(-t)v_2$$

where $a\colon \mathbb{R} \to \mathbb{R}$ is a bounded continuous function such that $a(t) \to a_+$ as $t \to +\infty$ and $a(t) \to a_-$ as $t \to -\infty$, for some constants $a_+ \neq a_-$.

Exercise 2.22. Show that the values of the exponents χ^+ and χ^- coincide (up to the change of sign), but the filtrations $\mathcal{V}^+$ and $\mathcal{V}^-$ are not coherent.

2.5. Lyapunov exponents for sequences of matrices

In this section we study Lyapunov exponents associated with sequences of matrices, which can be viewed as the discrete-time version of the matrix functions we dealt with in the previous sections. Consider a sequence of invertible $p \times p$ matrices $(A_m)_{m\in\mathbb{N}}$ with entries in $\mathbb{C}$ such that

$$\sup\{\|A_m\| : m \in \mathbb{N}\} < \infty.$$

Set

$$\mathcal{A}(m,n) = \begin{cases} A_{m-1}\cdots A_n, & m > n, \\ \mathrm{Id}, & m = n. \end{cases}$$

Consider the function $\chi^+\colon \mathbb{C}^p \to \mathbb{R} \cup \{-\infty\}$ defined by

$$\chi^+(v) = \limsup_{m\to+\infty} \frac{1}{m} \log \|\mathcal{A}(m,1)v\|.$$

Exercise 2.23. Show that χ^+ is a Lyapunov exponent in $\mathbb{C}^p$.

By Theorem 2.1, the function χ^+ attains only finitely many distinct values $\chi_1^+ < \cdots < \chi_s^+$ on $\mathbb{C}^p \setminus \{0\}$ where $s \leq p$. By (2.8), each number χ_i^+ is finite and occurs with some multiplicity k_i so that $\sum_{i=1}^s k_i = p$.

We consider now the sequence of dual matrices $B_m = (A_m^*)^{-1}$ and set

$$\mathcal{B}(m,n) = \begin{cases} (B_{m-1})^{-1}\cdots(B_n)^{-1}, & m > n, \\ \mathrm{Id}, & m = n. \end{cases}$$

The function $\chi^{*+}\colon \mathbb{C}^p \to \mathbb{R} \cup \{-\infty\}$ defined by

$$\chi^{*+}(v^*) = \limsup_{m\to+\infty} \frac{1}{m} \log \|\mathcal{B}(m,1)v^*\|$$

is a dual Lyapunov exponent. To see this, let us choose dual bases $(v_1, \ldots, v_p)$ and $(v_1^*, \ldots, v_p^*)$ of $\mathbb{C}^p$ and set $v_{im} = \mathcal{A}(m,1)v_i$ and $v_{im}^* = \mathcal{B}(m,1)v_i^*$. For every $m \in \mathbb{N}$ we have

$$\langle v_{im}, v_{im}^*\rangle = \langle \mathcal{A}(m,1)v_i, (\mathcal{A}(m,1)^*)^{-1}v_i^*\rangle = \langle v_i, v_i^*\rangle = 1.$$

Therefore, $1 \leq \|\mathcal{A}(m,1)v_i\| \cdot \|\mathcal{B}(m,1)v_i^*\|$ and the exponents χ^+ and χ^{*+} are dual.

We discuss the regularity of the pair of Lyapunov exponents (χ^+, χ^{*+}). Let $\mathbf{v} = (v_1, \ldots, v_p)$ be a basis of $\mathbb{C}^p$. Denote by $\Gamma_\ell(m) = \Gamma_\ell^{\mathbf{v}}(m)$ the volume of the ℓ-parallelepiped generated by the vectors $v_{im} = \mathcal{A}(m,1)v_i$, $i = 1, \ldots, \ell$. The following result is a discrete-time version of Theorem 2.12.

Theorem 2.24. *Assume that the pair of Lyapunov exponents* (χ^+, χ^{*+}) *is regular. Then:*

(1)

$$\lim_{m\to+\infty} \frac{1}{m} \log|\det \mathcal{A}(m,1)| = \sum_{i=1}^{s} k_i \chi_i^+; \tag{2.22}$$

(2) *for any basis* $\mathbf{v}$ *of* $\mathbb{C}^p$ *that is subordinate to the filtration* $\mathcal{V}_{\chi^+}$ *and any subset* $K \subset \{1, \ldots, n\}$,

$$\chi^+(\{v_i\}_{i\in K}) = \lim_{m\to+\infty} \frac{1}{m} \log \operatorname{vol}(\{\mathcal{A}(m,1)v_i\}_{i\in K}) = \sum_{i\in K} \chi^+(v_i)$$

and the angle σ_m *between the subspaces* $\operatorname{span}\{\mathcal{A}(m,1)v_i : i \in K\}$ *and* $\operatorname{span}\{\mathcal{A}(m,1)v_i : i \notin K\}$ *satisfies*

$$\lim_{m\to+\infty} \frac{1}{m} \log \sin \sigma_m = 0;$$

(3) *for any basis* $\mathbf{v}$ *of* $\mathbb{C}^p$ *that is subordinate to* $\mathcal{V}_{\chi^+}$ *and is ordered and for any* $1 \leq \ell \leq p$,

$$\lim_{m\to+\infty} \frac{1}{m} \log \Gamma_\ell^{\mathbf{v}}(m) = \sum_{i=1}^{\ell} \chi^+(v_i).$$

Proof. The following exercise shows that if the pair of Lyapunov exponents (χ^+, χ^{*+}) is regular, then the relation (2.22) holds.

Exercise 2.25. Show that for any basis $(v_1, \ldots, v_p)$ of $\mathbb{C}^p$ we have

$$-\sum_{i=1}^{p} \chi^{*+}(v_i) \leq \underline{\chi}(\det \mathcal{A}) \leq \overline{\chi}(\det \mathcal{A}) \leq \sum_{i=1}^{p} \chi^+(v_i),$$

where

$$\overline{\chi}(\det \mathcal{A}) = \limsup_{m\to+\infty} \frac{1}{m} \log|\det \mathcal{A}(m,1)|$$

and

$$\underline{\chi}(\det \mathcal{A}) = \liminf_{m\to+\infty} \frac{1}{m} \log|\det \mathcal{A}(m,1)|.$$

Use these inequalities to prove statement (1).

We now prove that statement (1) implies statement (2). Choose any basis $\{v_1, \dots, v_p\}$ of $\mathbb{C}^p$ and observe that

$$\begin{aligned} \operatorname{vol}(\mathcal{A}(m,1)v_1, \dots, \mathcal{A}(m,1)v_p) &= \operatorname{vol}(\{\mathcal{A}(m,1)v_i\}_{i\in K}) \\ &\quad \times \operatorname{vol}(\{\mathcal{A}(m,1)v_i\}_{i\notin K}) \sin\sigma_m. \end{aligned} \tag{2.23}$$

Given a collection of vectors $v_1, \dots, v_k$, we define

$$\chi^+(v_1, \dots, v_k) = \limsup_{m\to+\infty} \frac{1}{m} \log \operatorname{vol}(\mathcal{A}(m,1)v_1, \dots, \mathcal{A}(m,1)v_k).$$

Exercise 2.26. Show that the relation (2.22) implies that

$$\sum_{i=1}^{p} \chi^+(v_i) = \chi^+(v_1, \dots, v_p). \tag{2.24}$$

Since $\operatorname{vol}(v_1, \dots, v_k) \le \prod_{i=1}^{k} \|v_i\|$, we obtain that

$$\chi^+(v_1, \dots, v_k) \le \sum_{i=1}^{k} \chi^+(v_i).$$

This implies that

$$\chi^+(\{v_i\}_{i\in K}) \le \sum_{i\in K} \chi^+(v_i), \quad \chi^+(\{v_i\}_{i\notin K}) \le \sum_{i\notin K} \chi^+(v_i), \tag{2.25}$$

and hence,

$$\chi^+(v_1, \dots, v_p) \le \sum_{i=1}^{p} \chi^+(v_i) + \liminf_{m\to+\infty} \frac{1}{m} \log\sin\sigma_m.$$

Therefore, in view of (2.24) we have that

$$\liminf_{m\to+\infty} \frac{1}{m} \log\sin\sigma_m \ge 0.$$

Since $\sin\sigma_m$ is bounded, we conclude that

$$\lim_{m\to+\infty} \frac{1}{m} \log\sin\sigma_m = 0.$$

Therefore using (2.23), we find that

$$\begin{aligned} \sum_{i\in K} \chi^+(v_i) + \sum_{i\notin K} \chi^+(v_i) &= \chi^+(v_1, \dots, v_n) \\ &\le \chi^+(\{v_i\}_{i\in K}) + \chi^+(\{v_i\}_{i\notin K}). \end{aligned}$$

It follows from (2.25) that

$$\chi^+(\{v_i\}_{i\in K}) = \sum_{i\in K} \chi^+(v_i), \quad \chi^+(\{v_i\}_{i\notin K}) = \sum_{i\notin K} \chi^+(v_i).$$

This means that the relation (2.22) implies statement (2) of the theorem.

Exercise 2.27. Show that statement (2) of the theorem implies statement (3).

This exercise completes the proof of the theorem. □

One can show that the relation (2.22) as well as each of the other two statements of the theorem are indeed equivalent to the regularity of the pair of Lyapunov exponents (χ^+, χ^{*+}). Hence, the regularity is completely determined by the Lyapunov exponent χ^+ alone. Therefore, similarly to the continuous time case we call the Lyapunov exponent χ^+ *forward regular* if the pair of Lyapunov exponents (χ^+, χ^{*+}) is regular. Reversing the time, we introduce the Lyapunov exponent $\chi^- \colon \mathbb{C}^p \to \mathbb{R} \cup \{-\infty\}$ by

$$\chi^-(v) = \limsup_{m \to -\infty} \frac{1}{|m|} \log \|\mathcal{A}(m, 1)v\|.$$

The function χ^- attains finitely many values $\chi_1^- > \cdots > \chi_{s^-}^-$ where $s^- \le p$. Let $\mathcal{V}^-$ be the filtration of $\mathbb{C}^p$ associated to χ^-,

$$\mathbb{C}^p = V_0^- \supsetneqq V_1^- \supsetneqq \cdots \supsetneqq V_{s^-}^- \supsetneqq V_{s^-+1}^- = \{0\},$$

where

$$V_i^- = \{v \in \mathbb{C}^p : \chi^-(v) \le \chi_i^-\},$$

and let $k_i^- = \dim V_i^- - \dim V_{i+1}^-$ be the *multiplicity* of the value χ_i^- such that $\sum_{i=1}^{s^-} k_i^- = p$.

Consider the dual Lyapunov exponent $\chi^{*-} \colon \mathbb{C}^p \to \mathbb{R} \cup \{-\infty\}$ given by

$$\chi^{*-}(v^*) = \limsup_{m \infty} \frac{1}{|m|} \log \|\mathcal{A}(m, 1)^* v\|.$$

The regularity of the pair of Lyapunov exponents (χ^-, χ^{*-}) is completely determined by the Lyapunov exponent χ^-, thus justifying calling the Lyapunov exponent χ^- *backward regular* if the pair of Lyapunov exponents (χ^-, χ^{*-}) is regular.

We say that the Lyapunov exponent χ^+ is *Lyapunov–Perron regular* or simply *LP-regular* if the exponent χ^+ is forward regular, the exponent χ^- is backward regular, and the filtrations $\mathcal{V}^+$ and $\mathcal{V}^-$ are coherent (see Section 2.4 and Remark 2.20).

We stress that the pure existence of the limit in (2.22) (without the requirement that it be equal to the sum of the Lyapunov exponents) does *not* guarantee that the pair (χ^+, χ^{*+}) is regular.

Example 2.28. Let $\mathcal{A}^+ = A$ be the sequence of matrices where $A_0 = \left(\begin{smallmatrix} 1 & 0 \\ 2 & 4 \end{smallmatrix}\right)$ and $A_m = \left(\begin{smallmatrix} 1 & 0 \\ -2^{m+1} & 4 \end{smallmatrix}\right)$ for each $m \ge 1$ so that $\mathcal{A}(m, 1) = \left(\begin{smallmatrix} 1 & 0 \\ 2^m & 4^m \end{smallmatrix}\right)$ for every $m \ge 1$. Given a vector $v = (a, b) \ne (0, 0)$, we have $\chi^+(v) = \log 2$ if $b = 0$ and

$\chi^+(v) = \log 4$ if $b \neq 0$. Let $v_1 = (1,0)$ and $v_2 = (0,1)$. Then $\chi^+(v_1) = \log 2$ and $\chi^+(v_2) = \log 4$. Since $\det \mathcal{A}(m,1) = 4^m$, there exists the limit

$$\chi^+(v_1, v_2) = \lim_{m\to+\infty} \frac{1}{m} \log|\det \mathcal{A}(m,1)| = \log 4.$$

On the other hand,

$$\log 4 < \log 2 + \log 4 = \chi^+(v_1) + \chi^+(v_2)$$

and the pair (χ^+, χ^{*+}) is not regular.

In the one-dimensional case the situation is different.

Exercise 2.29. Consider a sequence of real numbers (i.e., a sequence of one-dimensional matrices) $(A_m)_{m\in\mathbb{N}} \subset GL(1,\mathbb{R}) = \mathbb{R} \setminus \{0\}$. Show that the pair of Lyapunov exponents (χ^+, χ^{*+}) associated to this sequence is regular if and only if the limit in (2.22) exists.

Chapter 3

Lyapunov Stability Theory of Nonautonomous Equations

The stability theory of differential equations is centered around the problem of whether a given solution $x(t) = x(t, x_0) \in \mathbb{C}^p$ of the equation

$$\dot{x} = f(t, x) \tag{3.1}$$

is stable under small perturbations of either the initial condition $x_0 = x(0)$ or the function f (in the topology of the space of C^r functions, $r \geq 0$; it is assumed that the solution $x(t)$ is well-defined for all $t \geq 0$). The latter is important in applications since the function f is usually known only up to a given precision.

Stability under small perturbations of initial conditions means that every solution $y(t)$, whose initial condition $y(0)$ lies in the δ-ball around x_0, stays within the ε-neighborhood of the solution $x(t)$ for all sufficiently small ε and some $\delta = \delta(\varepsilon)$ chosen appropriately (in particular, it is assumed that the solution $y(t)$ is well-defined for all $t \geq 0$). One says that the stability is exponential if the solutions $y(t)$ approach $x(t)$ with an exponential rate and that it is uniformly exponential if the convergence is uniform over the initial condition. The uniform exponential stability survives under small perturbations of the right-hand side function f and, therefore, is "observable in practical situations".

In order to study the stability of a solution $x(t)$ of the nonlinear system (3.1) under small perturbations of the initial conditions, one linearizes this system along $x(t)$, i.e., considers the linear system of differential equations known as the system of *variational equations*:

$$\dot{v} = A(t)v, \text{ where } A(t) = \frac{df}{dt}(t, x(t)). \tag{3.2}$$

Moving back from the linear system (3.2) to the nonlinear system (3.1) can be viewed as solving the problem of stability under small perturbations of the right-hand side function in (3.2), thus reducing the study of stability under small perturbations of the initial condition to the study of stability under small perturbations of the right-hand side function.

To characterize stability of the zero solution $v(t) = 0$ of the linear system (3.2), one introduces the Lyapunov exponent associated with the system (3.2), which is a function of the initial condition v_0 (and, hence, of the corresponding solution $v(t) = v(t, v_0)$), given by

$$\chi(v_0) = \limsup_{t \to +\infty} \frac{1}{t} \log \|v(t, v_0)\|,$$

with the convention that $\log 0 = -\infty$.

One can show that if all values of the Lyapunov exponent are negative, then the zero solution is exponentially (albeit possibly nonuniformly) stable. If only some of these values are negative, then the zero solution is conditionally exponentially stable, i.e., it is stable along some directions in the space. These directions form a subspace called *stable.*

One of the main results in the classical stability theory asserts that if the zero solution $v(t) = 0$ of the linear system (3.2) is uniformly exponentially stable, then so is the solution $x(t)$ of the nonlinear system (3.1). In the case when the zero solution is conditionally exponentially stable, the solution $x(t)$ is stable along a local smooth submanifold—called *local stable manifold*—which is tangent to the stable subspace. If the solution $v(t) = 0$ is only (nonuniformly) exponentially stable, the situation becomes much more subtle and requires the additional assumption that the solution is *LP-regular.*

In this chapter we present some core results in the Lyapunov–Perron absolute and conditional stability theory.

3.1. Stability of solutions of ordinary differential equations

We first introduce the notions of stability and asymptotic stability for solutions of ordinary differential equations considering the general case of nonautonomous equations.

Let $f\colon \mathbb{R}\times\mathbb{R}^n \to \mathbb{R}^n$ be a continuous function such that the differential equation

$$x' = f(t,x) \tag{3.3}$$

has unique solutions. For each pair $(t_0, x_0) \in \mathbb{R}\times\mathbb{R}^n$ let $x(t,t_0,x_0)$ be the solution of the initial value problem

$$\begin{cases} x' = f(t,x), \\ x(t_0) = x_0. \end{cases}$$

A solution $x(t,t_0,\bar{x}_0)$ of equation (3.3) defined for every $t > t_0$ is said to be *(Lyapunov) stable* if for each $\varepsilon > 0$ there exists $\delta > 0$ such that if $\|x_0 - \bar{x}_0\| < \delta$, then:

(1) $x(t,t_0,x_0)$ is defined for $t > t_0$;

(2) $\|x(t,t_0,x_0) - x(t,t_0,\bar{x}_0)\| < \varepsilon$ for $t > t_0$.

Otherwise, the solution $x(t,t_0,\bar{x}_0)$ is said to be *unstable*. We also introduce the stronger notion of asymptotic stability. A solution $x(t,t_0,\bar{x}_0)$ of equation (3.3) defined for every $t > t_0$ is said to be *asymptotically stable* if:

(1) $x(t,t_0,\bar{x}_0)$ is stable;

(2) there exists $\alpha > 0$ such that if $\|x_0 - \bar{x}_0\| < \alpha$, then

$$\|x(t,t_0,x_0) - x(t,t_0,\bar{x}_0)\| \to 0 \quad \text{when} \quad t \to +\infty.$$

An asymptotically stable solution $x(t,t_0,\bar{x}_0)$ is said to be *exponentially stable* if there exist $\beta, c, \lambda > 0$ such that if $\|x_0 - \bar{x}_0\| < \beta$, then

$$\|x(t,t_0,x_0) - x(t,t_0,\bar{x}_0)\| \le ce^{-\lambda t}, \quad t > t_0.$$

We note that in general a solution satisfying the second condition in the definition of asymptotic stability need not be stable.

Exercise 3.1. For the equation in polar coordinates

$$r' = r(1-r), \quad \theta' = \sin^2(\theta/2),$$

show that the fixed point $(1,0)$ satisfies the second condition in the notion of asymptotic stability but not the first one.

Now we consider an autonomous differential equation

$$x' = f(x), \tag{3.4}$$

where $f\colon \mathbb{R}^n \to \mathbb{R}^n$ is a continuous function such that equation (3.4) has unique and global solutions. It induces a flow φ_t on $\mathbb{R}^n$ such that the solution of equation (3.4) satisfying the initial condition $x(0) = x_0$ is given by $x(t) = \varphi_t(x_0)$. The vector field f can be recovered from the equation

$$f(x) = \frac{\partial}{\partial t}\varphi_t(x)|_{t=0}.$$

Given a trajectory $\{\varphi_t(x) : t \in \mathbb{R}\}$ of a point $x \in M$, we consider the *system of variational equations*

$$v' = A(x,t)v, \tag{3.5}$$

where

$$A(x,t) = d_{\varphi_t(x)}f.$$

In general this is a nonautonomous linear differential equation. It turns out that the stability of a given trajectory of the flow φ_t can be described by studying small perturbations $g(t,v)$ of its system of variational equations (3.5). We thus consider the nonautonomous linear equation

$$v' = A(t)v \tag{3.6}$$

in $\mathbb{R}^n$, where $A(t)$ is an $n \times n$ matrix with real entries depending continuously on $t \in \mathbb{R}$. For each $v_0 \in \mathbb{R}^n$ there exists a unique global solution $v(t)$ of equation (3.6) satisfying the initial condition $v(0) = v_0$. One can easily verify that for equation (3.6) a given solution is stable (respectively asymptotically stable, unstable) if and only if all solutions are stable (respectively asymptotically stable, unstable). Thus we only need to consider the trivial solution $v(t) = 0$.

Let $V(t)$ be a monodromy matrix, that is, an $n\times n$ matrix whose columns form a basis of the n-dimensional space of solutions of equation (3.6).

Exercise 3.2. Show that the trivial solution is:

(1) stable if and only if $\sup\{\|V(t)\| : t > 0\} < +\infty$;

(2) asymptotically stable if and only if

$$\|V(t)\| \to 0 \quad \text{when} \quad t \to +\infty;$$

(3) exponentially stable if and only if there exist $c, \lambda > 0$ such that

$$\|V(t)\| \le ce^{-\lambda t}, \quad t > 0,$$

or equivalently if

$$\limsup_{t\to+\infty} \frac{1}{t} \log \|V(t)\| < 0. \tag{3.7}$$

Let χ^+_{s+} be the largest value of the Lyapunov exponent χ^+ in (2.9). Condition (3.7) is equivalent to

$$\chi^+_{s+} < 0. \tag{3.8}$$

Exercise 3.3. Assume that the matrix function $A(t)$ is constant, that is, $A(t) = A$ for every $t \in \mathbb{R}$ and some $n \times n$ matrix A.

(1) Show that the trivial solution is stable if and only if A has no eigenvalues with positive real part and each eigenvalue with zero real part has a diagonal Jordan block (in the Jordan canonical form).

(2) Show that the trivial solution is asymptotically stable (in which case it is also exponentially stable) if and only if A has only eigenvalues with negative real part.

One can obtain a similar characterization of stability in the case when the matrix function $A(t)$ is periodic. Namely, assume that

$$A(t+T) = A(t), \quad t \in \mathbb{R},$$

for some constant $T > 0$ (this includes the constant case). In this case any monodromy matrix is of the form $P(t)E^{Bt}$ for some $n \times n$ matrices $P(t)$ and B with $P(t+T) = P(t)$ for $t \in \mathbb{R}$. The eigenvalues of the matrix B (which are well-defined $\operatorname{mod}(2\pi/T)$) are called *characteristic exponents.* Since the matrix $P(t)$ is periodic, the stability of solutions depends only on the characteristic exponents. Thus the trivial solution is stable if and only if there are no characteristic exponents with positive real part and each characteristic exponent with zero real part the corresponding Jordan block of B is diagonal. Moreover, the trivial solution is asymptotically stable (in which case it is also exponentially stable) if and only if there are only characteristic exponents with negative real part.

We consider the problem of the stability under nonlinear perturbations. Under condition (3.8) every solution of equation (3.6) is exponentially stable.

Now we consider a nonlinear differential equation

$$u' = A(t)u + f(t,u), \tag{3.9}$$

which is a *perturbation* of equation (3.6). We assume that $f(t,0) = 0$ and hence, $u(t) = 0$ is a solution of equation (3.9). We also assume that there exists a neighborhood U of 0 in $\mathbb{R}^n$ such that f is continuous on $[0,+\infty) \times U$ and for every u, $v \in U$ and $t \geq 0$,

$$\|f(t,u) - f(t,v)\| \leq K\|u-v\|(\|u\| + \|v\|)^{q-1} \tag{3.10}$$

for some constants $K > 0$ and $q > 1$. This means that the perturbation is sufficiently small in U. The number q is called the *order of the perturbation.*

One can ask whether condition (3.8) implies that the solution $u(t) = 0$ of the perturbed equation (3.9) is exponentially stable. Perron [**66**] showed that in general this may not be true.

Example 3.4. Consider the nonlinear system of differential equations in $\mathbb{R}^2$ given by

$$\begin{aligned} u_1' &= [-\omega - a(\sin\log t + \cos\log t)]u_1, \\ u_2' &= [-\omega + a(\sin\log t + \cos\log t)]u_2 + |u_1|^{\lambda+1}, \end{aligned} \tag{3.11}$$

for some positive constants ω, a, and λ. It is a perturbation of the system of linear equations

$$\begin{aligned} v_1' &= [-\omega - a(\sin\log t + \cos\log t)]v_1, \\ v_2' &= [-\omega + a(\sin\log t + \cos\log t)]v_2. \end{aligned} \tag{3.12}$$

We assume that

$$a < \omega < (2e^{-\pi} + 1)a \quad \text{and} \quad 0 < \lambda < \frac{2a}{\omega - a} - e^{\pi}. \tag{3.13}$$

Notice that the perturbation $f(t, (u_1, u_2)) = (0, |u_1|^{\lambda+1})$ satisfies condition (3.10) with $q = \lambda + 1 > 1$.

The general solution of (3.11) is given by

$$\begin{aligned} u_1(t) &= c_1 e^{-\omega t - at\sin\log t}, \\ u_2(t) &= c_2 e^{-\omega t + at\sin\log t} \\ &\quad + |c_1|^{\lambda+1} e^{-\omega t + at\sin\log t} \int_{t_0}^{t} e^{-(2+\lambda)a\tau\sin\log\tau - \omega\lambda\tau}\, d\tau, \end{aligned} \tag{3.14}$$

while the general solution of (3.12) is given by

$$v_1(t) = d_1 e^{-\omega t - at\sin\log t}, \quad v_2(t) = d_2 e^{-\omega t + at\sin\log t},$$

where c_1, c_2, d_1, d_2, and t_0 are arbitrary numbers.

Exercise 3.5. Show that the values of the Lyapunov exponent associated with (3.12) are $\chi_1^+ = \chi_2^+ = -\omega + a < 0$.

Let $u(t) = (u_1(t), u_2(t))$ be a solution of the system of nonlinear equations (3.11). In view of (3.14) it is also a solution of the system of linear equations

$$\begin{aligned} u_1' &= [-\omega - a(\sin\log t + \cos\log t)]u_1, \\ u_2' &= [-\omega + a(\sin\log t + \cos\log t)]u_2 + \delta(t)u_1, \end{aligned} \tag{3.15}$$

where

$$\delta(t) = \operatorname{sgn} c_1 |c_1|^{\lambda} e^{-\omega\lambda t - a\lambda t\sin\log t}.$$

We have

$$|\delta(t)| \le |c_1|^{\lambda} e^{(-\omega+a)\lambda t}$$

and hence, by (3.13), condition (2.8) holds for (3.15). Fix $0 < \varepsilon < \pi/4$ and for each $k \in \mathbb{N}$ set

$$t_k = e^{2k\pi - \frac{1}{2}\pi}, \quad t_k' = e^{2k\pi - \frac{1}{2}\pi - \varepsilon}.$$

Clearly, $0 < t_k' < t_k$ and $t_k \to \infty$ as $k \to \infty$. We have that

$$\int_{t_0}^{t_k} e^{-(2+\lambda)a\tau \sin \log \tau - \omega\lambda\tau}\, d\tau > \int_{t_k'}^{t_k} e^{-(2+\lambda)a\tau \sin \log \tau - \omega\lambda\tau}\, d\tau$$

and for every $\tau \in [t_k', t_k]$,

$$2k\pi - \frac{\pi}{2} - \varepsilon \le \log \tau \le 2k\pi - \frac{\pi}{2},$$

$$(2+\lambda)a\tau \cos \varepsilon \le -(2+\lambda)a\tau \sin \log \tau.$$

This implies that

$$\int_{t_k'}^{t_k} e^{-(2+\lambda)a\tau \sin \log \tau - \omega\lambda\tau}\, d\tau \ge \int_{t_k'}^{t_k} e^{(2+\lambda)a\tau \cos \varepsilon - \omega\lambda\tau}\, d\tau.$$

Set $r = (2+\lambda)a \cos \varepsilon - \omega\lambda$ and if necessary choose ε so small that $r > 0$. It follows that if $k \in \mathbb{N}$ is sufficiently large, then

$$\int_{t_0}^{t_k} e^{-(2+\lambda)a\tau \sin \log \tau - \omega\lambda\tau}\, d\tau > \int_{t_k'}^{t_k} e^{r\tau}\, d\tau > ce^{rt_k},$$

where $c = [1 - \exp(e^{-\varepsilon} - 1)]/r$. Set

$$t_k^* = t_k e^{\pi} = e^{2k\pi + \frac{1}{2}\pi}.$$

We obtain

$$\begin{aligned} e^{at_k^* \sin \log t_k^*} \int_{t_0}^{t_k^*} e^{-(2+\lambda)a\tau \sin \log \tau - \omega\lambda\tau}\, d\tau &> e^{at_k^*} \int_{t_0}^{t_k} e^{-(2+\lambda)a\tau \sin \log \tau - \omega\lambda\tau}\, d\tau \\ &> ce^{at_k^* + rt_k} = ce^{(a + re^{-\pi})t_k^*}. \end{aligned}$$

It follows from (3.13) that if $c_1 \ne 0$ and ε is sufficiently small, then the Lyapunov exponent of any solution $u(t)$ of (3.11) (that is also a solution of (3.15)) satisfies

$$\chi^+(u) \ge -\omega + a + re^{-\pi} = -\omega + a + [(2+\lambda)a\cos\varepsilon - \omega\lambda]e^{-\pi} > 0.$$

Therefore, the solution $u(t)$ is not asymptotically stable. This completes the construction of the example.

Lyapunov introduced regularity conditions which guarantee the exponential stability of the solution $u(t) = 0$ of the perturbed equation (3.9). Although there are many different ways to state the regularity conditions, for a given differential equation the regularity is often difficult to verify.

3.2. Lyapunov absolute stability theorem

We now state the *Lyapunov Stability Theorem.* It claims that under an additional assumption known as *forward regularity,* condition (3.8) indeed implies the exponential stability of the trivial solution $u(t) = 0$ of equation (3.9). The theorem was essentially established by Lyapunov.

Theorem 3.6. *Assume that the matrix function $A(t)$ satisfies condition (2.8). Assume also that the Lyapunov exponent χ^+ associated with equation (2.7) is forward regular and satisfies condition (3.8). Then the solution $u(t) = 0$ of the perturbed equation (3.9) is exponentially stable.*

We present some more general results on exponential stability of equation (3.9) of which the above theorem is an immediate corollary. Consider the perturbed differential equation (3.9) and assume that the function $f(t, u)$ satisfies (3.10). Let $V(t)$ be a monodromy matrix (the fundamental solution matrix) for which $V(0) = \mathrm{Id}$. Denote by $\mathcal{V}(t, s)$ the Cauchy matrix of equation (2.7) defined by $\mathcal{V}(t, s) = V(t)V(s)^{-1}$.

Theorem 3.7. *Assume that there are two continuous functions R and r on $[0, +\infty)$ such that for every $t \geq 0$,*

$$\int_0^t R(\tau)\, d\tau \leq D_1 < +\infty, \tag{3.16}$$

$$\int_0^\infty e^{(q-1)\int_0^s (R+r)(\tau)\, d\tau}\, ds = D_2 < +\infty, \tag{3.17}$$

and for every $0 \leq s \leq t$,

$$\|\mathcal{V}(t, s)\| \leq D_3 e^{\int_s^t R(\tau)\, d\tau + (q-1)\int_0^s r(\tau)\, d\tau}. \tag{3.18}$$

Then there exist $D > 0$ and $\delta > 0$ such that for every $u_0 \in \mathbb{R}^n$ with $\|u_0\| < \delta$, there is a unique solution u of equation (3.9) which satisfies:

1. *u is well-defined on the interval $[0, \infty)$, and $u(0) = u_0$;*
2. *for every t,*

$$\|u(t)\| \leq \|u_0\| D e^{\int_0^t R(\tau)\, d\tau}. \tag{3.19}$$

Proof. Equation (3.9) is equivalent to the integral equation

$$u(t) = \mathcal{V}(t, 0)u_0 + \int_0^t \mathcal{V}(t, s) f(s, u(s))\, ds. \tag{3.20}$$

Moreover, by (3.10), for every $u_0 \in \mathbb{R}^n$ with $u_0 \in U$ there exists a unique solution $u(t)$ satisfying the initial condition $u(0) = u_0$.

We consider the linear space $\mathcal{B}_C$ of $\mathbb{R}^n$-valued continuous functions $x = x(t)$ on the interval $[0, \infty)$ such that

$$\|x(t)\| \le Ce^{\int_0^t R(\tau)\, d\tau}$$

for every $t \ge 0$ where $C > 0$ is a constant. We endow this space with the norm

$$\|x\|_R = \sup\{\|x(t)\|e^{-\int_0^t R(\tau)\, d\tau} : t \ge 0\}.$$

It follows from (3.16) that $\|x(t)\| \le \|x\|_R e^{D_1}$. Therefore, if $\|x\|_R$ is sufficiently small, then $x(t) \in U$ for every $t \ge 0$. This means that the operator

$$(Jx)(t) = \int_0^t \mathcal{V}(t,s) f(s, x(s))\, ds$$

is well-defined on $\mathcal{B}_C$. Let

$$L = K(\|x_1\|_R + \|x_2\|_R)^{q-1}\|x_1 - x_2\|_R.$$

By (3.10), (3.17), and (3.18), we have

$$\begin{aligned}
\|(Jx_1)(t) - (Jx_2)(t)\| &\le \int_0^t D_3 e^{\int_s^t R(\tau)\, d\tau + (q-1)\int_0^s r(\tau)\, d\tau} L e^{q\int_0^s R(\tau)\, d\tau}\, ds \\
&= D_3 L e^{\int_0^t R(\tau)\, d\tau} \int_0^t e^{(q-1)\int_0^s (R+r)(\tau)\, d\tau}\, ds \\
&\le D_3 D_2 L e^{\int_0^t R(\tau)\, d\tau}.
\end{aligned}$$

It follows that

$$\|Jx_1 - Jx_2\|_R \le D_3 D_2 K(\|x_1\|_R + \|x_2\|_R)^{q-1}\|x_1 - x_2\|_R.$$

Now choose $\varepsilon > 0$ such that

$$\theta := D_3 D_2 K(2\varepsilon)^{q-1} < 1. \tag{3.21}$$

Whenever $x_1, x_2 \in \mathcal{B}_\varepsilon$, we have

$$\|Jx_1 - Jx_2\|_R \le \theta\|x_1 - x_2\|_R. \tag{3.22}$$

Choose a point $u_0 \in \mathbb{R}^n$ and set $\xi(t) = \mathcal{V}(t, 0)u_0$. It follows from (3.18) that

$$\|\xi(t)\| \le \|u_0\| D_3 e^{\int_0^t R(\tau)\, d\tau}$$

and hence, $\|\xi\|_R \le \|u_0\| D_3$. Consider the operator $\tilde{J}$ defined by

$$(\tilde{J}x)(t) = \xi(t) + (Jx)(t)$$

on the space $\mathcal{B}_\varepsilon$. Note that $\mathcal{B}_\varepsilon$ is a complete metric space with the distance $d(x, y) = \|x - y\|_R$. We have

$$\|\tilde{J}x\|_R \le D_3\|u_0\| + \theta\|x\|_R < D_3\delta + \theta\varepsilon < \varepsilon \tag{3.23}$$

provided that δ is sufficiently small. Therefore, $\tilde{J}(\mathcal{B}_\varepsilon) \subset \mathcal{B}_\varepsilon$. By (3.21) and (3.22), the operator $\tilde{J}$ is a contraction, and hence, there exists a unique function $u \in \mathcal{B}_\varepsilon$ which is a solution of (3.20). Inequality (3.19) now follows readily from (3.23). □

An immediate consequence of this theorem is the following statement due to Malkin (see [**60**]).

Theorem 3.8. *Assume that the Cauchy matrix of* (2.7) *admits the estimate*

$$\|\mathcal{V}(t,s)\| \leq De^{\alpha(t-s)+\beta s}, \tag{3.24}$$

for any $0 \leq s \leq t$ *and some constants* α, β, D, *such that*

$$(q-1)\alpha + \beta < 0. \tag{3.25}$$

Then the trivial solution of equation (3.9) *is exponentially stable.*

Proof. Set

$$R(t) = \alpha \quad \text{and} \quad r(t) = \frac{\beta}{q-1}.$$

It follows from (3.24) and (3.25) that conditions (3.17) and (3.18) hold. In order to verify condition (3.16), note that by setting $t = s$ in (3.18) we obtain

$$1 = \|\mathcal{V}(t,t)\| \leq D_3 e^{(q-1)\int_0^t r(\tau)\,d\tau}.$$

It follows that for every t,

$$\frac{\beta t}{q-1} = \int_0^t r(\tau)\,d\tau \geq -C, \tag{3.26}$$

where $C \geq 0$ is a constant. Therefore, by (3.25),

$$\alpha t = \int_0^t R(\tau)\,d\tau \leq C + \int_0^t (R(\tau) + r(\tau))\,d\tau \leq C \tag{3.27}$$

for all t. This implies condition (3.16). Therefore, by (3.19), any solution u of (3.9) satisfies

$$\|u(t)\| \leq \|u_0\| De^{\alpha t} \leq \|u_0\| De^{C}. \tag{3.28}$$

This shows that the trivial solution of (3.9) is asymptotically stable. By (3.26) and (3.27), we have $\beta \geq 0$ and $\alpha \leq 0$. In view of (3.25), $\alpha < 0$, and in view of (3.28), the solution $u(t) = 0$ is exponentially stable. The desired result follows. □

Another important consequence of Theorem 3.7 is the following statement due to Lyapunov (see [**59**]). It shows that exponential stability may hold even for some nonregular systems, provided that the regularity coefficient is sufficiently small.

Theorem 3.9. *Assume that the maximal value* $\chi_{\max}$ *of the Lyapunov exponent of* (2.7) *is strictly negative (see* (3.8)*) and that*

$$(q-1)\chi_{\max}+\gamma<0,$$

where γ *is the regularity coefficient. Then the trivial solution of equation* (3.9) *is exponentially stable.*

Proof. Consider a subordinate basis $(v_1(t),\dots,v_n(t))$ of the space of solutions of (2.7) such that the numbers

$$\chi_1'\le\cdots\le\chi_n'=\chi_{\max}$$

are the values of the Lyapunov exponent χ^+ of the vectors $v_1(t),\dots,v_n(t)$. We may assume that the matrix $V(t)$ whose columns are $v_1(t),\dots,v_n(t)$ is such that the columns $w_1(t),\dots,w_n(t)$ of the matrix $W(t)=[V(t)^*]^{-1}$ form a subordinate basis of the space of solutions of the dual equation $\dot{w}=-A(t)^*w$.

Let $\mu_1,\dots,\mu_n$ be the values of the Lyapunov exponent $\tilde{\chi}^+$ of the vectors $w_1(t),\dots,w_n(t)$. For every $\varepsilon>0$ there exists $D_\varepsilon>0$ such that

$$\|v_j(t)\|\le D_\varepsilon e^{(\chi_j'+\varepsilon)t}\quad\text{and}\quad\|w_j(t)\|\le D_\varepsilon e^{(\mu_j+\varepsilon)t}$$

for every t. Note that there is a pair of dual bases $\mathbf{v}(t)=(v_1(t),\dots,v_n(t))$ and $\mathbf{w}(t)=(w_1(t),\dots,w_n(t))$ for which

$$\gamma=\gamma(\chi,\chi^*)=\max\left\{\chi(v_j)+\chi^*(w_j):j=1,\dots,n\right\}.$$

Therefore,

$$\gamma=\max\{\chi_j'+\mu_j:j=1,\dots,n\}.$$

It follows that $\chi_j'+\mu_j\le\gamma$ for every $j=1,\dots,n$. Consider the Cauchy matrix

$$\mathcal{V}(t,s)=V(t)V(s)^{-1}=V(t)W(s)^*.$$

Its entries are

$$v_{ik}(t,s)=\sum_{j=1}^n v_{ij}(t)\overline{w_{kj}(s)},$$

where $v_{ij}(t)$ is the ith coordinate of the vector $v_j(t)$ and $w_{kj}(s)$ is the kth coordinate of the vector $w_j(s)$. It follows that

$$\begin{aligned}|v_{ik}(t,s)|&\le\sum_{j=1}^n|v_{ij}(t)|\cdot|\overline{w_{kj}(s)}|\\&\le\sum_{j=1}^n\|v_j(t)\|\cdot\|w_j(s)\|\\&\le nD_\varepsilon{}^2e^{(\chi_j'+\varepsilon)t+(\mu_j+\varepsilon)s}\\&=nD_\varepsilon{}^2e^{(\chi_j'+\varepsilon)(t-s)+(\chi_j'+\mu_j+2\varepsilon)s}.\end{aligned}$$

Therefore, there exists $D > 0$ such that

$$\|\mathcal{V}(t,s)\| \le De^{(\chi_{\max}+\varepsilon)(t-s)+(\gamma+2\varepsilon)s}. \tag{3.29}$$

Since ε can be chosen arbitrarily small, the desired result follows from Theorem 3.8 by setting $\alpha = \chi_{\max} + \varepsilon$ and $\beta = \gamma + 2\varepsilon$. □

We conclude by observing that the Lyapunov Stability Theorem 3.6 is an immediate corollary of Theorem 3.9. Moreover, if χ^+ is forward regular, setting $\gamma = 0$ in (3.29), we obtain that for every solution $u(t)$ of equation (3.9) and every $s \in \mathbb{R}$,

$$\begin{aligned}\|u(t)\| &\le De^{(\chi_{\max}+\varepsilon)(t-s)+2\varepsilon s}\|u(s)\| \\ &= De^{2\varepsilon s}e^{(\chi_{\max}+\varepsilon)(t-s)}\|u(s)\|.\end{aligned} \tag{3.30}$$

In other words the contraction constant $De^{2\varepsilon s}$ may deteriorate along the solution of equation (3.9). This implies that the "size" of the neighborhood at time s, where the exponential stability of the trivial solution is guaranteed, may decay with subexponential rate. In Section 4.1 we will establish an analog of (3.30) for dynamical systems with discrete time (see statement (L7) of Theorem 4.3) and illustrate the crucial role that it plays in the nonuniform hyperbolicity theory.

3.3. Lyapunov conditional stability theorem

In this section we discuss the concept of conditional stability that is a foundation for hyperbolicity theory and we present a version of Theorem 3.6 that deals with conditional stability.

A solution $x(t, t_0, \bar{x}_0)$ of equation (3.3) defined for every $t > t_0$ is said to be *(conditionally) stable* along a set $V \subset \mathbb{R}^n$ if for each $\varepsilon > 0$ there exists $\delta > 0$ such that if $x_0, \bar{x}_0 \in V$ and $\|x_0 - \bar{x}_0\| < \delta$, then

(1) $x(t, t_0, x_0)$ is defined for $t > t_0$;

(2) $\|x(t, t_0, x_0) - x(t, t_0, \bar{x}_0)\| < \varepsilon$ for $t > t_0$.

A solution $x(t, t_0, \bar{x}_0)$ of equation (3.3) defined for every $t > t_0$ is said to be *(conditionally) asymptotically stable* along a set $V \subset \mathbb{R}^n$ if

(1) $x(t, t_0, \bar{x}_0)$ is stable along V;

(2) there exists $\alpha > 0$ such that if $x_0, \bar{x}_0 \in V$ and $\|x_0 - \bar{x}_0\| < \alpha$, then

$$\|x(t, t_0, x_0) - x(t, t_0, \bar{x}_0)\| \to 0 \quad \text{when} \quad t \to +\infty.$$

An asymptotically stable solution $x(t, t_0, \bar{x}_0)$ along V is said to be *(conditionally) exponentially stable* along V if there exist $\beta, c, \lambda > 0$ such that if $x_0, \bar{x}_0 \in V$ and $\|x_0 - \bar{x}_0\| < \beta$, then

$$\|x(t, t_0, x_0) - x(t, t_0, \bar{x}_0)\| \le ce^{-\lambda t}, \quad t > t_0.$$

The classical Hadamard–Perron Theorem claims that a hyperbolic fixed point is exponentially stable along its stable manifold.

We now describe a generalization of the Lyapunov Stability Theorem when *not all* Lyapunov exponents are negative. Namely, let $\chi_1^+ < \cdots < \chi_s^+$ be the distinct values of the Lyapunov exponent χ^+ for equation (2.7). Assume that

$$\chi_k^+ < 0, \tag{3.31}$$

where $1 \le k < s$ and the number χ_{k+1}^+ can be negative, positive, or equal to 0 (compare to (3.8)).

We recall that a *local smooth submanifold* of $\mathbb{R}^n$ is the graph a smooth function $\psi\colon B \to \mathbb{R}^n$, where B is the open unit ball in $\mathbb{R}^k$ for $k \le n$.

Theorem 3.10. *Assume that the Lyapunov exponent χ^+ for equation* (2.7) *with the matrix function $A(t)$ satisfying* (2.8) *is forward regular and satisfies condition* (3.31). *Then there exists a local smooth submanifold $V^s \subset \mathbb{R}^n$ which passes through* 0, *is tangent at* 0 *to the linear subspace V_k (defined by* (2.1)*), and is such that the trivial solution of* (3.9) *is exponentially stable along V^s.*

This statement is an extension of the classical Hadamard–Perron Theorem to the case of conditional stability and is a particular case of a general stable manifold theorem that we discuss in Section 7.1.

Following [**11**], we give an outline of the proof of the theorem restricting ourselves to showing the existence of *Lipschitz* stable manifolds along which the trivial solution is exponentially stable. The stable manifolds are obtained as graphs of Lipschitz functions. Let

$$E(t) = V(t) \bigoplus_{i=1}^{k} E_i \quad \text{and} \quad F(t) = V(t) \bigoplus_{i=k+1}^{s} E_i,$$

where $V(t)$ is a monodromy matrix of equation (3.6). Take $\varepsilon > 0$ such that

$$\chi_k + \alpha + \varepsilon < 0 \quad \text{and} \quad \chi_k + 2\varepsilon < \chi_{k+1},$$

where $\alpha = 2\varepsilon(1 + 2/q)$. Given $\delta > 0$, consider the set

$$X_\alpha = \{(s, \xi) : s \ge 0 \text{ and } \xi \in B(\delta e^{-\alpha s})\},$$

where $B(\delta)$ is the open ball in $E(s)$ of radius $\delta > 0$ centered at zero. Let $\mathfrak{X}_\alpha$ be the space of continuous functions $\varphi\colon X_\alpha \to \mathbb{R}^n$ such that for each $s \ge 0$,

$$\varphi(s, B(\delta e^{-\alpha s})) \subset F(s), \quad \varphi(s, 0) = 0,$$

and

$$\|\varphi(s, x) - \varphi(s, y)\| \le \|x - y\| \quad \text{for} \quad x, y \in B(\delta e^{-\alpha s}).$$

The graph of a function $\varphi \in \mathfrak{X}_\alpha$ is given by

$$\mathcal{V} = \{(s,\xi,\varphi(s,\xi)) : (s,\xi) \in X_\alpha\} \subset \mathbb{R}_0^+ \times \mathbb{R}^n. \tag{3.32}$$

We will show that there exists $\varphi \in \mathfrak{X}_\alpha$ such that for every initial condition $(s,\xi) \in X_\beta \subset X_\alpha$, with $\beta = \alpha + 2\varepsilon$, the corresponding solution is entirely contained in $\mathcal{V}$. In order for $\mathcal{V}$ to be forward invariant, any solution with initial condition in $\mathcal{V}$ must remain in $\mathcal{V}$ for all time and thus must be of the form

$$(x(t), y(t)) = (x(t), \varphi(t, x(t)))$$

with components in $E(t)$ and $F(t)$, respectively. Equation (3.9) is then equivalent to

$$x(t) = U(t,s)x(s) + \int_s^t U(t,\tau) f(\tau, x(\tau), \varphi(\tau, x(\tau)))\, d\tau, \tag{3.33}$$

$$\varphi(t, x(t)) = V(t,s)\varphi(s, x(s)) + \int_s^t V(t,\tau) f(\tau, x(\tau), \varphi(\tau, x(\tau)))\, d\tau, \tag{3.34}$$

where

$$U(t,s) = V(t)V(s)^{-1}|E(s) \quad \text{and} \quad V(t,s) = V(t)V(s)^{-1}|F(s).$$

Let

$$X_\alpha^* = \{(s,\xi) : s \geq 0 \text{ and } \xi \in E(s)\}.$$

Consider the space $\mathfrak{X}_\alpha^*$ of continuous functions $\varphi\colon X_\alpha^* \to \mathbb{R}^n$ for which $\varphi|X_\alpha \in \mathfrak{X}_\alpha$ and

$$\varphi(s,\xi) = \varphi(s, \delta e^{-\alpha s}\xi/\|\xi\|), \quad s \geq 0,\ \xi \notin B(\delta e^{-\alpha s}).$$

Arguing similarly as in the proof of Theorem 3.7, one can show that for each $\varphi \in \mathfrak{X}_\alpha^*$ there exists a unique function $x(t) = x_\varphi(t)$ satisfying (3.33).

Lemma 3.11. *There exists $R > 0$ such that for all sufficiently small $\delta > 0$, $\varphi \in \mathfrak{X}_\alpha^*$, and $(s,\xi) \in X_\alpha$, one can find a unique continuous function $x = x_\varphi\colon [s,+\infty) \to \mathbb{R}^n$ satisfying:*

(1) $x_\varphi(s) = \xi$, $x_\varphi(t) \in E(t)$, *and* (3.33) *holds for* $t \geq s$;

(2)
$$\|x_\varphi(t)\| \leq Re^{(\chi_k+\varepsilon)(t-s)+2\varepsilon s}\|\xi\|, \quad t \geq s.$$

It remains to find a function φ satisfying identity (3.34) with $x = x_\varphi$. We first observe that this identity can be reduced to the identity

$$\varphi(s,\xi) = -\int_s^\infty V(\tau,s)^{-1} f(\tau, x_\varphi(\tau), \varphi(\tau, x_\varphi(\tau)))\, d\tau \tag{3.35}$$

for every $(s,\xi) \in X_\alpha$. This means that we need to find a fixed point of the operator Φ defined for each $\varphi \in \mathfrak{X}_\alpha^*$ by

$$(\Phi\varphi)(s,\xi) = -\int_s^\infty V(\tau,s)^{-1} f(\tau, x_\varphi(\tau), \varphi(\tau, x_\varphi(\tau)))\, d\tau$$

with $(s, \xi) \in X_\alpha$. Here x_φ is the unique continuous function given by Lemma 3.11 if $x_\varphi(s) = \xi$ and by

$$(\Phi\varphi)(s, \xi) = (\Phi\varphi)(s, \delta e^{-\alpha s}\xi/\|\xi\|)$$

otherwise. One can show that Φ is a contraction in the space $\mathfrak{X}_\alpha^*$ and thus, given a sufficiently small $\delta > 0$, there exists a unique function $\varphi \in \mathfrak{X}_\alpha^*$ such that (3.35) holds for every $(s, \xi) \in X_\alpha$. The desired Lipschitz stable manifold $\mathcal{V}$ in (3.32) is the restriction of this function φ to X_α.

Chapter 4

Elements of the Nonuniform Hyperbolicity Theory

Hyperbolicity theory is a fundamental new step in developing the Lyapunov stability theory for dynamical systems with either continuous or discrete time. It deals almost exclusively with situations where stability is conditional, i.e., appears only along some but not all directions in space. Indeed, along other directions the system is unstable. Trajectories with such conditional stability are said to be hyperbolic, either uniformly or nonuniformly according to how the rates of contraction and expansion vary from point to point. While trajectories that are completely stable (i.e., stable along all directions in space) are usually isolated, hyperbolic trajectories can occupy a "large" part of the phase space (for example, they may form a set of positive or full Lebesgue measure) and in the case of Anosov systems every trajectory is hyperbolic.

In this chapter we are mostly concerned with the concept of nonuniform hyperbolicity. In this case the rates of contraction and expansion may vary from point to point in a "wild" and, in general, uncontrollable way. It turns out, however, that for trajectories that are LP-regular one can obtain some important estimates on how the rates of contraction and expansion vary along the trajectories. Although at first glance these estimates seem to be technical, they play a crucial role in studying ergodic properties of hyperbolic systems.

In this connection one may wonder whether a given dynamical system admits LP-regular trajectories and if so how large the set of such trajectories is. A celebrated result by Oseledets known as the Multiplicative Ergodic Theorem (the name was coined by Oseledets) claims that LP-regular trajectories form a set of full measure with respect to any invariant Borel measure on M. Thus, from the measure-theoretic point of view, there is no need to impose the requirement that a given trajectory is LP-regular.

4.1. Dynamical systems with nonzero Lyapunov exponents

Consider a flow φ_t on a smooth compact Riemannian p-manifold M which is generated by the vector field $\mathfrak{X}$ on M such that

$$\mathfrak{X}(x) = \frac{d\varphi_t(x)}{dt}|_{t=0}.$$

We will always assume that the vector field $\mathfrak{X}$ depends smoothly on x and we refer to φ_t as a smooth flow on M. For every $x_0 \in M$ the trajectory $\{x(x_0,t) = \varphi_t(x_0) : t \in \mathbb{R}\}$ represents a solution of the nonlinear differential equation

$$\dot{v} = \mathfrak{X}(v) \tag{4.1}$$

on the manifold M. This solution is uniquely defined by the initial condition $x(x_0, 0) = x_0$.

With the flow φ_t one can associate a certain collection of single linear differential equations (4.1) "along" each trajectory of the flow. The stability of a given trajectory can be described by studying small perturbations of the system of variational equations. The perturbation term is of type (3.9) and since the flow is smooth, it satisfies condition (3.10). The results of the previous section apply and allow one to study the stability of trajectories via the Lyapunov exponents. Although it is still a very difficult problem to verify whether a given trajectory is Lyapunov regular, it turns out that "most" trajectories (in the sense of measure theory) have this property.

A similar but technically simpler approach can be used to establish the stability of trajectories of dynamical systems with discrete time.

Let $f: M \to M$ be a diffeomorphism of a compact smooth Riemannian p-dimensional manifold M. Given $x \in M$, consider the trajectory $\{f^m(x)\}_{m\in\mathbb{Z}}$. The family of maps $\{d_{f^m(x)}f\}_{m\in\mathbb{Z}}$ can be viewed as an analog of the system of variational equations in the continuous time case. Given $x \in M$ and $v \in T_xM$, the formula

$$\chi^+(x, v) = \limsup_{m\to+\infty} \frac{1}{m} \log \|d_x f^m v\| \tag{4.2}$$

defines the *Lyapunov exponent* specified by the diffeomorphism f at the point x. By the general theory of Lyapunov exponents (see Section 2.1), the function $\chi^+(x,\cdot)$ attains only finitely many values on $T_xM\setminus\{0\}$, which we denote by

$$\chi_1^+(x)<\cdots<\chi_{s^+(x)}^+(x),$$

where $s^+(x)\le p$. We also denote by $\mathcal{V}_x^+$ the filtration of T_xM associated to $\chi^+(x,\cdot)$:

$$\{0\}=V_0^+(x)\subsetneqq V_1^+(x)\subsetneqq\cdots\subsetneqq V_{s^+(x)}^+(x)=T_xM,$$

where $V_i^+(x)=\{v\in T_xM:\chi^+(x,v)\le\chi_i^+(x)\}$. The number

$$k_i^+(x)=\dim V_i^+(x)-\dim V_{i-1}^+(x)$$

is the *multiplicity* of the value $\chi_i^+(x)$. We have

$$\sum_{i=1}^{s^+(x)}k_i^+(x)=p.$$

Finally, the collection of pairs

$$\operatorname{Sp}\chi^+(x)=\{(\chi_i^+(x),k_i^+(x)):1\le i\le s^+(x)\}$$

is called the *Lyapunov spectrum* of the exponent $\chi^+(x,\cdot)$.

We note that the functions $\chi_i^+(x)$, $s^+(x)$, and $k_i^+(x)$ are *invariant* under f and are Borel *measurable* (but not necessarily continuous).

For every $x\in M$ and $v\in T_xM$ we set

$$\chi^-(x,v)=\limsup_{m\to-\infty}\frac{1}{|m|}\log\|d_xf^mv\|.$$

Again, by the general theory of Lyapunov exponents, the function $\chi^-(x,\cdot)$ takes on finitely many values on $T_xM\setminus\{0\}$:

$$\chi_1^-(x)>\cdots>\chi_{s^-(x)}^-(x),$$

where $s^-(x)\le p$. We denote by $\mathcal{V}_x^-$ the filtration of T_xM associated to $\chi^-(x,\cdot)$:

$$T_xM=V_1^-(x)\supsetneqq\cdots\supsetneqq V_{s^-(x)}^-(x)\supsetneqq V_{s^-(x)+1}^-(x)=\{0\},$$

where $V_i^-(x)=\{v\in T_xM:\chi^-(x,v)\le\chi_i^-(x)\}$. The number

$$k_i^-(x)=\dim V_i^-(x)-\dim V_{i+1}^-(x)$$

is the *multiplicity* of the value $\chi_i^-(x)$. The collection of pairs

$$\operatorname{Sp}\chi^-(x)=\{(\chi_i^-(x),k_i^-(x))\colon i=1,\dots,s^-(x)\}$$

is called the *Lyapunov spectrum* of the exponent $\chi^-(x,\cdot)$.

In order to simplify our notation, in what follows, we will often drop the superscript + from the notation of the forward Lyapunov exponents and the associated quantities if it does not cause any confusion.

We say that a point $x \in M$ is

(1) *forward regular* if the Lyapunov exponent χ^+ is forward regular;

(2) *backward regular* if the Lyapunov exponent χ^- is backward regular;

(3) *Lyapunov–Perron regular* or simply *LP-regular* if the Lyapunov exponent χ^+ is forward regular, the Lyapunov exponent χ^- is backward regular, and the filtrations $\mathcal{V}_x^-$ and $\mathcal{V}_x^-$ are coherent (see Sections 2.4 and 2.5).

Note that if x is LP-regular, then so is the point $f(x)$ and thus, one can speak of the whole trajectory $\{f^m(x)\}$ as being forward, backward, or LP-regular.

We state a result characterizing the LP-regularity, which is an analog of Theorem 2.21 for the discrete time case. Recall that the cotangent bundle T^*M consists of 1-forms on M. The diffeomorphism f acts on T^*M by its codifferential

$$d_x^* f \colon T_{f(x)}^* M \to T_x^* M$$

defined by

$$d_x^* f \varphi(v) = \varphi(d_x f v), \quad v \in T_x M, \quad \varphi \in T_{f(x)}^* M.$$

We denote the inverse map by

$$d_x' f = (d_x^* f)^{-1} \colon T_x^* M \to T_{f(x)}^* M$$

and define the Lyapunov exponent χ^* on T^*M by the formula

$$\chi^*(x, \varphi) = \limsup_{m \to +\infty} \frac{1}{m} \log \|d_x' f^m \varphi\|.$$

Given $\mathbf{v} = (v_1, \dots, v_k)$, we denote by $\Gamma_k^{\mathbf{v}}(m)$ the volume of the parallelepiped generated by the vectors $d_x f^m v_1, \dots, d_x f^m v_k$.

Theorem 4.1. *The point $x \in M$ is LP-regular if and only if there exist a decomposition*

$$T_x M = \bigoplus_{i=1}^{s(x)} E_i(x) \tag{4.3}$$

into subspaces $E_i(x)$ and numbers $\chi_1(x) < \dots < \chi_{s(x)}(x)$ such that:

(1) *$E_i(x)$ is invariant under $d_x f$, i.e.,*

$$d_x f E_i(x) = E_i(f(x)),$$

and depends Borel measurably on x;

(2) *for $v \in E_i(x) \setminus \{0\}$,*

$$\lim_{m \to \pm\infty} \frac{1}{m} \log \|d_x f^m v\| = \chi_i(x)$$

with uniform convergence on $\{v \in E_i(x) : \|v\| = 1\}$;

(3) *if* $\mathbf{v} = (v_1, \ldots, v_{k_i(x)})$ *is a basis of* $E_i(x)$, *then*

$$\lim_{m\to\pm\infty} \frac{1}{m} \log \Gamma^{\mathbf{v}}_{k_i(x)}(m) = \chi_i(x) k_i(x);$$

(4) *for any* $v, w \in T_xM \setminus \{0\}$ *with* $\angle(v,w) \neq 0$,

$$\lim_{m\to\pm\infty} \frac{1}{m} \log \angle(d_x f^m v, d_x f^m w) = 0;$$

(5) *the Lyapunov exponent* $\chi(x,\cdot)$ *is exact, that is,*

$$\liminf_{m\to\pm\infty} \frac{1}{m} \log \Gamma^{\mathbf{v}}_k(m) = \limsup_{m\to\pm\infty} \frac{1}{m} \log \Gamma^{\mathbf{v}}_k(m)$$

for any $1 \leq k \leq p$ *and any vectors* $v_1, \ldots, v_k$;

(6) *there exists a decomposition of the cotangent bundle*

$$T^*_x M = \bigoplus_{i=1}^{s(x)} E^*_i(x)$$

into subspaces $E^*_i(x)$ *associated with the Lyapunov exponent* χ^*; *the subspaces* $E^*_i(x)$ *are invariant under* $d'_x f$, *i.e.*,

$$d'_x f E^*_i(x) = E^*_i(f(x)),$$

and depend (Borel) measurably on x; *moreover, if* $\{v_i(x) : i = 1, \ldots, p\}$ *is a subordinate basis with* $v_i(x) \in E_j(x)$ *for* $n_{j-1}(x) < i \leq n_j(x)$ *and if* $\{v^*_i(x) : i = 1, \ldots, p\}$ *is a dual basis, then* $v^*_i(x) \in E^*_j(x)$ *for* $n_{j-1}(x) < i \leq n_j(x)$.

The decomposition (4.3) is called the *Oseledets decomposition* associated with the Lyapunov exponent (4.2). The following theorem of Oseledets (see [**65**]) is a key result in studying the LP-regularity of trajectories of dynamical systems. It shows that LP-regularity is "typical" from the measure-theoretic point of view.

Theorem 4.2 (Multiplicative Ergodic Theorem). *If* f *is a* C^1 *diffeomorphism of a compact smooth Riemannian manifold* M, *then the set of Lyapunov–Perron regular points has full measure with respect to any* f*-invariant Borel probability measure on* M.

The proof of this theorem is given in Chapter 6.

We denote by $\mathcal{R} \subset M$ the set of points that are LP-regular. Although the notion of LP-regularity does not require any invariant measure to be present, the crucial consequence of Theorem 4.2 is that $\mathcal{R}$ is nonempty and indeed has full measure with respect to any f-invariant Borel probability measure on M.

We stress that there may exist trajectories which are both forward and backward regular but *not* LP-regular. However, such trajectories form a negligible set with respect to any f-invariant Borel probability measure. Note that only in some exceptional situations is *every* point in M LP-regular.[1]

Let ν be an *ergodic* f-invariant Borel measure. Since the values of Lyapunov exponents are invariant Borel functions, there exist numbers $s = s^\nu$, $\chi_i = \chi_i^\nu$, and $k_i = k_i^\nu$ for $i = 1, \dots, s$ such that

$$s(x) = s, \quad \chi_i(x) = \chi_i, \quad k_i(x) = k_i \tag{4.4}$$

for ν-almost every x. The collection of pairs

$$\operatorname{Sp}\chi(\nu) = \{(\chi_i, k_i) : 1 \le i \le s\}$$

is called the *Lyapunov spectrum* of the measure ν.

We now consider dynamical systems whose spectrum of the Lyapunov exponent does not vanish on some subset of M. More precisely, let

$$\begin{aligned}\mathcal{E} = \big\{x \in \mathcal{R} :\ &\text{there exists } 1 \le k(x) < s(x) \\ &\text{with } \chi_{k(x)}(x) < 0 \text{ and } \chi_{k(x)+1}(x) > 0\big\}.\end{aligned} \tag{4.5}$$

This set is f-invariant. We say that f is a *dynamical system with nonzero Lyapunov exponents* if there exists an f-invariant Borel probability measure ν on M such that $\nu(\mathcal{E}) = 1$. The measure ν is called a *hyperbolic measure* for f.

Observe that every point $x \in \mathcal{E}$ is LP-regular and satisfies (4.4). Therefore, by the Multiplicative Ergodic Theorem, for every $x \in \mathcal{E}$,

$$E^s(x) = \bigoplus_{i=1}^{k} E_i(x) \quad \text{and} \quad E^u(x) = \bigoplus_{i=k+1}^{s} E_i(x).$$

The following theorem from [**68**] describes the properties of these subspaces.

Theorem 4.3. *The subspaces $E^s(x)$ and $E^u(x)$, $x \in \mathcal{E}$, have the following properties:*

(L1) *they depend Borel measurably on x;*

(L2) *they form a splitting of the tangent space, i.e. $T_xM = E^s(x) \oplus E^u(x)$;*

(L3) *they are invariant:*

$$d_xfE^s(x) = E^s(f(x)) \quad \text{and} \quad d_xfE^u(x) = E^u(f(x)).$$

Furthermore, given $x \in \mathcal{E}$, there exist $\varepsilon_0 = \varepsilon_0(x) > 0$ and functions $C(x, \varepsilon) > 0$ and $K(x, \varepsilon) > 0$ with $0 < \varepsilon \le \varepsilon_0$ such that:

[1] This is true, for example, when f is a linear hyperbolic toral automorphism (see Section 1.1).

(L4) *the subspace* $E^s(x)$ *is* stable*: if* $v \in E^s(x)$ *and* $n > 0$*, then*

$$\|d_x f^n v\| \le C(x,\varepsilon) e^{(\chi_{k(x)}(x)+\varepsilon)n} \|v\|;$$

(L5) *the subspace* $E^u(x)$ *is* unstable*: if* $v \in E^u(x)$ *and* $n < 0$*, then*

$$\|d_x f^n v\| \le C(x,\varepsilon) e^{(\chi_{k(x)+1}(x)-\varepsilon)n} \|v\|;$$

(L6)

$$\angle(E^s(x), E^u(x)) \ge K(x,\varepsilon);$$

(L7) *the functions* $C(x,\varepsilon)$ *and* $K(x,\varepsilon)$ *are Borel measurable in* x *and for every* $m \in \mathbb{Z}$*,*

$$C(f^m(x),\varepsilon) \le C(x,\varepsilon) e^{\varepsilon|m|} \quad \text{and} \quad K(f^m(x),\varepsilon) \ge K(x,\varepsilon) e^{-\varepsilon|m|}.$$

We remark that condition (L7) is crucial and is a manifestation of the LP-regularity (it is an analog of condition (3.30) in the discrete time case). Roughly speaking, it means that the estimates (L4), (L5), and (L6) may deteriorate as $|m| \to \infty$ but only with subexponential rate.[2] We stress that the rates of contraction along stable subspaces and expansion along unstable subspaces are substantially stronger.

Proof of Theorem 4.3. The first three statements are immediate consequence of Theorem 4.1.

Given $q > 0$, consider the sets

$$\mathcal{E}_q = \left\{ x \in \mathcal{E} : \chi_{k(x)}(x) < -\frac{1}{q},\ \chi_{k(x)+1}(x) > \frac{1}{q} \right\},$$

which are f-invariant, nested, i.e, $\mathcal{E}_q \subset \mathcal{E}_{q+1}$ for each $q > 0$, and exhaust $\mathcal{E}$, i.e., $\bigcup_{q>0} \mathcal{E}_q = \mathcal{E}$. It suffices to prove statements (L4)–(L7) of the theorem for every nonempty set $\mathcal{E}_q$. In what follows, we fix such a $q > 0$ and choose a sufficiently small number $\varepsilon_0 = \varepsilon_0(q)$. We need the following lemma.

Lemma 4.4. *Let* $X \subset M$ *be an* f*-invariant Borel set and let* $A(x,\varepsilon)$ *be a positive Borel function on* $X \times [0,\varepsilon_0)$*,* $0 < \varepsilon_0 < 1$*, such that for every* $\varepsilon_0 \ge \varepsilon > 0$*,* $x \in X$*, and* $m \in \mathbb{Z}$*,*

$$M_1(x,\varepsilon) e^{-\varepsilon|m|} \le A(f^m(x),\varepsilon) \le M_2(x,\varepsilon) e^{\varepsilon|m|},$$

where $M_1(x,\varepsilon)$ *and* $M_2(x,\varepsilon)$ *are Borel functions. Then one can find positive Borel functions* $B_1(x,\varepsilon)$ *and* $B_2(x,\varepsilon)$ *such that*

$$B_1(x,\varepsilon) \le A(x,\varepsilon) \le B_2(x,\varepsilon), \tag{4.6}$$

and for $m \in \mathbb{Z}$*,*

$$B_1(x,\varepsilon) e^{-2\varepsilon|m|} \le B_1(f^m(x),\varepsilon), \quad B_2(x,\varepsilon) e^{2\varepsilon|m|} \ge B_2(f^m(x),\varepsilon). \tag{4.7}$$

[2] More precisely, this means that the estimates (L4), (L5), and (L6) may deteriorate as $|m| \to \infty$ with exponential rate $e^{\varepsilon|m|}$ for *arbitrarily* small ε. However, $C(x,\varepsilon)$ may increase to ∞ and $K(x,\varepsilon)$ may decrease to zero as $\varepsilon \to 0$.

Proof of the lemma. It follows from the conditions of the lemma that there exists $m(x,\varepsilon) > 0$ such that if $m \in \mathbb{Z}$ and $|m| > m(x,\varepsilon)$, then

$$-2\varepsilon \leq \frac{1}{|m|} \log A(f^m(x),\varepsilon) \leq 2\varepsilon.$$

Set

$$B_1(x,\varepsilon) = \min_{-m(x,\varepsilon)\leq i\leq m(x,\varepsilon)} \left\{1, A(f^i(x),\varepsilon)e^{2\varepsilon|i|}\right\},$$

$$B_2(x,\varepsilon) = \max_{-m(x,\varepsilon)\leq i\leq m(x,\varepsilon)} \left\{1, A(f^i(x),\varepsilon)e^{-2\varepsilon|i|}\right\}.$$

The functions $B_1(x,\varepsilon)$ and $B_2(x,\varepsilon)$ are Borel functions. Moreover, if $n \in \mathbb{Z}$, then

$$B_1(x,\varepsilon)e^{-2\varepsilon|n|} \leq A(f^n(x),\varepsilon) \leq B_2(x,\varepsilon)e^{2\varepsilon|n|}. \tag{4.8}$$

Furthermore, if $b_1 \leq 1 \leq b_2$ are such that

$$b_1 e^{-2\varepsilon|n|} \leq A(f^n(x),\varepsilon) \tag{4.9}$$

and

$$b_2 e^{2\varepsilon|n|} \geq A(f^n(x),\varepsilon) \tag{4.10}$$

for all $n \in \mathbb{Z}$, then $b_1 \leq B_1(x,\varepsilon)$ and $b_2 \geq B_1(x,\varepsilon)$. In other words,

$$\begin{aligned} B_1(x,\varepsilon) &= \sup\{b \leq 1 : \text{inequality (4.9) holds for all } n \in \mathbb{Z}\}, \\ B_2(x,\varepsilon) &= \inf\{b \geq 1 : \text{inequality (4.10) holds for all } n \in \mathbb{Z}\}. \end{aligned} \tag{4.11}$$

Inequalities (4.6) follow from (4.8) (with $n = 0$). We also have

$$A(f^{n+m}(x),\varepsilon) \leq B_2(x,\varepsilon)e^{2\varepsilon|n+m|} \leq B_2(x,\varepsilon)e^{2\varepsilon|n|+2\varepsilon|m|},$$

$$A(f^{n+m}(x),\varepsilon) \geq B_1(x,\varepsilon)e^{-2\varepsilon|n+m|} \geq B_1(x,\varepsilon)e^{-2\varepsilon|n|-2\varepsilon|m|}.$$

Comparing these inequalities with (4.8) written at the point $f^m(x)$ and taking (4.11) into account, we obtain (4.7). The proof of the lemma is complete. □

We apply Lemma 4.4 to construct the function $K(x,\varepsilon)$. Let $K\colon \mathcal{E}_q \to \mathbb{R}$ be a Borel function. It is said to be *tempered* at the point x if

$$\lim_{m\to\pm\infty} \frac{1}{m} \log K(f^m(x)) = 0. \tag{4.12}$$

It follows that the function $A(x,\varepsilon) = K(x)$ satisfies all the conditions of Lemma 4.4.

Fix $0 < \varepsilon < \varepsilon_0(q)$ and for $x \in \mathcal{E}_q$ consider the function

$$\gamma(x) = \angle(E^s(x), E^u(x)).$$

Exercise 4.5. Show that $\gamma(x)$ is a tempered function.

Therefore, applying Lemma 4.4, we conclude that the function $K(x,\varepsilon) = B_1(x, \frac{1}{2}\varepsilon)$ satisfies conditions (L6) and (L7) of the theorem provided the number $\varepsilon_0(q)$ is sufficiently small.

We will now show how to construct the function $C(x,\varepsilon)$. The proof is an elaboration for the discrete time case of arguments in the proof of Theorem 3.9 (see (3.30)).

Lemma 4.6. *There exist $\varepsilon_0 = \varepsilon_0(q)$ and a positive Borel function $D(x,\varepsilon)$ (where $x \in \mathcal{E}_q$ and $0 < \varepsilon < \varepsilon_0$) such that for $m \in \mathbb{Z}$ and $1 \le i \le s$,*

$$D(f^m(x),\varepsilon) \le D(x,\varepsilon)^2 e^{2\varepsilon|m|} \tag{4.13}$$

and for any $n \ge 0$,

$$\|df_{ix}^n\| \le D(x,\varepsilon)e^{(\chi_i+\varepsilon)n}, \quad \|df_{ix}^{-n}\| \ge D(x,\varepsilon)^{-1}e^{-(\chi_i+\varepsilon)n}$$

where $\chi_i = \chi_i(x)$ and $df_{ix}^n = d_x f^n|E_i(x)$.

Proof of the lemma. Let $x \in \mathcal{E}_q$. By Theorem 4.1 (we use the notation of that theorem) there exists $\varepsilon_0 = \varepsilon_0(q)$ such that for every $0 < \varepsilon < \varepsilon_0$ one can find a number $n(x,\varepsilon) \in \mathbb{N}$ such that for $n \ge n(x,\varepsilon)$,

$$\chi_i - \varepsilon \le \frac{1}{n}\log\|df_{ix}^n\| \le \chi_i + \varepsilon, \quad -\chi_i - \varepsilon \le \frac{1}{n}\log\|df_{ix}^{-n}\| \le -\chi_i + \varepsilon,$$

and

$$-\chi_i - \varepsilon \le \frac{1}{n}\log\|d'f_{ix}^n\| \le -\chi_i + \varepsilon, \quad \chi_i - \varepsilon \le \frac{1}{n}\log\|d'f_{ix}^{-n}\| \le \chi_i + \varepsilon,$$

where $d'f_{ix}^n = d'_x f^n|E_i^*(x)$ (recall that $E_i^*(x) \subset T_x^*M$ is the dual space to $E_i(x)$ and $d'_x f$ is the inverse of the codifferential). Set

$$D_1^+(x,\varepsilon) = \min_{1\le i\le s}\ \min_{0\le j\le n(x,\varepsilon)} \left\{1, \|df_{ix}^j\|e^{(-\chi_i+\varepsilon)j}, \|d'f_{ix}^j\|e^{(\chi_i+\varepsilon)j}\right\},$$

$$D_1^-(x,\varepsilon) = \min_{1\le i\le s}\ \min_{-n(x,\varepsilon)\le j\le 0} \left\{1, \|df_{ix}^j\|e^{(-\chi_i-\varepsilon)j}, \|d'f_{ix}^j\|e^{(\chi_i-\varepsilon)j}\right\},$$

$$D_2^+(x,\varepsilon) = \max_{1\le i\le s}\ \max_{0\le j\le n(x,\varepsilon)} \left\{1, \|df_{ix}^j\|e^{(-\chi_i-\varepsilon)j}, \|d'f_{ix}^j\|e^{(\chi_i-\varepsilon)j}\right\},$$

$$D_2^-(x,\varepsilon) = \max_{1\le i\le s}\ \max_{-n(x,\varepsilon)\le j\le 0} \left\{1, \|df_{ix}^j\|e^{(-\chi_i+\varepsilon)j}, \|d'f_{ix}^j\|e^{(\chi_i+\varepsilon)j}\right\},$$

and

$$D_1(x,\varepsilon) = \min\{D_1^+(x,\varepsilon), D_1^-(x,\varepsilon)\},$$

$$D_2(x,\varepsilon) = \max\{D_2^+(x,\varepsilon), D_2^-(x,\varepsilon)\},$$

$$D(x,\varepsilon) = \max\{D_1(x,\varepsilon)^{-1}, D_2(x,\varepsilon)\}.$$

The function $D(x,\varepsilon)$ is measurable, and if $n \geq 0$ and $1 \leq i \leq s$, then

$$\begin{aligned} D(x,\varepsilon)^{-1}e^{(\chi_i-\varepsilon)n} &\leq \|df_{ix}^n\| \leq D(x,\varepsilon)e^{(\chi_i+\varepsilon)n}, \\ D(x,\varepsilon)^{-1}e^{(-\chi_i-\varepsilon)n} &\leq \|df_{ix}^{-n}\| \leq D(x,\varepsilon)e^{(-\chi_i+\varepsilon)n}, \\ D(x,\varepsilon)^{-1}e^{(-\chi_i-\varepsilon)n} &\leq \|d'f_{ix}^n\| \leq D(x,\varepsilon)e^{(-\chi_i+\varepsilon)n}, \\ D(x,\varepsilon)^{-1}e^{(\chi_i-\varepsilon)n} &\leq \|d'f_{ix}^{-n}\| \leq D(x,\varepsilon)e^{(\chi_i+\varepsilon)n}. \end{aligned} \tag{4.14}$$

Moreover, if $d \geq 1$ is a number for which inequalities (4.14) hold for all $n \geq 0$ and $1 \leq i \leq s$ with $D(x,\varepsilon)$ replaced by d, then $d \geq D(x,\varepsilon)$. Therefore,

$$\begin{aligned} D(x,\varepsilon) = \inf\{d \geq 1 : &\text{ the inequalities (4.14) hold for all } n \geq 0 \\ &\text{and } 1 \leq i \leq s \text{ with } D(x,\varepsilon) \text{ replaced by } d\}. \end{aligned} \tag{4.15}$$

We wish to compare the values of the function $D(x,\varepsilon)$ at the points x and $f^m x$ for $m \in \mathbb{Z}$. Notice that for every $x \in M$, $v \in T_xM$, and $\varphi \in T_x^*M$ with $\varphi(v) = 1$ we have

$$(d'_x f\varphi)(d_x fv) = \varphi((d_x f)^{-1} d_x fv) = \varphi(v) = 1. \tag{4.16}$$

Using the Riemannian metric on the manifold M, we introduce the identification map $\tau_x \colon T_x^*M \to T_xM$ such that $\langle \tau_x(\varphi), v\rangle = \varphi(v)$ where $v \in T_xM$ and $\varphi \in T_x^*M$.

Let $\{v_k^n : k = 1, \dots, p\}$ be a basis of $E_i(f^n(x))$ and let $\{w_k^n : k = 1, \dots, p\}$ be the dual basis of $E_i^*(f^n(x))$. We have $\tau_{f^n(x)}(w_k^n) = v_k^n$. Denote by $A_{n,m}^i$ and $B_{n,m}^i$ the matrices corresponding to the linear maps $df_{if^m(x)}^n$ and $d'f_{if^m(x)}^n$ with respect to the above bases. It follows from (4.16) that

$$A_{m,0}^i(B_{m,0}^i)^* = \mathrm{Id}$$

where * stands for matrix transposition. Hence, for every $n > 0$ the matrix corresponding to the map $df_{if^m(x)}^n$ is

$$A_{n,m}^i = A_{n+m,0}^i(A_{m,0}^i)^{-1} = A_{n+m,0}^i(B_{m,0}^i)^*.$$

Therefore, in view of (4.14), we obtain the following:

(1) if $n > 0$, then

$$\begin{aligned} \|df_{if^m(x)}^n\| &\leq D(x,\varepsilon)^2 e^{(\chi_i+\varepsilon)(n+m)+(-\chi_i+\varepsilon)m} \\ &= D(x,\varepsilon)^2 e^{2\varepsilon m} e^{(\chi_i+\varepsilon)n}, \\ \|df_{if^m(x)}^n\| &\geq D(x,\varepsilon)^{-2} e^{(\chi_i-\varepsilon)(n+m)+(-\chi_i-\varepsilon)m} \\ &= D(x,\varepsilon)^{-2} e^{-2\varepsilon m} e^{(\chi_i-\varepsilon)n}, \end{aligned}$$

(2) if $n > 0$ and $m - n \geq 0$, then

$$\begin{aligned}\|df_{if^m(x)}^{-n}\| &\leq D(x,\varepsilon)^2 e^{(\chi_i+\varepsilon)(m-n)+(-\chi_i+\varepsilon)m}\\ &= D(x,\varepsilon)^2 e^{2\varepsilon m} e^{(-\chi_i+\varepsilon)n},\\ \|df_{if^m(x)}^{-n}\| &\geq D(x,\varepsilon)^{-2} e^{(\chi_i-\varepsilon)(m-n)+(-\chi_i-\varepsilon)m}\\ &= D(x,\varepsilon)^{-2} e^{-2\varepsilon m} e^{(-\chi_i-\varepsilon)n},\end{aligned}$$

(3) if $n > 0$ and $n - m \geq 0$, then

$$\begin{aligned}\|df_{if^m(x)}^{-n}\| &\leq D(x,\varepsilon)^2 e^{(\chi_i+\varepsilon)(n-m)+(-\chi_i+\varepsilon)m}\\ &= D(x,\varepsilon)^2 e^{2\varepsilon m} e^{(-\chi_i+\varepsilon)n},\\ \|df_{if^m(x)}^{-n}\| &\geq D(x,\varepsilon)^{-2} e^{(\chi_i-\varepsilon)(n-m)+(-\chi_i-\varepsilon)m}\\ &= D(x,\varepsilon)^{-2} e^{-2\varepsilon m} e^{(-\chi_i-\varepsilon)n}.\end{aligned}$$

Similar inequalities hold for the maps $d'f_{if^m(x)}^n$ for each $n, m \in \mathbb{Z}$. Comparing this with the inequalities (4.14) applied to the point $f^m(x)$ and using (4.15), we conclude that if $m \geq 0$, then

$$D(f^m(x),\varepsilon) \leq D(x,\varepsilon)^2 e^{2\varepsilon m}. \tag{4.17}$$

Similar arguments show that if $m \leq 0$, then

$$D(f^{-m}(x),\varepsilon) \leq D(x,\varepsilon)^2 e^{-2\varepsilon m}. \tag{4.18}$$

It follows from (4.17) and (4.18) that if $m \in \mathbb{Z}$, then

$$D(f^m(x),\varepsilon) \leq D(x,\varepsilon)^2 e^{2\varepsilon|m|}.$$

This completes the proof of the lemma. □

We now proceed with the proof of the theorem. Replacing in (4.13) m by $-m$ and x by $f^m(x)$, we obtain

$$D(f^m(x),\varepsilon) \geq \sqrt{D(x,\varepsilon)}e^{-\varepsilon|m|}. \tag{4.19}$$

Consider two disjoint subsets $\sigma_1, \sigma_2 \subset [1,s] \cap \mathbb{N}$ and set

$$L_1(x) = \bigoplus_{i\in\sigma_1} E_i(x), \quad L_2(x) = \bigoplus_{i\in\sigma_2} E_i(x)$$

and $\gamma_{\sigma_1\sigma_2}(x) = \angle(L_1(x), L_2(x))$. By Theorem 4.1 the function $\gamma_{\sigma_1\sigma_2}$ is tempered and hence, in view of Lemma 4.4 one can find a function $K_{\sigma_1\sigma_2}(x)$ satisfying condition (L7) such that

$$\gamma_{\sigma_1\sigma_2}(x) \geq K_{\sigma_1\sigma_2}(x).$$

Set

$$T(x,\varepsilon) = \min K_{\sigma_1\sigma_2}(x),$$

where the minimum is taken over all pairs of disjoint subsets $\sigma_1, \sigma_2 \subset [1, s] \cap \mathbb{N}$. The function $T(x, \varepsilon)$ satisfies condition (L7).

Let $v \in E^s(x)$. Write $v = \sum_{i=1}^k v_i$ where $v_i \in E_i(x)$. We have

$$\|v\| \le \sum_{i=1}^{k} \|v_i\| \le LT^{-1}(x, \varepsilon)\|v\|,$$

where $L > 1$ is a constant. Let us set

$$C'(x, \varepsilon) = LD(x, \varepsilon)T(x, \varepsilon)^{-1}.$$

It follows from (4.13) and (4.19) that the function $C'(x, \varepsilon)$ satisfies the condition of Lemma 4.4 with

$$M_1(x, \varepsilon) = \frac{2}{\pi} L\sqrt{D(x, \varepsilon)} \quad \text{and} \quad M_2(x, \varepsilon) = LD(x, \varepsilon)^2 T(x, \varepsilon)^{-1}.$$

Therefore, there exists a function $C_1(x, \varepsilon) \ge C'(x, \varepsilon)$ for which the statements of Lemma 4.4 hold.

Applying the above arguments to the inverse map f^{-1} and the subspace $E^u(x)$, one can construct a function $C_2(x, \varepsilon)$ for which the statements of Lemma 4.4 hold. The desired function $C(x, \varepsilon)$ is defined by

$$C(x, \varepsilon) = \max\{C_1(x, \varepsilon/2), C_2(x, \varepsilon/2)\}.$$

This completes the proof of Theorem 4.3. □

4.2. Nonuniform complete hyperbolicity

In this section we introduce one of the principal concepts of smooth ergodic theory—the notion of nonuniform hyperbolicity—and we discuss its relation to dynamical systems with nonzero Lyapunov exponents introduced in the previous section.

Let $f \colon M \to M$ be a diffeomorphism of a compact smooth Riemannian manifold M of dimension p and let $Y \subset M$ be an f-invariant nonempty measurable subset. Also let $\lambda, \mu \colon Y \to (0, \infty)$ and $\varepsilon \colon Y \to [0, \varepsilon_0]$ with $\varepsilon_0 > 0$ be measurable functions satisfying

$$\lambda(f(x)) = \lambda(x), \quad \mu(f(x)) = \mu(x), \quad \varepsilon(f(x)) = \varepsilon(x) \tag{4.20}$$

(i.e., these functions are f-invariant), and

$$\lambda(x)e^{\varepsilon(x)} < 1 < \mu(x)e^{-\varepsilon(x)}, \quad x \in Y. \tag{4.21}$$

We say that the set Y is *nonuniformly (completely) hyperbolic* if there exist measurable functions $C, K \colon Y \to (0, \infty)$ such that for every $x \in Y$:

(H1) there exists a decomposition $T_x M = E^1(x) \oplus E^2(x)$, depending measurably on $x \in Y$, such that

$$d_x f E^1(x) = E^1(f(x)) \quad \text{and} \quad d_x f E^2(x) = E^2(f(x)); \tag{4.22}$$

(H2) for $v \in E^1(x)$ and $m > 0$,

$$\|d_x f^m v\| \le C(x)\lambda(x)^m e^{\varepsilon(x)m}\|v\|;$$

(H3) for $v \in E^2(x)$ and $m < 0$,

$$\|d_x f^m v\| \le C(x)\mu(x)^m e^{-\varepsilon(x)m}\|v\|;$$

(H4) $\angle(E^1(x), E^2(x)) \ge K(x)$;

(H5) for $m \in \mathbb{Z}$,

$$C(f^m(x)) \le C(x)e^{\varepsilon(x)|m|}, \quad K(f^m(x)) \ge K(x)e^{-\varepsilon(x)|m|}.$$

We stress that, in general, one should expect the functions C and K to be only (Borel) measurable but not continuous. This means that these functions may jump arbitrarily near a given point $x \in Y$ in an uncontrollable way. Condition (H5), however, provides some control over these functions along the trajectory $\{f^n(x)\}$ for $x \in Y$: the function C can increase and the function K can decrease with a small exponential rate. If ν is an invariant Borel probability measure, for which $\nu(Y) > 0$, then given $\varepsilon > 0$, there exists a subset $A \subset Y$ with $\nu(A) \ge \nu(Y) - \varepsilon > 0$ such that the function C is bounded from above on A. Moreover, due to the Poincaré recurrence theorem almost every point $x \in A$ returns to A infinitely often. Therefore, the function C indeed oscillates along the trajectory $f^n(x)$, for almost every $x \in A$, but may still become arbitrarily large. A similar observation holds for the function K.

Note that the dimensions of E^1 and E^2 are measurable f-invariant functions and hence the set Y can be decomposed into finitely many disjoint invariant measurable subsets on which the dimensions of E^1 and E^2 are constant.

Exercise 4.7. Show that if Y is nonuniformly (completely) hyperbolic, then for every $x \in Y$:

(1) $d_x f^m E^1(x) = E^1(f^m(x))$ and $d_x f^m E^2(x) = E^2(f^m(x))$;

(2) for $v \in E^a(x)$ and $m < 0$,

$$\|d_x f^m v\| \ge C(f^m(x))^{-1}\lambda(x)^m e^{\varepsilon(x)m}\|v\|;$$

(3) for $v \in E^2(x)$ and $m > 0$,

$$\|d_x f^m v\| \ge C(f^m(x))^{-1}\mu(x)^m e^{-\varepsilon(x)m}\|v\|.$$

We summarize the discussion in the previous section by saying that the set $\mathcal{E}$ (see (4.5)) of LP-regular points with nonzero Lyapunov exponents is nonuniformly (completely) hyperbolic with

$$\lambda(x) = e^{\chi_{k(x)}(x)}, \quad \mu(x) = e^{\chi_{k(x)+1}(x)},$$
$$C(x) = C(x,\varepsilon), \quad K(x) = K(x,\varepsilon)$$

for *any* fixed $0 < \varepsilon \le \varepsilon_0(x)$ with sufficiently small $\varepsilon_0(x)$ (see conditions (L1)–(L7) in Section 4.1). In fact, finding trajectories with nonzero Lyapunov exponents seems to be a universal approach in establishing nonuniform hyperbolicity.

We emphasize that the set of points with nonzero Lyapunov exponents whose regularity coefficient is sufficiently small (but may not necessarily be zero) is nonuniformly hyperbolic for some $\varepsilon > 0$.

We now introduce the notion of uniform hyperbolicity. Let $0 < \lambda < 1 < \mu$ be some numbers and let $K \subset M$ be a measurable subset. We stress that K need not be f-invariant. The set K is said to be *uniformly hyperbolic* if there exist $c > 0$ and $\gamma > 0$ such that for every $x \in K$:

(1) there exists a decomposition $T_xM = E^1(x) \oplus E^2(x)$, depending measurably on $x \in K$ and satisfying (4.22) whenever $f(x) \in K$;

(2) (a) for $v \in E^1(x)$ and $m > 0$,

$$\|d_xf^m v\| \le c\lambda^m \|v\|;$$

(b) for $v \in E^2(x)$ and $m < 0$,

$$\|d_xf^m v\| \le c\mu^m \|v\|;$$

(c) $\angle(E^1(x), E^2(x)) \ge \gamma$.

We will show below that for a nonuniformly hyperbolic set K of full measure with respect to an invariant measure there are in fact uniformly hyperbolic (noninvariant) sets $K_\delta \subset K$ of measure at least $1 - \delta$ for arbitrarily small $\delta > 0$. This observation is crucial in studying the topological and measure-theoretic properties. We stress that the "parameters" of uniform hyperbolicity, i.e., the numbers c and γ, may depend on δ approaching ∞ and 0, respectively. We will then show a crucial fact: this can occur only with a small exponential rate.

We introduce the notion of nonuniform (complete) hyperbolicity for dynamical systems with continuous time. Consider a smooth flow φ_t on a compact smooth Riemannian manifold M which is generated by a vector field $X(x)$. A measurable φ_t-invariant subset $Y \subset M$ is said to be *nonuniformly (completely) hyperbolic* if there exist measurable functions $\lambda, \mu\colon Y \to (0,\infty)$ and $\varepsilon\colon Y \to [0,\varepsilon_0]$ with $\varepsilon_0 > 0$ satisfying (4.20) and (4.21), measurable functions C, $K\colon Y \times (0,1) \to (0,\infty)$, and subspaces $E^s(x)$ and $E^u(x)$ for each $x \in Y$, which satisfy conditions (H2)–(H5) and the following condition:

(H1′) the subspaces $E^s(x)$ and $E^u(x)$ depend measurably on x and together with the subspace $E^0(x) = \{\alpha X(x) : \alpha \in \mathbb{R}\}$ form an invariant splitting of the tangent space, i.e.,

$$T_xM = E^s(x) \oplus E^u(x) \oplus E^0(x),$$

with

$$d_x\varphi_t E^s(x) = E^s(\varphi_t(x)) \quad \text{and} \quad d_x\varphi_t E^u(x) = E^u(\varphi_t(x)).$$

We say that a dynamical system (with discrete or continuous time) is *nonuniformly (completely) hyperbolic* if it possesses an invariant nonuniformly hyperbolic subset.

4.3. Regular sets

By Luzin's theorem every measurable function on a measurable space X is "nearly" continuous with respect to a finite measure μ; that is, it is continuous outside a set of arbitrarily small measure. In other words, X can be exhausted by an increasing sequence of measurable subsets on which the function is continuous. In line with this idea, the regular sets are built to exhaust an invariant nonuniformly (completely) hyperbolic set Y by an increasing sequence of (not necessarily invariant) *uniformly* (completely) hyperbolic subsets, demonstrating that nonuniform (complete) hyperbolicity is "nearly" uniform.

Let f be a diffeomorphism of a compact smooth Riemannian manifold M and let $\lambda, \mu, \varepsilon$ be positive numbers satisfying

$$0 < \lambda e^{\varepsilon} < 1 < \mu e^{-\varepsilon}. \tag{4.23}$$

Given an integer j, $1 \leq j < n$, and $\ell \geq 1$, we denote by $\Lambda^{\ell}_{\lambda\mu\varepsilon j}$ the set of points $x \in M$ for which there exists a decomposition $T_xM = E^1_x \oplus E^2_x$ such that for every $k \in \mathbb{Z}$ and $m > 0$ the following properties hold:

(1) $\dim E^a_x = j$ (and hence, $\dim E^2_x = n - j$);

(2) if $v \in d_x f^k E^a_x$, then

$$\|d_{f^k(x)} f^m v\| \leq \ell \lambda^m e^{\varepsilon(m+|k|)} \|v\|$$

and

$$\|d_{f^k(x)} f^{-m} v\| \geq \ell^{-1} \lambda^{-m} e^{-\varepsilon(|k-m|+m)} \|v\|;$$

(3) if $v \in d_x f^k E^2_x$, then

$$\|d_{f^k(x)} f^{-m} v\| \leq \ell \mu^{-m} e^{\varepsilon(m+|k|)} \|v\|$$

and

$$\|d_{f^k} f^m v\| \geq \ell^{-1} \mu^m e^{-\varepsilon(|k+m|+m)} \|v\|;$$

(4)

$$\angle(d_x f^k E^1_x, d_x f^k E^2_x) \geq \ell^{-1} e^{-\varepsilon|k|}.$$

The set $\Lambda^{\ell}_{\lambda\mu\varepsilon j}$ is called a *regular set* (or a *Pesin set*). It is a (not necessarily invariant) uniformly hyperbolic set for f. We also introduce the *level set*

$$\Lambda_{\lambda\mu\varepsilon j} := \bigcup_{\ell\geq 1} \Lambda^{\ell}_{\lambda\mu\varepsilon j}.$$

Exercise 4.8. Show that:

(1) $\Lambda^{\ell}_{\lambda\mu\varepsilon j} \subset \Lambda^{\ell+1}_{\lambda\mu\varepsilon j}$, i.e, regular sets are nested;

(2) if $m \in \mathbb{Z}$, then $f^m(\Lambda^{\ell}_{\lambda\mu\varepsilon j}) \subset \Lambda^{\ell'}_{\lambda\mu\varepsilon j}$, where $\ell' = \ell \exp(|m|\varepsilon)$;

(3) the set $\Lambda_{\lambda\mu\varepsilon j}$ is f-invariant;

(4) if $\varepsilon < \log(1 + 1/\ell)$, then

$$\Lambda^{\ell-1}_{\lambda\mu\varepsilon j} \subset f(\Lambda^{\ell}_{\lambda\mu\varepsilon j}) \subset \Lambda^{\ell+1}_{\lambda\mu\varepsilon j} \quad \text{and} \quad \Lambda^{\ell-1}_{\lambda\mu\varepsilon j} \subset f^{-1}(\Lambda^{\ell}_{\lambda\mu\varepsilon j}) \subset \Lambda^{\ell+1}_{\lambda\mu\varepsilon j};$$

(5) the regular sets $\Lambda^{\ell} = \Lambda^{\ell}_{\lambda\mu\varepsilon j}$ are closed (and hence compact);

(6) the subspaces E^1_x and E^2_x vary continuously with $x \in \Lambda^{\ell}$ (with respect to the distance in the Grassmannian bundle).

It follows that every regular set $\Lambda^{\ell}_{\lambda\mu\varepsilon j}$ is a (not necessarily invariant) uniformly hyperbolic set for f.

Consider a nonuniformly (completely) hyperbolic set Y for f. Given positive numbers $\lambda, \mu, \varepsilon$ satisfying (4.23) and an integer j, $1 \leq j < n$, consider the measurable set

$$Y_{\lambda\mu\varepsilon j} = \{x \in Y : \lambda(x) \leq \lambda < \mu \leq \mu(x),\, \varepsilon(x) \leq \varepsilon,\, E^1(x) = j\}.$$

Clearly, $Y_{\lambda\mu\varepsilon j}$ is invariant under f and is nonempty if the numbers $\lambda, \mu, \varepsilon$, and j are chosen appropriately. For each integer $\ell \geq 1$, consider the measurable subset

$$Y^{\ell}_{\lambda\mu\varepsilon j} = \{x \in Y_{\lambda\mu\varepsilon j} : C(x) \leq \ell,\, K(x) \geq \ell^{-1}\}.$$

We have

$$Y^{\ell}_{\lambda\mu\varepsilon j} \subset Y^{\ell+1}_{\lambda\mu\varepsilon j} \quad \text{and} \quad Y_{\lambda\mu\varepsilon j} = \bigcup_{\ell\geq 1} Y^{\ell}_{\lambda\mu\varepsilon j}.$$

Note that $Y^{\ell}_{\lambda\mu\varepsilon j}$ is a uniformly hyperbolic set for f but need not be invariant nor compact.

Exercise 4.9. Show that:

(1) $Y^{\ell}_{\lambda\mu\varepsilon j} \subset \Lambda^{\ell}_{\lambda\mu\varepsilon j}$ for every $\ell \geq 1$;

(2) $E^1_x = E^1(x)$ and $E^2_x = E^2(x)$ for every $x \in \Lambda$.

It follows that every nonuniformly (completely) hyperbolic set Y can be exhausted by a nested sequence of (not necessarily invariant) uniformly (completely) hyperbolic sets $Y^\ell_{\lambda\mu\varepsilon j}$. Moreover, to the set Y one can associate the family of nonempty f-invariant level sets

$$\{\Lambda_{\lambda\mu\varepsilon j} : \lambda, \mu, \varepsilon \text{ satisfy (4.23)}\} \tag{4.24}$$

and for each $\lambda, \mu, \varepsilon, j$ the collection of nonempty *compact* regular sets

$$\{\Lambda^\ell = \Lambda^\ell_{\lambda\mu\varepsilon j} : \ell \geq 1\}. \tag{4.25}$$

Note that f is nonuniformly (completely) hyperbolic on each level set $\Lambda_{\lambda\mu\varepsilon j}$ as well as on the set $\Lambda = \bigcup \Lambda_{\lambda\mu\varepsilon j}$ (here the union is taken over all numbers $\lambda, \mu, \varepsilon$ satisfying (4.23) and $1 \leq j < p$) that can be viewed as an "extension" of the "original" nonuniformly (completely) hyperbolic set Y. We stress that the rates of exponential contraction $\lambda(x)$ and of exponential expansion $\mu(x)$ are uniformly bounded away from 1 on each level set $\Lambda_{\lambda\mu\varepsilon j}$ but may be arbitrarily close to 1 on Λ.

4.4. Nonuniform partial hyperbolicity

In Section 4.1 we studied diffeomorphisms whose values of the Lyapunov exponent are *all* nonzero on a nonempty set $\mathcal{E}$ (with some of the values being negative and the remaining ones being positive; see (4.5)). As we saw in Section 4.2, the set $\mathcal{E}$ is nonuniformly hyperbolic and the hyperbolicity is complete. In this section we discuss the more general case of partial hyperbolicity. It deals with the situation when *some* of the values of the Lyapunov exponent are negative and some among the remaining ones may be zero.

While for dynamical systems that are nonuniformly completely hyperbolic one can obtain a sufficiently complete description of their ergodic properties (with respect to smooth invariant measures; see Chapter 9), dynamical systems that are nonuniformly partially hyperbolic may not possess "nice" ergodic properties. However, some principal results describing local behavior of systems that are nonuniformly completely hyperbolic can be extended without much extra work to systems that are only nonuniformly partially hyperbolic.[3]

Let $Z \subset M$ be an f-invariant nonempty measurable subset and let $\lambda, \mu \colon Z \to (0, \infty)$ and $\varepsilon \colon Z \to [0, \varepsilon_0]$ for some $\varepsilon_0 > 0$ be measurable functions satisfying

$$\lambda(f(x)) = \lambda(x), \quad \mu(f(x)) = \mu(x), \quad \varepsilon(f(x)) = \varepsilon(x)$$

(i.e., these functions are f-invariant), and

$$\lambda(x)e^{\varepsilon(x)} < \mu(x)e^{-\varepsilon(x)}.$$

[3]This includes constructing local stable manifolds and establishing their absolute continuity (see Remarks 7.2 and 8.13).

We say that an invariant measurable set Z is *nonuniformly partially hyperbolic in the broad sense* if there exist measurable functions $C, K \colon \Lambda \to (0,\infty)$ such that conditions (H2)–(H5) hold.

As in Section 4.3, to each set Z that is nonuniformly partially hyperbolic in the broad sense one can associate a collection of level sets $Z_{\lambda\mu\varepsilon j}$, and for each λ, μ, and ε, a collection of regular sets $Z^{\ell}_{\lambda\mu\varepsilon j}$ over $\ell > 1$. Here λ, μ, and ε are positive numbers satisfying

$$0 < \lambda e^{\varepsilon} < \mu e^{-\varepsilon}.$$

The level sets are invariant and the regular sets are nested and exhaust the set Z. The set $\Lambda = \bigcup Z_{\lambda\mu\varepsilon j}$ is nonuniformly partially hyperbolic in the broad sense and $Z \subset \Lambda$. Observe that each regular set is *uniformly* partially hyperbolic in the broad sense but not necessarily invariant.

Let $f \colon M \to M$ be a diffeomorphism of a compact smooth Riemannian manifold M of dimension p. Consider the f-invariant set

$$\mathcal{F} = \{x \in \mathcal{R} : \text{there exists } 1 \le k(x) < s(x) \text{ with } \chi_{k(x)}(x) < 0\}.$$

Repeating the arguments in the proof of Theorem 4.3, one can show that f is nonuniformly partially hyperbolic in the broad sense on $\mathcal{F}$.

4.5. Hölder continuity of invariant distributions

As we saw in Section 1.1, the stable and unstable subspaces of an Anosov diffeomorphism f depend continuously on the point in the manifold. Since these subspaces at a point x are determined by the whole positive and, respectively, negative semitrajectory through x, their dependence on the point may not be differentiable even if f is real analytic. However, one can show that they depend Hölder continuously on the point.[4]

We remind the reader of the definition of Hölder continuous distribution. A k-dimensional *distribution* E on a smooth manifold M is a family of k-dimensional subspaces $E(x) \subset T_xM$. A Riemannian metric on M naturally induces distances in TM and in the space of k-dimensional distributions on TM. The Hölder continuity of a distribution E can be defined using these distances. More precisely, for a subspace $A \subset \mathbb{R}^p$ (where $p = \dim M$) and a vector $v \in \mathbb{R}^p$, set

$$d(v, A) = \min_{w \in A} \|v - w\|.$$

In other words, $d(v, A)$ is the distance from v to its orthogonal projection on A. For subspaces A and B in $\mathbb{R}^p$, define

$$d(A, B) = \max\left\{ \max_{v \in A, \|v\|=1} d(v, B), \max_{w \in B, \|w\|=1} d(w, A) \right\}.$$

[4]This result was proved by Anosov in [**2**] and is a corollary of Theorem 4.11 below.

Let $D \subset \mathbb{R}^p$ be a subset and let E be a k-dimensional distribution. The distribution E is said to be *Hölder continuous* with *Hölder exponent* $\alpha \in (0,1]$ and *Hölder constant* $L > 0$ if there exists $\varepsilon_0 > 0$ such that

$$d(E(x), E(y)) \leq L\|x-y\|^\alpha$$

for every $x, y \in D$ with $\|x-y\| \leq \varepsilon_0$.

Now let E be a continuous distribution on M. Choose a small number $\varepsilon > 0$ and an atlas $\{U_i\}$ of M. We say that E is *Hölder continuous* if the restriction $E|U_i$ is Hölder continuous for every i.

Exercise 4.10. (1) Show that if a distribution E on M is Hölder continuous with respect to an atlas $\{U_i\}$ of M, then it is also Hölder continuous with respect to any other atlas of M with the same Hölder exponent (but the Hölder constant may be different).

(2) Show that if a distribution E on M is Hölder continuous, then it remains Hölder continuous with the same Hölder exponent if the Riemannian metric is replaced by an equivalent smooth metric.

(3) Show that a distribution E on M is Hölder continuous if and only if there are positive constants C, α, and ε such that for every two points x and y with $\rho(x,y) \leq \varepsilon$ we have

$$d(E(x), \tilde{E}(x)) \leq C\rho(x,y)^\alpha,$$

where $\tilde{E}(x)$ is the subspace of T_xM that is the parallel transport of the subspace $E(y) \subset T_yM$ along the unique geodesic connecting x and y.[5]

Finally, given a subset $\Lambda \subset M$, we say that a distribution $E(x)$ on Λ is Hölder continuous if for an atlas $\{U_i\}$ of M, the restriction $E|U_i \cap \Lambda$ is Hölder continuous for every i.

If f is a diffeomorphism that is nonuniformly completely hyperbolic on an invariant subset Λ, then the stable and unstable subspaces depend only (Borel) measurably on the point in Λ. However, as we saw in Section 4.2, these subspaces vary continuously on the point in a regular set, and in this section we show that they depend Hölder continuously on the point. It should be stressed that in the case of nonuniform hyperbolicity, the Hölder continuity property requires higher regularity of the system, i.e., that f is of class $C^{1+\alpha}$. In what follows, we consider only the stable subspaces; the Hölder continuity of the unstable subspaces follows by reversing the time.

Let f be a $C^{1+\alpha}$ diffeomorphism of a compact smooth manifold M and let Y be a nonuniformly (completely) hyperbolic set for f. Consider the

[5] The geodesic connecting x and y is unique if the number ε is sufficiently small.

corresponding collection $\Lambda_{\lambda\mu\varepsilon j}$ of level sets and for each λ, μ, and ε the corresponding collection of regular sets $\Lambda^\ell = \Lambda^\ell_{\lambda\mu\varepsilon j}$, $\ell \geq 1$.

Theorem 4.11. *The stable and unstable distributions $E^s(x)$ and $E^u(x)$ depend Hölder continuously on $x \in \Lambda^\ell$.*

We shall prove a more general statement of which Theorem 4.11 is an easy corollary. It applies to the cases of complete as well as of partial hyperbolicity. By the Whitney Embedding Theorem, every manifold M can be embedded in the Euclidean space $\mathbb{R}^N$ for a sufficiently large N. If M is compact, the Riemannian metric on M is equivalent to the distance $\|x-y\|$ induced by the embedding. We assume in Theorem 4.13, without loss of generality, that the manifold is embedded in $\mathbb{R}^N$.

Given a number $\kappa > 0$, we say that two subspaces $E_1, E_2 \subset \mathbb{R}^N$ are *κ-transverse* if $\|v_1 - v_2\| \geq \kappa$ for all unit vectors $v_1 \in E_1$ and $v_2 \in E_2$.

Exercise 4.12. Show that the subspaces $E^s(x)$ and $E^u(x)$ are κ-transverse for $x \in \Lambda^\ell$ and some $\kappa > 0$ which is independent of x.

Theorem 4.13. *Let M be a compact m-dimensional C^2 submanifold of $\mathbb{R}^N$ for some $m < N$, and let $f\colon M \to M$ be a $C^{1+\beta}$ map for some $\beta \in (0,1)$. Assume that there exist a set $D \subset M$ and real numbers $0 < \lambda < \mu$, $c > 0$, and $\kappa > 0$ such that for each $x \in D$ there are κ-transverse subspaces $E_1(x)$, $E_2(x) \subset T_xM$ such that:*

(1) *$T_xM = E_1(x) \oplus E_2(x)$;*

(2) *for every $n > 0$ and every $v_1 \in E_1(x)$, $v_2 \in E_2(x)$ we have*

$$\|d_xf^n v_1\| \leq c\lambda^n \|v_1\| \text{ and } \|d_xf^n v_2\| \geq c^{-1}\mu^n \|v_2\|.$$

Then for every $a > \max_{z\in M} \|d_zf\|^{1+\beta}$, the distribution E_1 is Hölder continuous with exponent

$$\alpha = \frac{\log\mu - \log\lambda}{\log a - \log\lambda}\beta.$$

Proof. We follow the argument in **[18]** and we begin with two technical lemmas.

Lemma 4.14. *Let A_n and B_n, for $n = 0, 1, \ldots$, be two sequences of real $N \times N$ matrices such that for some $\Delta \in (0,1)$,*

$$\|A_n - B_n\| \leq \Delta a^n$$

for every positive integer n. Assume that there exist subspaces E_A, $E_B \subset \mathbb{R}^N$ and numbers $0 < \lambda < \mu$ and $C > 1$ such that $\lambda < a$ and for each $n \geq 0$,

$$\|A_nv\| \leq C\lambda^n\|v\| \text{ if } v \in E_A; \qquad \|A_nw\| \geq C^{-1}\mu^n\|w\| \text{ if } w \in E_A{}^\perp;$$

$$\|B_nv\| \leq C\lambda^n\|v\| \text{ if } v \in E_B; \qquad \|B_nw\| \geq C^{-1}\mu^n\|w\| \text{ if } w \in E_B{}^\perp.$$

Then

$$\operatorname{dist}(E_A, E_B) \leq 3C^2 \frac{\mu}{\lambda} \Delta^{\frac{\log \mu - \log \lambda}{\log a - \log \lambda}}.$$

Proof of the lemma. Set

$$Q_A^n = \{v \in \mathbb{R}^N : \|A_n v\| \leq 2C\lambda^n \|v\|\}$$

and

$$Q_B^n = \{v \in \mathbb{R}^N : \|B_n v\| \leq 2C\lambda^n \|v\|\}.$$

For each $v \in \mathbb{R}^N$, write $v = v_1 + v_2$, where $v_1 \in E_A$ and $v_2 \in {E_A}^{\perp}$. If $v \in Q_A^n$, then

$$\|A_n v\| = \|A_n(v_1 + v_2)\| \geq \|A_n v_2\| - \|A_n v_1\| \geq C^{-1}\mu^n \|v_2\| - C\lambda^n \|v_1\|,$$

and hence,

$$\|v_2\| \leq C\mu^{-n}(\|A_n v\| + C\lambda^n \|v_1\|) \leq 3C^2 \left(\frac{\lambda}{\mu}\right)^n \|v\|.$$

Therefore,

$$\operatorname{dist}(v, E_A) \leq 3C^2 \left(\frac{\lambda}{\mu}\right)^n \|v\|. \tag{4.26}$$

Set $\gamma = \lambda/a < 1$. There exists a unique nonnegative integer n such that $\gamma^{n+1} < \Delta \leq \gamma^n$. If $w \in E_B$, then

$$\begin{aligned}
\|A_n w\| &\leq \|B_n w\| + \|A_n - B_n\| \cdot \|w\| \\
&\leq C\lambda^n \|w\| + \Delta a^n \|w\| \\
&\leq (C\lambda^n + (\gamma a)^n)\|w\| \leq 2C\lambda^n \|w\|.
\end{aligned}$$

It follows that $w \in Q_A^n$ and hence, $E_B \subset Q_A^n$. By symmetry, $E_A \subset Q_B^n$. By (4.26) and the choice of n, we obtain

$$\operatorname{dist}(E_A, E_B) \leq 3C^2 \left(\frac{\lambda}{\mu}\right)^n \leq 3C^2 \frac{\mu}{\lambda} \Delta^{\frac{\log \mu - \log \lambda}{\log a - \log \lambda}}.$$

This completes the proof of the lemma. □

For a diffeomorphism $f\colon M \to M$ of class $C^{1+\beta}$ the following result holds.

Exercise 4.15. There are positive constants $L > 0$ and $r > 0$ such that for any two points $x, y \in M$ for which $\|x - y\| < r$ we have that

$$\|d_x f - d_y f\| \leq L\|x - y\|^\beta. \tag{4.27}$$

Since M is compact, covering M by finitely many balls of radius r, we obtain that (4.27) holds for any $x, y \in M$ (with an appropriately chosen constant L). The following result extends the estimate (4.27) to powers of the map f.

Lemma 4.16. *Let $f\colon M \to M$ be a $C^{1+\beta}$ map of a compact m-dimensional C^2 submanifold $M \subset \mathbb{R}^N$. Then for every $a > \max_{z\in M} \|d_z f\|^{1+\beta}$ there exists $D > 1$ such that for every $n \in \mathbb{N}$ and every $x, y \in M$ we have*

$$\|d_x f^n - d_y f^n\| \le D a^n \|x - y\|^\beta.$$

Proof of the lemma. Let D' be such that

$$\|d_x f - d_y f\| \le D' \|x - y\|^\beta.$$

Set $b = \max_{z\in M} \|d_z f\| \ge 1$ and observe that for every x, $y \in M$,

$$\|f^n(x) - f^n(y)\| \le b^n \|x - y\|.$$

Fix $a > b$. Then the lemma holds true for $n = 1$ and any $D \ge D'$. For the inductive step we note that

$$\begin{aligned}
\|d_x f^{n+1} - d_y f^{n+1}\| &\le \|d_{f^n(x)} f\| \cdot \|d_x f^n - d_y f^n\| \\
&\quad + \|d_{f^n(x)} f - d_{f^n(y)} f\| \cdot \|d_y f^n\| \\
&\le b D a^n \|x - y\|^\beta + D' \big(b^n \|x - y\|\big)^\beta b^n \\
&\le D a^{n+1} \|x - y\|^\beta \left(\frac{b}{a} + \frac{D'}{D} \frac{(b^{1+\beta})^n}{a^{n+1}}\right).
\end{aligned}$$

If $a > b^{1+\beta}$, then there exists $D \ge D'$ for which the factor in parentheses is less than 1. □

We proceed with the proof of the theorem.

For $x \in M$, let $(T_x M)^\perp$ denote the orthogonal complement to the tangent plane $T_x M$ in $\mathbb{R}^N$. Since the distribution $(TM)^\perp$ is smooth, it is sufficient to prove that the distribution $F = E_1 \oplus (TM)^\perp$ is Hölder continuous.

Since $E_1(x)$ and $E_2(x)$ are κ-transverse and of complementary dimensions in $T_x M$, there exists $d > 1$ such that $\|d_x f^n w\| \ge d^{-1} \mu^n \|w\|$ for every $x \in D$ and $w \perp E_1(x)$.

For x, $y \in D$ and a positive integer n, let A_n and B_n be $N \times N$ matrices such that

$$\begin{aligned}
A_n v &= d_x f^n v \text{ if } v \in T_x M \quad \text{and} \quad A_n w = 0 \text{ if } w \in (T_x M)^\perp, \\
B_n v &= d_y f^n v \text{ if } v \in T_y M \quad \text{and} \quad B_n w = 0 \text{ if } w \in (T_y M)^\perp.
\end{aligned}$$

By Lemma 4.16,

$$\|A_n - B_n\| \le D a^n \|x - y\|^\beta.$$

Now Theorem 4.13 follows from Lemma 4.14 with $\Delta = D\|x - y\|^\beta$, $E_A = F(x)$, $E_B = F(y)$, and $C = \max\{c, d\}$. □

Chapter 5

Cocycles over Dynamical Systems

A sequence of matrices $(A_m)_{m\in\mathbb{Z}}$ in $\mathbb{R}^n$ can be viewed as a matrix function $\mathcal{A}: \mathbb{Z}\times\mathbb{Z}\to GL(n,\mathbb{R})$ satisfying the particular property that

$$\mathcal{A}(m,\ell+k)=\mathcal{A}(\sigma^k(m),\ell)\mathcal{A}(m,k),$$

where

$$\mathcal{A}(m,\ell)=A_{m+\ell-1}\cdots A_m$$

and $\sigma\colon \mathbb{Z}\to\mathbb{Z}$ is the shift, i.e., $\sigma(m)=m+1$. This property identifies the function $\mathcal{A}$ as a (linear) cocycle over the shift σ. A general concept of cocycle is a far-reaching extension of this simple example in which the shift is replaced by an invertible map f acting on an abstract space X. One can then build the theory of Lyapunov exponents for cocycles extending the concepts of LP-regularity, nonuniform hyperbolicity, etc. This allows one to substantially broaden the applications of the theory. We stress that while LP-regularity and nonuniform hyperbolicity do not require the presence of any invariant measure, we shall see below that the study of cocycles over measurable transformations preserving finite measures provides many interesting new results.

Consider a diffeomorphism $f\colon M\to M$ of a compact smooth Riemannian n-manifold M. Given a trajectory $\{f^m(x)\}_{m\in\mathbb{Z}}$, $x\in M$, we can identify each tangent space $T_{f^m(x)}M$ with $\mathbb{R}^n$ and thus obtain a sequence of matrices

$$(A_m(x))_{m\in\mathbb{Z}}=\{d_{f^m(x)}f\}_{m\in\mathbb{Z}}.$$

This generates a cocycle $\mathcal{A}_x$ over the transformation $f\colon X\to X$, where $X=\{f^m(x)\}_{m\in\mathbb{Z}}$. This cocycle can be viewed as the *individual derivative cocycle* associated with the trajectory of x. Dealing with such cocycles allows one to study LP-regularity and nonuniform hyperbolicity of individual trajectories and, as we shall see below, construct stable and unstable local and global manifolds along such trajectories.

While individual derivative cocycles depend on the choice of the individual trajectory, one can build the "global" cocycle associated with the diffeomorphism f. To this end we represent the manifold M as a finite union $\bigcup_i \Delta_i$ of differentiable copies Δ_i of the n-simplex such that:

(1) in each Δ_i one can introduce local coordinates in such a way that $T(\Delta_i)$ can be identified with $\Delta_i \times \mathbb{R}^n$;

(2) all nonempty intersections $\Delta_i \cap \Delta_j$, for $i \neq j$, are $(n-1)$-dimensional submanifolds.

In each Δ_i the derivative of f can be interpreted as a linear cocycle. This implies that $df\colon M \to \mathbb{R}^n$ can be interpreted as a measurable cocycle $\mathcal{A}$ with $df(x)$ the matrix representation of df in local coordinates. We call $\mathcal{A}$ the *derivative cocycle* specified by the diffeomorphism f. One can show that it does not depend on the choice of the decomposition $\{\Delta_i\}$; more precisely, if $\{\Delta_i'\}$ is another decomposition of M into copies of the n-simplex satisfying requirements (1) and (2) above, then the corresponding cocycle $\mathcal{A}'$ is equivalent to the cocycle $\mathcal{A}$ (see Section 5.1.3 for the definition of equivalent cocycles).

The derivative cocycle allows one to apply the results about general cocycles to smooth dynamical systems. A remarkable manifestation of this fact is the Multiplicative Ergodic Theorem for smooth dynamical systems (see Theorem 4.2) that claims that a "typical" individual cocycle with respect to a finite invariant measure is LP-regular. This theorem is an immediate corollary of the corresponding result for cocycles in Section 6 (see Theorem 6.1).

5.1. Cocycles and linear extensions

5.1.1. Linear multiplicative cocycles. Consider an invertible measurable transformation $f\colon X \to X$ of a measure space X. We call the function $\mathcal{A}\colon X \times \mathbb{Z} \to GL(n,\mathbb{R})$ a *linear multiplicative cocycle over* f or simply a *cocycle* if it has the following properties:

(1) for each $x \in X$ we have $\mathcal{A}(x,0) = \mathrm{Id}$, and given $m,k \in \mathbb{Z}$,

$$\mathcal{A}(x,m+k) = \mathcal{A}(f^k(x),m)\mathcal{A}(x,k);$$

(2) for every $m \in \mathbb{Z}$ the function $\mathcal{A}(\cdot,m)\colon X \to GL(n,\mathbb{R})$ is measurable.

Every cocycle is generated by a measurable function $A\colon X \to GL(n,\mathbb{R})$ called the *generator*. More precisely, every such function determines a cocycle by the formula

$$\mathcal{A}(x,m) = \begin{cases} A(f^{m-1}(x)) \cdots A(f(x))A(x) & \text{if } m > 0, \\ \mathrm{Id} & \text{if } m = 0, \\ A(f^m(x))^{-1} \cdots A(f^{-2}(x))^{-1} A(f^{-1}(x))^{-1} & \text{if } m < 0. \end{cases}$$

On the other hand, a cocycle $\mathcal{A}$ is generated by the matrix function $A = \mathcal{A}(\cdot, 1)$.

A cocycle $\mathcal{A}$ over f induces a *linear extension* $F\colon X \times \mathbb{R}^n \to X \times \mathbb{R}^n$ of f to $X \times \mathbb{R}^n$ (also known as a *linear skew product*). It is given by the formula

$$F(x,v) = (f(x), A(x)v).$$

In other words, the action of F on the fiber over x to the fiber over $f(x)$ is given by the linear map $A(x)$. If $\pi\colon X \times \mathbb{R}^n \to X$ is the projection defined by $\pi(x,v) = x$, then the diagram

$$\begin{array}{ccc} X \times \mathbb{R}^n & \xrightarrow{F} & X \times \mathbb{R}^n \\ {\scriptstyle\pi}\big\downarrow & & \big\downarrow{\scriptstyle\pi} \\ X & \xrightarrow{f} & X \end{array}$$

is commutative and for each $m \in \mathbb{Z}$ we obtain

$$F^m(x,v) = (f^m(x), \mathcal{A}(x,m)v). \tag{5.1}$$

Linear extensions are particular cases of bundle maps of measurable vector bundles that we now consider. Let E and X be measure spaces and let $\pi\colon E \to X$ be a measurable transformation. E is called a *measurable vector bundle* over X if there exists a countable collection of measurable subsets $Y_i \subset X$, which cover X, and measurable maps $\pi_i\colon X \times \mathbb{R}^n \to X$ with $\pi_i^{-1}(Y_i) = Y_i \times \mathbb{R}^n$. A bundle map $F\colon E \to E$ over a measurable transformation $f\colon X \to X$ is a measurable map, which makes the following diagram commutative:

$$\begin{array}{ccc} E & \xrightarrow{F} & E \\ {\scriptstyle\pi}\big\downarrow & & \big\downarrow{\scriptstyle\pi} \\ X & \xrightarrow{f} & X. \end{array}$$

The following proposition shows that from the measure theory point of view every vector bundle over a compact metric space is trivial, and hence without loss of generality, one may always assume that $E = X \times \mathbb{R}^n$. In other words, every bundle map of E is essentially a linear extension provided that the base space X is a compact metric space.

Proposition 5.1. *If E is a measurable vector bundle over a compact metric space (X, ν), then there is a subset $Y \subset X$ such that $\nu(Y) = 1$ and $\pi^{-1}(Y)$ is (isomorphic to) a trivial vector bundle.*

Exercise 5.2. Prove Proposition 5.1. Hint: Using the compactness of the space X, construct a finite cover by balls $\{B(x_1, r_1), \ldots, B(x_k, r_k)\}$ such that $\nu(\partial B(x_i, r_i)) = 0$ and $\pi | B(x_i, r_i)$ is measurably isomorphic to $B(x_i, r_i) \times \mathbb{R}^p$ for each $i = 1, \ldots, k$.

5.1.2. Operations with cocycles. Starting from a given cocycle, one can build other cocycles using some basic constructions in ergodic theory and algebra (see [**9**] for more details).

Let $\mathcal{A}$ be a cocycle over a measurable transformation f of a Lebesgue space X. For each $m \geq 1$ consider the transformation $f^m : X \to X$ and the measurable cocycle $\mathcal{A}^m$ over f^m with the generator

$$A^m(x) := A(f^{m-1}(x)) \cdots A(x).$$

The cocycle $\mathcal{A}^m$ is called the *mth power cocycle* of $\mathcal{A}$.

Let $f : X \to X$ be a measure-preserving transformation of a Lebesgue space (X, ν) and let $Y \subset X$ be a measurable subset of positive ν-measure. By Poincaré's Recurrence Theorem the set $Z \subset Y$ of points $x \in Y$ such that $f^n(x) \in Y$ for infinitely many positive integers n has measure $\nu(Z) = \nu(Y)$. We define the transformation $f_Y : Y \to Y$ (mod 0) as follows: for each $x \in Z$, set

$$k_Y(x) = \min\{k \geq 1 : f^k(x) \in Y\} \quad \text{and} \quad f_Y(x) = f^{k_Y(x)}(x).$$

One can easily verify that the function k_Y and the map f_Y are measurable on Z. We call k_Y the *(first) return time* to Y and f_Y the *(first) return map* or *induced transformation* on Y.

Lemma 5.3 (See, for example, [**27**]).

(1) *We have that $\int_Y k_Y \, d\nu = \nu(\bigcup_{n \geq 0} f^n(Y))$ and $k_Y \in L^1(X, \nu)$.*
(2) *The measure ν is invariant under f_Y.*

Since $k_Y \in L^1(X, \nu)$, by Birkhoff's Ergodic Theorem, the function

$$\tau_Y(x) = \lim_{k \to +\infty} \frac{1}{k} \sum_{i=0}^{k-1} k_Y(f_Y^i(x))$$

is well-defined for ν-almost all $x \in Y$ and $\tau_Y \in L^1(X, \nu)$.

If $\mathcal{A}$ is a measurable linear cocycle over f with generator A, we define the *induced cocycle* $\mathcal{A}_Y$ over f_Y to be the cocycle with the generator

$$A_Y(x) = A^{k_Y(x)}(x).$$

5.1.3. Cohomology and tempered equivalence. In this section we introduce the concept of two cocycles with the same base transformation being equivalent. Since cocycles act on fibers by linear transformations, we should first require that these linear actions be equivalent, i.e., that the corresponding matrices be conjugate by a linear coordinate change. We will then impose some requirements on how the coordinate change depends on the base point.

Following this line of thinking, consider two cocycles $\mathcal{A}$ and $\mathcal{B}$ over an invertible measurable transformation $f\colon X \to X$ and let $A, B\colon X \to GL(n, \mathbb{R})$ be their generators, respectively. The cocycles act on the fiber $\{x\} \times \mathbb{R}^n$ of $X \times \mathbb{R}^n$ by matrices $A(x)$ and $B(x)$, respectively. Now let $L\colon X \to GL(n, \mathbb{R})$ be a measurable matrix function such that for every $x \in X$,

$$A(x) = L(f(x))^{-1}B(x)L(x).$$

In other words, the actions of the cocycles $\mathcal{A}$ and $\mathcal{B}$ on the fibers are conjugate via the matrix function $L(x)$.

Exercise 5.4. Let $\mathcal{A}$ be a cocycle over an invertible measurable transformation $f\colon X \to X$ and let A be its generator. Also let $L\colon X \to GL(n, \mathbb{R})$ be a measurable matrix function. Given $x \in X$, let $v_{f(x)} = A(x)v_x$ and $u_x = L(x)v_x$. Show that the matrix function $B\colon X \to GL(n, \mathbb{R})$, for which $u_{f(x)} = B(x)u_x$ satisfies

$$A(x) = L(f(x))^{-1}B(x)L(x)$$

and generates a measurable cocycle $\mathcal{B}$ over f.

One can naturally think of the cocycles $\mathcal{A}$ and $\mathcal{B}$ as being equivalent. However, since the function L is in general only measurable, without any additional assumption on L the measure-theoretic properties of the cocycles $\mathcal{A}$ and $\mathcal{B}$ can be very different. We now introduce a sufficiently general class of coordinate changes, which have the important property of being tempered and which make the notion of equivalence productive.

Let $Y \subset X$ be an f-invariant nonempty measurable set. A measurable function $L\colon X \to GL(n, \mathbb{R})$ is said to be *tempered on Y with respect to f* or simply *tempered on Y* if for every $x \in Y$ we have

$$\lim_{m\to\pm\infty} \frac{1}{m} \log\|L(f^m(x))\| = \lim_{m\to\pm\infty} \frac{1}{m} \log\|L(f^m(x))^{-1}\| = 0.$$

A cocycle over f is said to be *tempered on Y* if its generator is tempered on Y.

If the real functions $x \mapsto \|L(x)\|$, $\|L(x)^{-1}\|$ are bounded or, more generally, have finite essential supremum, then the function L is tempered with respect to any invertible transformation $f\colon X \to X$ on any f-invariant nonempty measurable subset $Y \subset X$. The following statement provides a more general criterion for a function L to be tempered.

Proposition 5.5. *Let $f\colon X \to X$ be an invertible transformation preserving a probability measure ν, and let $L\colon X \to GL(n, \mathbb{R})$ be a measurable function. If*

$$\log\|L\|, \log\|L^{-1}\| \in L^1(X, \nu),$$

then L is tempered on some set of full ν-measure.

Proof. We need the following result.

Exercise 5.6. Let $f\colon X \to X$ be a measurable transformation preserving a probability measure ν. Show that if the function $\varphi\colon X \to \mathbb{R}$ is in $L^1(X, \nu)$, then

$$\lim_{m \to +\infty} \frac{\varphi(f^m(x))}{m} = 0$$

for ν-almost every $x \in X$.

The proposition follows by applying Exercise 5.6 to the functions $x \mapsto \log\|L(x)\|$ and $x \mapsto \log\|L(x)^{-1}\|$. □

We now proceed with the notion of equivalence for cocycles. Let $\mathcal{A}$ and $\mathcal{B}$ be two cocycles over an invertible measurable transformation f and let A, B be their generators, respectively. Given a measurable subset $Y \subset X$, we say that the cocycles are *equivalent on* Y or *cohomologous on* Y if there exists a measurable function $L\colon X \to GL(n, \mathbb{R})$, which is tempered on Y, such that for every $x \in Y$ we have

$$A(x) = L(f(x))^{-1} B(x) L(x). \tag{5.2}$$

This is clearly an equivalence relation and if two cocycles $\mathcal{A}$ and $\mathcal{B}$ are equivalent, we write $\mathcal{A} \sim_Y \mathcal{B}$. Equation (5.2) is called the *cohomological equation.*

It follows from (5.2) that for any $x \in Y$ and $m \in \mathbb{Z}$,

$$\mathcal{A}(x, m) = L(f^m(x))^{-1} \mathcal{B}(x, m) L(x). \tag{5.3}$$

Proposition 5.5 immediately implies the following result.

Corollary 5.7. *If $L\colon X \to \mathbb{R}$ is a measurable function such that* $\log\|L\|$, $\log\|L^{-1}\| \in L^1(X, \nu)$, *then any two cocycles $\mathcal{A}$ and $\mathcal{B}$ satisfying* (5.3) *are equivalent cocycles.*

Notice that if two cocycles $\mathcal{A}$ and $\mathcal{B}$ are equivalent and $x = f^m(x)$ is a periodic point, then, by (5.3), the matrices $\mathcal{A}(x, m)$ and $\mathcal{B}(x, m)$ are conjugate. In a similar manner to that in Livshitz's theorem, one can ask whether the converse holds. This is indeed true for some classes of cocycles with hyperbolic behavior. We refer the reader to [**49**] for a detailed discussion.

5.2. Lyapunov exponents and Lyapunov–Perron regularity for cocycles

Let $\mathcal{A}$ be a cocycle over an invertible measurable transformation f of a measure space X. For every $x \in X$, the cocycle $\mathcal{A}$ generates the sequence of matrices $\{A_m\}_{m\in\mathbb{Z}} = \{A(f^m(x))\}_{m\in\mathbb{Z}}$. Using the notion of the Lyapunov exponent for these sequences of matrices (see Section 2.5), we will introduce the notion of the Lyapunov exponent for the cocycle. In doing so, one should carefully examine the dependence of the Lyapunov exponent when moving from one such sequence of matrices to another one, i.e., moving from one trajectory of f to another one. This is a crucial new element in studying Lyapunov exponents for cocycles over dynamical systems versus studying Lyapunov exponents for sequences of matrices.

Given a point $(x, v) \in X \times \mathbb{R}^n$, define the *(forward) Lyapunov exponent of* (x, v) *(with respect to the cocycle* $\mathcal{A}$*)* by

$$\chi^+(x, v) = \chi^+(x, v, \mathcal{A}) = \limsup_{m\to+\infty} \frac{1}{m} \log\|\mathcal{A}(x, m)v\|.$$

Exercise 5.8. Show that for each $x \in X$ the function $\chi^+(x, \cdot)$ is a Lyapunov exponent in $\mathbb{R}^n$ and that it does not depend on the choice of the inner product on $\mathbb{R}^n$.

With the convention that $\log 0 = -\infty$, we obtain $\chi^+(x, 0) = -\infty$ for every $x \in X$.

Fix $x \in X$. By Theorem 2.1, there exist a positive integer $p^+(x) \leq n$, a collection of numbers

$$\chi_1^+(x) < \chi_2^+(x) < \cdots < \chi_{p^+(x)}^+(x),$$

and a filtration $\mathcal{V}_x^+$ of linear subspaces

$$\{0\} = E_0^+(x) \subsetneq E_1^+(x) \subsetneq \cdots \subsetneq E_{p^+(x)}^+(x) = \mathbb{R}^n,$$

such that:

(1) $E_i^+(x) = \{v \in \mathbb{R}^n : \chi^+(x, v) \leq \chi_i^+(x)\}$;

(2) $\chi^+(x, v) = \chi_i^+(x)$ for $v \in E_i^+(x) \setminus E_{i-1}^+(x)$.

The numbers $\chi_i^+(x)$ are the values of the Lyapunov exponent χ^+ at x. The collection of linear spaces $E_i^+(x, \mathcal{A}) = E_i^+(x)$ is the filtration $\mathcal{V}_x^+$ of $\mathbb{R}^n$ associated to χ^+ at x. The number

$$k_i^+(x) = \dim E_i^+(x) - \dim E_{i-1}^+(x)$$

is the multiplicity of the value $\chi_i^+(x)$. The Lyapunov spectrum of χ^+ at x is the collection of pairs

$$\operatorname{Sp}_x^+ \mathcal{A} = \{(\chi_i^+(x), k_i^+(x)) : i = 1, \ldots, p^+(x)\}.$$

Observe that $k_i^+(f(x)) = k_i^+(x)$ for every $x \in X$ and $1 \le i \le p^+(x)$, and hence, $\mathrm{Sp}_{f(x)}^+ \mathcal{A} = \mathrm{Sp}_x^+ \mathcal{A}$ for any $x \in X$.

Exercise 5.9. Show that the following properties hold:

(1) the functions χ^+ and p^+ are Borel measurable;

(2) $\chi^+ \circ F = \chi^+$ and $p^+ \circ f = p^+$, where F is the linear extension given by (5.1);

(3) $A(x)E_i^+(x) = E_i^+(f(x))$ and $\chi_i^+(f(x)) = \chi_i^+(x)$ for every $x \in X$ and $1 \le i \le p^+(x)$.

The Lyapunov exponent, its spectrum, and associated filtrations are invariant under tempered coordinate changes.

Proposition 5.10. *Let $\mathcal{A}$ and $\mathcal{B}$ be equivalent cocycles over a measurable transformation $f\colon X \to X$ and let $L\colon X \to GL(n,\mathbb{R})$ be a measurable function that is tempered (see (5.2)). For any $x \in X$:*

(1) *the forward and backward Lyapunov spectra at x coincide; i.e.,*

$$\mathrm{Sp}_x^+ \mathcal{A} = \mathrm{Sp}_x^+ \mathcal{B} \quad \text{and} \quad \mathrm{Sp}_x^- \mathcal{A} = \mathrm{Sp}_x^- \mathcal{B};$$

(2) *$L(x)$ preserves the forward and backward filtrations of $\mathcal{A}$ and $\mathcal{B}$; i.e., for each $i = 1, \dots, p^+(x)$,*

$$L(x)E_i^+(x,\mathcal{A}) = E_i^+(x,\mathcal{B}),$$

and for each $i = 1 \dots, p^-(x)$,

$$L(x)E_i^-(x,\mathcal{A}) = E_i^-(x,\mathcal{B}).$$

Proof. Since the function L satisfies (5.3), for each $v \in \mathbb{R}^n$ we obtain

$$\begin{aligned}\limsup_{m\to\pm\infty} \frac{1}{|m|} \log\|\mathcal{A}(x,m)v\| \le{} & \limsup_{m\to\pm\infty} \frac{1}{|m|} \log\|L(f^m(x))^{-1}\| \\ & + \limsup_{m\to\pm\infty} \frac{1}{|m|} \|\mathcal{B}(x,m)L(x)v\|.\end{aligned}$$

Since L is tempered, this implies that

$$\chi^+(x,v,\mathcal{A}) \le \chi^+(x,L(x)v,\mathcal{B}) \quad \text{and} \quad \chi^-(x,v,\mathcal{A}) \le \chi^-(x,L(x)v,\mathcal{B}).$$

Writing $\mathcal{B}(x,m)L(x) = L(f^m(x))\mathcal{A}(x,m)$, we conclude in a similar way that

$$\chi^+(x,L(x)v,\mathcal{B}) \le \chi^+(x,v,\mathcal{A}) \quad \text{and} \quad \chi^-(x,L(x)v,\mathcal{B}) \le \chi^-(x,v,\mathcal{A}).$$

The desired result now follows. $\square$

We say that x is a *forward* (respectively, *backward*) regular point for $\mathcal{A}$ if the sequence of matrices $(A(f^m(x)))_{m\in\mathbb{Z}}$ is forward (respectively, backward) regular.

One can easily check that if x is a forward (respectively, backward) regular point for $\mathcal{A}$, then for every $m \in \mathbb{Z}$ the point $f^m(x)$ is also forward (respectively, backward) regular for $\mathcal{A}$. Furthermore, if $\mathcal{A}$ and $\mathcal{B}$ are equivalent cocycles on Y, then the point $y \in Y$ is forward (respectively, backward) regular for $\mathcal{A}$ if and only if it is forward (respectively, backward) regular for $\mathcal{B}$.

Consider the families of filtrations $\mathcal{V}^+ = \{\mathcal{V}_x^+\}_{x\in X}$ and $\mathcal{V}^- = \{\mathcal{V}_x^-\}_{x\in X}$ of the cocycle $\mathcal{A}$. Recall that for each $x \in X$, $\mathcal{V}^+(x)$ and $\mathcal{V}^-(x)$ are filtrations associated with the Lyapunov exponents χ^+ and χ^- of the cocycle $\mathcal{A}$. For each $x \in X$ these filtrations determine filtrations $\mathcal{V}_x^+$ and $\mathcal{V}_x^-$ of the Lyapunov exponents $\chi^+(x,\cdot)$ and $\chi^-(x,\cdot)$ for the sequence of matrices $\{A_m\}_{m\in\mathbb{Z}} = \{A(f^m(x))\}_{m\in\mathbb{Z}}$.

We say that the families of filtrations $\mathcal{V}^+$ and $\mathcal{V}^-$ are *coherent* at a point $x \in X$ if the following properties hold:

(1) $p^+(x) = p^-(x) =: p(x)$;

(2) there exists a decomposition

$$\mathbb{R}^n = \bigoplus_{j=1}^{p(x)} H_j(x) \tag{5.4}$$

into subspaces $H_j(x)$ such that $A(x)H_j(x) = H_j(f(x))$ and

$$E_i^+(x) = \bigoplus_{j=1}^{i} H_j(x) \quad \text{and} \quad E_i^-(x) = \bigoplus_{j=i}^{p(x)} H_j(x);$$

(3) for $v \in H_i(x) \setminus \{0\}$, $i = 1, \ldots, p(x)$,

$$\lim_{m\to\pm\infty} \frac{1}{m} \log\|\mathcal{A}(x,m)v\| = \chi_i^+(x) = -\chi_i^-(x) =: \chi_i(x), \tag{5.5}$$

with uniform convergence on $\{v \in H_i(x) : \|v\| = 1\}$.

We call the decomposition (5.4) the *Oseledets decomposition*, and we call the subspaces $H_i(x)$ the *Oseledets subspaces* at x.

We remark that property (2) is equivalent to the following property: for $i = 1, \ldots, p(x)$ the spaces

$$H_i(x) = E_i^+(x) \cap E_i^-(x) \tag{5.6}$$

satisfy (5.4).

We say that a point $x \in X$ is *Lyapunov–Perron regular* or simply *LP-regular* for $\mathcal{A}$ if the following conditions hold:

(1) x is simultaneously forward and backward regular for $\mathcal{A}$;

(2) the families of filtrations $\mathcal{V}^+$ and $\mathcal{V}^-$ are coherent at x.

One can easily see that the set of LP-regular points is f-invariant. Condition (2) requires some degree of compatibility between forward and backward regularity. We remark that if $\mathcal{A}$ and $\mathcal{B}$ are equivalent cocycles on Y, then the point $y \in Y$ is LP-regular for $\mathcal{A}$ if and only if it is LP-regular for $\mathcal{B}$. We will show in Theorem 6.1 that under fairly general assumptions the set of LP-regular points has full measure with respect to any invariant measure.

Proposition 5.11. *If $x \in X$ is an LP-regular point for $\mathcal{A}$, then the following properties hold:*

(1) *the exponents $\chi^+(x, \cdot)$ and $\chi^-(x, \cdot)$ are exact; that is, the limits*

$$\lim_{m\to\pm\infty} \frac{1}{m} \log \operatorname{vol}(\mathcal{A}(x,m)v_1, \dots, \mathcal{A}(x,m)v_k)$$

exist for any $1 \le k \le n$ and any vectors $v_1, \dots, v_k$;

(2) $\dim H_i(x) = k_i^+(x) = k_i^-(x) =: k_i(x)$;

(3) *for any vectors $v_1, \dots, v_{k_i(x)} \in H_i(x)$ with $\operatorname{vol}(v_1, \dots, v_{k_i(x)}) \neq 0$, we have*

$$\lim_{m\to\pm\infty} \frac{1}{m} \log \operatorname{vol}(\mathcal{A}(x,m)v_1, \dots, \mathcal{A}(x,m)v_{k_i(x)}) = \chi_i(x)k_i(x);$$

(recall that $\operatorname{vol}(v_1, \dots, v_n)$ is the volume of the parallelepiped formed by the vectors $v_1, \dots, v_n$).

Identifying the space $H_i(f^m(x))$ with $\mathbb{R}^{k_i(x)}$ for every $m \in \mathbb{Z}$, one can rewrite property (3) in the following way: if $i = 1, \dots, p(x)$, then

$$\lim_{m\to\pm\infty} \frac{1}{m} \log|\det(\mathcal{A}(x,m)|H_i(x))| = \chi_i(x)k_i(x).$$

It also follows from Theorem 2.24 that for every LP-regular point $x \in X$, $1 \le i, j \le p(x)$ with $i \neq j$, and every distinct vectors $v, w \in H_i(x)$, we have

$$\lim_{m\to\pm\infty} \frac{1}{m} \log \sin \angle(H_i(f^m(x)), H_j(f^m(x))) = 0,$$

i.e., the angles between any two spaces $H_i(x)$ and $H_j(x)$ can grow at most subexponentially along the orbit of x, and

$$\lim_{m\to\pm\infty} \frac{1}{m} \log \sin \angle(\mathcal{A}(x,m)v, \mathcal{A}(x,m)w) = 0.$$

Remark 5.12. One can extend the notion of nonuniform hyperbolicity (see Section 4.2) to cocycles over dynamical systems and, in particular, prove an analog of Theorem 4.3 for cocycles. We refer the reader to [**9**] for details.

5.3. Examples of measurable cocycles over dynamical systems

Measurable cocycles over dynamical systems appear naturally in many areas of mathematics. In this section we describe two rather simple situations.

5.3.1. Reducible cocycles. Let $\mathcal{A}$ be a cocycle over a measurable transformation $f\colon X \to X$. We say that $\mathcal{A}$ is *reducible* if it is equivalent to a cocycle whose generator is constant along trajectories. More precisely, if $A\colon X \to GL(n, \mathbb{R})$ is the generator of $\mathcal{A}$, then there exist measurable functions $B, L\colon X \to GL(n, \mathbb{R})$ such that:

(1) B is f-invariant and L is tempered on X;

(2) $A(x) = L(f(x))^{-1}B(x)L(x)$ for every $x \in X$.

In particular,

$$\mathcal{A}(x, m) = L(f^m(x))^{-1}B(x)^m L(x).$$

We note that if $\mathcal{A}$ is a reducible cocycle, then every point $x \in X$ is LP-regular and the values of the Lyapunov exponent at a point $x \in X$ are equal to $\log|\lambda_i(x)|$, where $\lambda_i(x)$ are the eigenvalues of the matrix $B(x)$.

Reducible cocycles are natural objects in the classical Floquet theory (where the dynamical system in the base is a periodic flow). Consider an autonomous differential equation $x' = F(x)$, and let $\varphi_t(x)$ be a T-periodic solution. The corresponding system of variational equations is given by

$$z' = C_x(t)z, \tag{5.7}$$

where $C_x(t) = d_{\varphi_t(x)}F$ is a T-periodic matrix function. By Floquet theory, any fundamental solution of equation (5.7) is of the form $X(t) = P(t)e^{Dt}$ for some T-periodic matrix function $P(t)$ and a constant matrix D. Setting $L(\varphi_t(x)) = P(t)^{-1}$, the Cauchy matrix can be written in the form

$$X(t)X(0)^{-1} = L(\varphi_t(x))^{-1}e^{Dt}L(x).$$

Since $P(t)$ is T-periodic and thus bounded, the function L is tempered. It follows that the cocycle $X(t)X(0)^{-1}$ over the flow restricted to the orbit of x is reducible. More generally, if the flow defined by the equation $x' = F(x)$ has only constant or periodic trajectories (possibly with different periods), then the corresponding cocycle defined by the systems of variational equations is reducible.

We describe another example of reducible cocycles associated with translations on the torus. Given a translation vector $\omega \in \mathbb{R}^n$, let $f\colon \mathbb{T}^n \to \mathbb{T}^n$ be the translation $f(x) = x + \omega \bmod 1$. Consider the cocycle $\mathcal{A}(\omega)$ over f generated by the time-1 map obtained from the solutions of a linear differential equation

$$y' = B(x + t\omega)y,$$

where $B\colon T^n \to GL(n,\mathbb{R})$ is a given function. It was shown by Floquet that if $\omega/\pi \in \mathbb{Q}^n$, then the cocycle $\mathcal{A}(\omega)$ is reducible.

The case $\omega/\pi \notin \mathbb{Q}^n$ is more subtle and whether the cocycle $\mathcal{A}(\omega)$ is reducible depends on ω. One can show (see [**43**]) that there are translation vectors ω for which some points $x \in T^n$ are not LP-regular for the cocycle $\mathcal{A}$. This implies that for these ω the cocycle is not reducible.

We shall describe a condition on ω that (along with some other requirements on the cocycle) guarantees that $\mathcal{A}(\omega)$ is reducible. If $\omega/\pi \notin \mathbb{Q}^n$, then the vector $\tilde{\omega} = (\omega, 2\pi)$ is rationally independent, that is, $\langle k, \tilde{\omega}\rangle \neq 0$ whenever $k \in \mathbb{Z}^{n+1} \setminus \{0\}$. We say that ω satisfies a *Diophantine condition* if the vector $\tilde{\omega}$ is Diophantine; that is,

$$|\langle k, \tilde{\omega}\rangle| \geq \frac{c}{\|k\|^\tau}$$

for every $k \in \mathbb{Z}^{n+1} \setminus \{0\}$ and some $c, \tau > 0$. One can show that the set of Diophantine vectors has full Lebesgue measure. Johnson and Sell [**44**] obtained the following criterion for reducibility.

Theorem 5.13. *If ω satisfies a Diophantine condition and all Lyapunov exponents are exact and simple everywhere, then the cocycle $\mathcal{A}(\omega)$ is reducible.*

We refer the reader to [**34, 55**] for further results on the reducibility of cocycles when ω satisfies a Diophantine condition.

5.3.2. Cocycles associated with Schrödinger operators. The Schrödinger equation is a linear partial differential equation that is designed to describe how the quantum state of a physical system changes in time. Its discrete-time version is an equation of the form

$$-(u_{n+1} + u_{n-1}) + V(f^n(x))u_n = Eu_n, \tag{5.8}$$

where $f(x) = x + \alpha \pmod 1$ is an irrational rotation of the circle S^1, $E \in \mathbb{R}$ is the total energy of the system, and $V\colon S^1 \to \mathbb{R}$ is the potential energy.

Exercise 5.14. Show that equation (5.8) is equivalent to

$$\begin{pmatrix} u_{n+1} \\ u_n \end{pmatrix} = A(f^n(x)) \begin{pmatrix} u_n \\ u_{n-1} \end{pmatrix},$$

where $A\colon S^1 \to SL(2,\mathbb{R})$ is given by

$$A(\theta) = \begin{pmatrix} V(\theta) - E & -1 \\ 1 & 0 \end{pmatrix}.$$

Let $\lambda_1(V,x)$ and $\lambda_2(V,x)$ be the Lyapunov exponents of the cocycle generated by A at a point $x \in S^1$, and set

$$\Lambda_i(V) = \int_{S^1} \lambda_i(V,x)\,dx, \quad i = 1,2.$$

One can show (see [**14**]) that the map

$$C^0(S^1,\mathbb{R}) \ni V \mapsto (\Lambda_1(V), \Lambda_2(V))$$

is discontinuous at a potential V if and only if $\Lambda_1(V)$ and $\Lambda_2(V)$ are nonzero and E lies in the spectrum of the associated Schrödinger operator

$$(H_x\Psi)(n) = -(\Psi(n+1) + \Psi(n-1)) + V(f^n(x))\Psi(n)$$

in the space $\ell^2(\mathbb{Z})$ (we recall that the spectrum of a linear operator B is the complement of the set of numbers $E \in \mathbb{C}$ such that the inverse $(B-E)^{-1}$ is well-defined and bounded). This is due to the fact that E lies in the complement of the spectrum if and only if the cocycle is uniformly hyperbolic; see also Ruelle [**78**], Bourgain [**16**], and Bourgain and Jitomirskaya [**17**].

We present a result due to Ávila and Krikorian [**6**] that establishes nonuniform hyperbolicity for Schrödinger cocycles. We say that a number α satisfies the *recurrent Diophantine condition* if there are infinitely many $n > 0$ for which the nth image of α under the Gauss map satisfies the Diophantine condition with fixed constant c and power τ.

Theorem 5.15. *Assume that $V \in C^r(S^1,\mathbb{R})$ with $r = \omega$ or ∞ and that α satisfies the recurrent Diophantine condition. Then for almost every E the Schrödinger cocycle over the rotation by α either has nonzero Lyapunov exponents or is C^r-equivalent to a constant cocycle.*

Chapter 6

The Multiplicative Ergodic Theorem

The goal of this chapter is to present one of the principal results of this book, known as Oseledets's Multiplicative Ergodic Theorem (see [**65**]). Consider an invertible measure-preserving transformation f of a Lebesgue space (X, ν) and a measurable cocycle $\mathcal{A}$ over f. We stress that verifying nonuniform hyperbolicity of the cocycle $\mathcal{A}$ on X (or on a subset $Y \subset X$ of positive measure; see Remark 5.12) amounts to showing that (1) Lyapunov exponents on Y are nonzero and (2) every point $x \in Y$ is LP-regular. There are certain methods that allow one to estimate Lyapunov exponents and, in particular, to show that they are not equal to zero. However, establishing LP-regularity of a given trajectory is a daunting task. This is where the Multiplicative Ergodic Theorem comes into play: it provides a condition on the cocycle $\mathcal{A}$ and its invariant measure ν that guarantees that almost every trajectory of f is LP-regular.

Theorem 6.1. *Assume that the generator A of the cocycle $\mathcal{A}$ satisfies*

$$\log^+\|A\|, \log^+\|A^{-1}\| \in L^1(X, \nu), \tag{6.1}$$

where $\log^+ a = \max\{\log a, 0\}$. Then the set of LP-regular points for $\mathcal{A}$ has full ν-measure.

In the next two sections we present a proof of this theorem following the original Oseledets approach (see [**65**]). Its idea is to first reduce the general case to the case of triangular cocycles and then to prove the theorem for such cocycles.

6.1. Lyapunov–Perron regularity for sequences of triangular matrices

We begin by considering sequences of lower triangular matrices and we present a useful criterion of LP-regularity, which we exploit in the proof of Theorem 6.1.

Theorem 6.2. *Let* $\mathcal{A}^+ = (a_{ij}^m)_{m\in\mathbb{Z}} \subset GL(n,\mathbb{R})$ *be a sequence of lower triangular matrices such that:*

(1) *the limit*

$$\lim_{m\to+\infty}\frac{1}{m}\sum_{k=0}^{m}\log|a_{ii}^k| =: \lambda_i, \quad i=1,\dots,n,$$

exists and is finite;

(2) *we have*

$$\limsup_{m\to+\infty}\frac{1}{m}\log^+|a_{ij}^m| = 0, \quad i,j=1,\dots,n.$$

Then the sequence $\mathcal{A}^+$ *is forward regular, and the numbers* λ_i *are the values of the Lyapunov exponent* χ^+ *(counted with their multiplicities but possibly not ordered).*

Proof. Before going into the detailed proof, let us explain the main point. For the sake of this discussion let us count each exponents according to its multiplicity; thus we have exactly n exponents. To verify the forward regularity, we will produce a basis $\{v_1,\dots,v_n\}$, which is subordinate to the standard filtration (i.e., related to the standard basis by an upper triangular coordinate change) and such that $\chi^+(v_i)=\lambda_i$.

If the exponents are ordered so that $\lambda_1\geq\lambda_2\geq\cdots\geq\lambda_n$, then the standard basis is in fact forward regular. To see this, notice that while multiplying lower triangular matrices one obtains a matrix whose off-diagonal entries contain a polynomially growing number of terms each of which can be estimated by the growth of the product of diagonal terms below.

However, if the exponents are not ordered that way, then an element e_i of the standard basis will grow according to the maximal of the exponents λ_j for $j\geq i$. In order to produce the right growth, one has to compensate for the growth caused by off-diagonal terms by subtracting from the vector e_i a certain linear combination of vectors e_j for which $\lambda_j>\lambda_i$. This can be done in a unique fashion. The proof proceeds by induction in n.

For $n=1$ the result follows immediately from condition (1). Given $n>1$, we assume that the sequence of lower triangular matrices $A_m\in GL(n+1,\mathbb{R})$ satisfies conditions (1) and (2). For each $m\geq1$ let A'_m be the triangular $n \times n$ matrix obtained from A_m by deleting its first row and first column. The matrices A'_m satisfy conditions (1) and (2). Consider the Lyapunov exponent

$$\chi'(v) = \limsup_{m \to +\infty} \frac{1}{m} \log\|\mathcal{A}'_m v\|,$$

where $\mathcal{A}'_m = A'_{m-1} \cdots A'_0$ and the filtration

$$\{0\} = E_0 \subsetneq E_1 \subsetneq \cdots \subsetneq E_p = \mathbb{R}^n$$

associated to the Lyapunov exponent χ'. By the induction hypothesis, χ' is forward regular. Write $\mathbb{R}^n = \bigoplus_{i=1}^p H_i$ where $H_i = E_{i-1}{}^{\perp} \cap E_i$ and let $\lambda'_2 \leq \cdots \leq \lambda'_{n+1}$ be the numbers $\lambda_2, \ldots, \lambda_{n+1}$ in condition (1) written in nondecreasing order. Since $\mathcal{A}'_m$ is forward regular, conditions (1) and (2) guarantee that for $i \geq 2$ and $v \in H_i \setminus \{0\}$,

$$\lim_{m \to +\infty} \frac{1}{m} \log\|\mathcal{A}'_m v\| = \lambda'_i. \tag{6.2}$$

It remains only to show that λ_1 is a value of the Lyapunov exponent χ for some vector $v \in \mathbb{R}^{n+1} \setminus \bigoplus_{i=1}^p H_i$. Indeed, since $\mathcal{A}'_m$ is forward regular, we obtain that

$$\begin{aligned}\lim_{m \to +\infty} \frac{1}{m} \log|\det \mathcal{A}_m| &= \lim_{m \to +\infty} \frac{1}{m} \sum_{k=0}^{m} \log|a_{11}^k| \\ &+ \lim_{m \to +\infty} \frac{1}{m} \log|\det \mathcal{A}'_m| = \sum_{i=1}^{n+1} \lambda_i\end{aligned}$$

and thus, $\mathcal{A}_m$ is forward regular. By Theorem 2.24, we have

$$\lim_{m \to +\infty} \frac{1}{m} \log|\sin \angle(H_i^m, \widehat{H}_i^m)| = 0, \tag{6.3}$$

where

$$H_i^m = \mathcal{A}'_m H_i, \quad \widehat{H}_i^m = \bigoplus_{j \neq i} \mathcal{A}'_m H_i.$$

Let $\{v_1^m, \ldots, v_n^m\}$ be a basis of $\mathbb{R}^n$ such that $\{v_{n_{i-1}+1}^m, \ldots, v_{n_i}^m\}$ is an orthonormal basis of H_i^m, where $n_i = \dim E_i$, $i = 1, \ldots, p$. We denote by C_m the coordinate change from the standard basis $\{e_0, \ldots, e_n\}$ of $\mathbb{R}^{n+1}$ to the basis $\{e_0, v_1^m, \ldots, v_n^m\}$. It follows from (6.3) that

$$\lim_{m \to +\infty} \frac{1}{m} \log\|C_m\| = \lim_{m \to +\infty} \frac{1}{m} \log\|C_m{}^{-1}\| = 0. \tag{6.4}$$

Consider the sequence of matrices $B_m = C_{m+1}{}^{-1} A_m C_m$. Note that for each i,

$$B_m \operatorname{span}\{e_{n_{i-1}+1}, \ldots, e_{n_i}\} = \operatorname{span}\{e_{n_{i-1}+1}, \ldots, e_{n_i}\}.$$

Hence the matrix B_m has the form

$$B_m = \begin{pmatrix} a_{11}^m & & & & \\ g_1^m & B_1^m & & & \\ g_2^m & 0 & B_2^m & & \\ \vdots & \vdots & & \ddots & \\ g_p^m & 0 & \cdots & 0 & B_p^m \end{pmatrix},$$

where each B_i^m is a $k_i \times k_i$ matrix and each $g_i^m \in \mathbb{R}^{k_i}$ is a column vector. Set $\mathcal{B}_m = B_{m-1} \cdots B_0$ and observe that

$$\mathcal{B}_m = {C_m}^{-1} \mathcal{A}_m C_0. \tag{6.5}$$

The following lemma establishes a crucial property of the column vector g_i^m.

Lemma 6.3. *For each $i = 1, \dots, p$ we have*

$$\limsup_{m \to +\infty} \frac{1}{m} \log^+ \|g_i^m\| = 0.$$

Proof of the lemma. Let $g^m \in \mathbb{R}^n$ be the column vector composed of the components of $g_1^m, \dots, g_p^m$. We have

$$\|g^m\| \leq \|{C_{m+1}}^{-1}\| \cdot \|a_{21}^m e_1 + \cdots + a_{n+1,1}^m e_n\| \cdot \|C_m\|.$$

Therefore using (6.4) and condition (2), we obtain

$$\limsup_{m \to +\infty} \frac{1}{m} \log^+ \|g^m\| \leq \limsup_{m \to +\infty} \frac{1}{m} \log^+ \max\{|a_{i1}^m| : 2 \leq i \leq n+1\} = 0. \tag{6.6}$$

Moreover, since g_i^m is the projection of g^m onto H_i^{m+1} along $\widehat{E}_i^{m+1}$, we have

$$\|g_i^m\| \leq \|g^m\| / \sin \angle(H_i^{m+1}, \widehat{H}_i^{m+1})$$

and hence,

$$\frac{1}{m} \log \|g_i^m\| \leq \frac{1}{m} \log \|g^m\| - \frac{1}{m} \log \sin \angle(H_i^{m+1}, \widehat{H}_i^{m+1}).$$

By (6.3) and (6.6),

$$0 \leq \limsup_{m \to +\infty} \frac{1}{m} \log^+ \|g_i^m\| \leq \limsup_{m \to +\infty} \frac{1}{m} \log^+ \|g^m\| = 0$$

and the lemma follows. □

We proceed with the proof of the theorem.

Case 1: $\lambda_1 \geq \lambda'_{n+1}$ or $\lambda_1 \geq \lambda_j$ for all $2 \leq j \leq n+1$. This is an easy situation since the off-diagonal elements will grow sufficiently slowly compared to the first element of the basis. Observe that

$$\mathcal{B}_m e_0 = \prod_{i=0}^{m-1} a_{11}^i e_0 + \sum_{j=1}^{p} \sum_{i=1}^{m-1} a_{11}^{i-1} \cdots a_{11}^0 B_j^{m-1} \cdots B_j^{i+1} g_j^i. \tag{6.7}$$

Moreover, $\mathcal{B}_m v = B_j^{m-1} \cdots B_j^0 v$ for each $v \in H_j$. Given j, $1 \le j \le p$, set

$$c_{mj}(v) = \frac{\|B_j^m \cdots B_j^0 v\|}{\|B_j^{m-1} \cdots B_j^0 v\|}, \quad v \in C_0{}^{-1} H_j \setminus \{0\}, \quad m \ge 0.$$

By (6.4), (6.5), and (6.2), we obtain for such v,

$$\begin{aligned}\lambda_j' &= \lim_{m\to+\infty} \frac{1}{m} \log\|\mathcal{A}_m' C_0 v\| = \lim_{m\to+\infty} \frac{1}{m} \log \frac{\|\mathcal{B}_m v\|}{\|v\|} \\ &= \lim_{m\to+\infty} \frac{1}{m} \log(c_{0j}(v) \cdots c_{m-1j}(v)) = \lim_{m\to+\infty} \frac{1}{m} \sum_{l=0}^{m-1} \log c_{lj}(v).\end{aligned} \tag{6.8}$$

Fix i, $0 \le i < m$, and set

$$K_{mij} = \log\|a_{11}^{i-1} \cdots a_{11}^0 B_j^{m-1} \cdots B_j^{i+1} g_j^i\|.$$

Note that

$$\begin{aligned}K_{mij} &= \sum_{\ell=0}^{i-1} \log|a_{11}^\ell| + \log\|B_j^{m-1} \cdots B_j^{i+1} g_j^i\| \\ &= \sum_{\ell=0}^{i-1} \log|a_{11}^\ell| + \sum_{r=i+1}^{m-1} \log c_{rj}(v) + \log\|g_j^i\|,\end{aligned}$$

where $v = (B_j^i \cdots B_j^0)^{-1} g_j^i$. By Lemma 6.3, for each $\varepsilon > 0$ there exists i_0 independent of m such that $\log^+\|g_j^i\| < \varepsilon i < \varepsilon m$ for every $i_0 \le i < m$. By condition (1) and (6.8), we may assume (choosing a larger i_0 if necessary) that

$$(\lambda_1 - \varepsilon) i < \sum_{\ell=0}^{i-1} \log|a_{11}^\ell| < (\lambda_1 + \varepsilon) i$$

and

$$(\lambda_j' - \varepsilon)(m - i) < \sum_{r=i+1}^{m-1} \log c_{rj}(v) < (\lambda_j' + \varepsilon)(m - i).$$

Since $\lambda_1 \ge \lambda_j'$, it follows that

$$K_{mij} < (\lambda_1 + \varepsilon) i + (\lambda_j' + \varepsilon)(m - i) + \varepsilon m \le (\lambda_1 + 2\varepsilon) m.$$

By (6.7), we conclude that

$$\begin{aligned}\exp[(\lambda_1 - \varepsilon) m] &\le \|\mathcal{B}_m e_0\| \\ &\le \exp[(\lambda_1 + \varepsilon) m] + \sum_{j=1}^{p} \sum_{i=0}^{m-1} \exp K_{mij} \\ &\le \exp[(\lambda_1 + \varepsilon) m] + pm \exp[(\lambda_1 + 2\varepsilon) m] \\ &\le (p+1) \exp[(\lambda_1 + 3\varepsilon) m]\end{aligned}$$

for all sufficiently large m for which $m \le \exp(\varepsilon m)$. It follows from (6.4) and (6.5) that

$$\chi'(e_0) = \lim_{m \to +\infty} \frac{1}{m} \log \|\mathcal{B}_m e_0\| = \lambda_1.$$

This completes the proof of the theorem in the case $\lambda_1 \ge \lambda'_{n+1}$.

Case 2: $\lambda_1 < \lambda'_j$ for some j such that $2 \le j \le n+1$. Using again Lemma 6.3, given $\varepsilon > 0$, we find that $\|g_j^m\| < \exp(\varepsilon m)$. Furthermore,

$$\|(B_j^m \cdots B_j^0)^{-1} g_j^m\| \le \exp[(-\lambda'_j + \varepsilon)m]$$

for all sufficiently large m. Therefore,

$$\begin{aligned}
\sum_{m=1}^{\infty} \|a_{11}^{m-1} \cdots a_{11}^0 (B_j^m \cdots B_j^0)^{-1} g_j^m\| \\
&\le \sum_{m=1}^{\infty} \exp[(\lambda_1 + \varepsilon)m + (-\lambda'_j + \varepsilon)m] \\
&\le \sum_{m=1}^{\infty} \exp[(\lambda_1 - \lambda'_j + 2\varepsilon)m] < +\infty
\end{aligned}$$

for all $m \ge 1$ and all sufficiently small $\varepsilon > 0$. This shows that the formula

$$h_j = -\sum_{m=1}^{\infty} a_{11}^{m-1} \cdots a_{11}^0 (B_j^m \cdots B_j^0)^{-1} g_j^m$$

defines a vector in H_j. This is exactly where the compensation needed to correct the excessive growth takes place. Set

$$v = e_0 + \sum_{j : \lambda_1 < \lambda'_j} h_j$$

and denote by $\mathrm{proj}_{H_j^m}$ the projection onto H_j^m along $\widehat{E}_j^m$. We have

$$\begin{aligned}
\mathrm{proj}_{H_j^m} \mathcal{B}_m v &= \mathrm{proj}_{H_j^m} \mathcal{B}_m (e_0 + h_j) \\
&= \sum_{i=1}^{m-1} a_{11}^{i-1} \cdots a_{11}^0 B_j^{m-1} \cdots B_j^{i+1} g_j^i + \mathcal{B}_m h_j \\
&= \sum_{i=1}^{m-1} a_{11}^{i-1} \cdots a_{11}^0 B_j^{m-1} \cdots B_j^{i+1} g_j^i + B_j^{m-1} \cdots B_j^0 h_j \\
&= -\sum_{i=m}^{\infty} a_{11}^{i-1} \cdots a_{11}^0 (B_j^i \cdots B_j^m)^{-1} g_j^i.
\end{aligned}$$

Proceeding as before, we obtain

$$
\begin{aligned}
\|\mathrm{proj}_{E_j^m} \mathcal{B}_m v\| &\le \sum_{i=m}^{\infty} \|a_{11}^{i-1} \cdots a_{11}^{0} (B_j^i \cdots B_j^m)^{-1} g_j^i\| \\
&\le \sum_{i=m}^{\infty} \exp[(\lambda_1+\varepsilon)i + (-\lambda_j' + \varepsilon)(i-m) + \varepsilon i] \\
&\le \exp[(\lambda_1+\varepsilon)m] \sum_{i=m}^{\infty} \exp[(\lambda_1 - \lambda_j' + 2\varepsilon)(i-m) + \varepsilon i] \\
&= D(\varepsilon) \exp[(\lambda_1 + 2\varepsilon)m]
\end{aligned}
\tag{6.9}
$$

for all sufficiently large $m \ge 1$. Here

$$
D(\varepsilon) = \sum_{i=1}^{\infty} \exp[(\lambda_1 - \lambda_j' + 3\varepsilon)i] < +\infty,
$$

provided $\varepsilon > 0$ is chosen such that $\lambda_1 - \lambda_j' + 3\varepsilon < 0$ (that is always possible since $\lambda_1 < \lambda_j'$).

Denote by proj_{e_0} the projection on e_0 along $\mathbb{R}^n = \bigoplus_{i=1}^p H_i$. Exploiting the special form of the matrix B_m described above, we obtain

$$
\|\mathrm{proj}_{e_0} \mathcal{B}_m v\| = \prod_{i=0}^{m-1} a_{11}^i.
$$

By Theorem 2.24, for each j,

$$
\lim_{m\to+\infty} \frac{1}{m} \log|\sin \angle(E_j^m, \widehat{E}_j^m)| = 0.
$$

Since e_0 is orthogonal to $\bigoplus_{i=1}^p H_i$, there exist $c_1 > 0$ and $c_2 > 0$ such that

$$
c_1 \|\mathcal{B}_m v\| \le \prod_{i=0}^{m-1} a_{11}^i + \sum_{j:\lambda_j > \lambda_1} \|\mathrm{proj}_{E_j^m} \mathcal{B}_m v\| \le c_2 \|\mathcal{B}_m v\|
$$

for every $m \ge 1$. By (6.4), (6.9), and condition (1), we have

$$
\chi'(C_0{}^{-1} v) = \lim_{m\to+\infty} \frac{1}{m} \log\|\mathcal{B}_m v\| = \lambda_1.
$$

The above discussion implies that the Lyapunov exponent χ' is exact with respect to the vector

$$
v_{n+1} = e_0 + C_0{}^{-1} \sum_{j:\lambda_1 < \lambda_j'} h_j
$$

and that $\chi'(v_{n+1}) = \lambda_1$. This completes the proof of Theorem 6.2. □

An analogous criterion for forward regularity holds for sequences of upper triangular matrices.

Using the correspondence between forward and backward sequences of matrices, we immediately obtain the corresponding criterion for backward regularity.

Theorem 6.4. *Let $\mathcal{A}^- = (a_{ij}^m)_{m\in\mathbb{Z}} \subset GL(n,\mathbb{R})$ be a sequence of lower triangular matrices such that:*

(1) *the limit*

$$\lim_{m\to-\infty} \frac{1}{|m|} \sum_{k=1}^{m} \log|a_{ii}^k| =: \lambda_i^-, \quad i = 1,\dots,n,$$

exists and is finite;

(2) *we have*

$$\limsup_{m\to-\infty} \frac{1}{|m|} \log^+|a_{ij}^m| = 0, \quad i,j = 1,\dots,n.$$

Then the sequence $\mathcal{A}^-$ is backward regular, and the numbers λ_i^- are the values of the Lyapunov exponent χ^- (counted with multiplicities but possibly not ordered).

6.2. Proof of the Multiplicative Ergodic Theorem

We split the proof of the theorem into two steps. First we show how to reduce the general case to the case of triangular cocycles. Then we present the proof of the latter case.

Step I. Reduction to triangular cocycles.

We construct an extension of the transformation f,

$$F\colon X \times SO(n,\mathbb{R}) \to X \times SO(n,\mathbb{R}),$$

where $SO(n,\mathbb{R})$ is the group of orthogonal $n \times n$ matrices. Given $(x,U) \in X \times SO(n,\mathbb{R})$, one can apply the Gram-Schmidt orthogonalization procedure to the columns of the matrix $A(x)U$ and write

$$A(x)U = R(x,U)T(x,U), \tag{6.10}$$

where $R(x,U)$ is orthogonal and $T(x,U)$ is lower triangular (with positive entries on the diagonal). The two matrices $R(x,U)$ and $T(x,U)$ are uniquely defined, and their entries are linear combinations of the entries of U. Set

$$F(x,U) = (f(x), R(x,U)).$$

Consider the projection $\pi\colon (x,U) \mapsto U$. By (6.10), we obtain

$$T(x,U) = ((\pi \circ F)(x,U))^{-1} A(x)\pi(x,U). \tag{6.11}$$

Let $\tilde{\mathcal{A}}$ and $\mathcal{T}$ be two cocycles over F defined, respectively, by $\tilde{A}(x,U) = A(x)$ and $\mathcal{T}(x,U) = T(x,U)$. Since $\|U\| = 1$ for every $U \in SO(n,\mathbb{R})$, it follows from (6.11) that the cocycles $\tilde{\mathcal{A}}$ and $\mathcal{T}$ are equivalent on $X \times SO(n,\mathbb{R})$. Therefore a point $(x,U) \in X \times SO(n,\mathbb{R})$ is LP-regular for $\tilde{\mathcal{A}}$ if and only if it is LP-regular for $\mathcal{T}$.

Without loss of generality, since X is a Lebesgue space, we may assume that it is a compact metric space. Let $\mathcal{M}$ be the set of all Borel probability measures $\tilde{\nu}$ on $X \times SO(n,\mathbb{R})$ that satisfy

$$\tilde{\nu}(B \times SO(n,\mathbb{R})) = \nu(B) \tag{6.12}$$

for all measurable sets $B \subset X$. Then $\mathcal{M}$ is a compact convex subset of a locally convex topological vector space. The map $F_*\colon \mathcal{M} \to \mathcal{M}$ defined by

$$(F_*\tilde{\nu})(B) = \tilde{\nu}(F^{-1}(B))$$

is a bounded linear operator. By the Tychonoff Fixed Point Theorem there exists a fixed point $\tilde{\nu}_0 \in \mathcal{M}$ for the operator F_*, i.e., a measure $\tilde{\nu}_0$ such that $\tilde{\nu}_0(F^{-1}(B)) = \tilde{\nu}_0(B)$ for every measurable set $B \subset X \times SO(n,\mathbb{R})$. By (6.12), we conclude that the set of LP-regular points for $\mathcal{A}$ has full ν-measure if and only if the set of LP-regular points for $\tilde{\mathcal{A}}$ has full $\tilde{\nu}_0$-measure, and hence if and only if the set of LP-regular points for $\mathcal{T}$ has full $\tilde{\nu}_0$-measure.

Step II. Proof for triangular cocycles.

From now on we assume that $A(x) = (a_{ij}(x))$ is a lower triangular matrix (i.e., $a_{ij}(x) = 0$ if $i < j$). Write $A(x)^{-1} = (b_{ij}(x))$ and note that $b_{ii}(x) = 1/a_{ii}(x)$ for each i. By (6.1) we obtain $\log^+|a_{ij}|, \log^+|b_{ij}| \in L^1(X,\nu)$. It follows from Exercise 5.6 that

$$\lim_{m\to+\infty} \frac{1}{m}\log^+|a_{ij}(f^m(x))| = \lim_{m\to-\infty} \frac{1}{m}\log^+|b_{ij}(f^m(x))| = 0 \tag{6.13}$$

for every $1 \le i,j \le n$ and ν-almost every $x \in X$. Note that

$$|\log|a_{ii}|| = \log^+|a_{ii}| - \log^-|a_{ii}| = \log^+|a_{ii}| - \log^+|b_{ii}|. \tag{6.14}$$

By (6.1) and (6.14) we have $\log|a_{ii}| = -\log|b_{ii}| \in L^1(X,\nu)$. Birkhoff's Ergodic Theorem guarantees the existence of measurable functions $\lambda_i \in L^1(X,\nu)$ for $i = 1,\dots,n$, such that

$$\lim_{m\to+\infty} \frac{1}{m}\sum_{k=0}^{m-1} \log|a_{ii}(f^k(x))| = \lim_{m\to-\infty} \frac{1}{m}\sum_{k=m}^{-1} \log|b_{ii}(f^k(x))| = \lambda_i(x) \tag{6.15}$$

for each $i = 1,\dots,n$ and ν-almost every $x \in X$. Let $Y \subset X$ be the set of points $x \in X$ for which (6.13) and (6.15) hold. Notice that Y is a set of full ν-measure. We will show that Y consists of LP-regular points for $\mathcal{A}$, and thus the set of LP-regular points for $\mathcal{A}$ has full ν-measure.

By Theorems 6.2 and 6.4, the sequence $A(f^m(x))_{m\in\mathbb{Z}}$ is simultaneously forward and backward regular for every $x \in Y$. Moreover, the numbers $\lambda_i(x)$ are the forward Lyapunov exponents counted with their multiplicities (but possibly not ordered) and are symmetric to the backward Lyapunov exponents counted with their multiplicities (but possibly not ordered). We conclude that $p^+(x) = p^-(x) =: p(x)$ and $\chi_i^-(x) = -\chi_i^+(x)$ for $i = 1, \ldots, p(x)$. We now show that the spaces $H_1(x), \ldots, H_{p(x)}(x)$ defined by (5.6) satisfy (5.4).

Lemma 6.5. *If $x \in Y$ and $v \in \mathbb{R}^n \setminus \{0\}$, then $\chi^+(x,v) + \chi^-(x,v) \geq 0$.*

Proof of the lemma. For every $v \in \mathbb{R}^n \setminus \{0\}$, there exists i such that $v = \alpha_i e_i + \cdots + \alpha_n e_n$ with $\alpha_i \neq 0$, where $e_1, \ldots, e_n$ is the standard basis of $\mathbb{R}^n$. For any $m > 0$ the projection of $\mathcal{A}(x,m)v$ over $\operatorname{span}\{e_i\}$ along its orthogonal complement is

$$\prod_{k=0}^{m-1} a_{ii}(f^k(x))\alpha_i e_i.$$

Thus,

$$\log\|\mathcal{A}(x,m)v\| \geq \sum_{k=0}^{m-1} \log|a_{ii}(f^k(x))| + \log|\alpha_i|,$$

and $\chi^+(x,v) \geq \lambda_i(x)$. In a similar way, for every $m < 0$,

$$\log\|\mathcal{A}(x,m)v\| \geq \sum_{k=-m}^{-1} \log|b_{ii}(f^k(x))| + \log|\alpha_i|,$$

and $\chi^-(x,v) \geq -\lambda_i(x)$. Hence, $\chi^+(x,v) + \chi^-(x,v) \geq 0$. □

Lemma 6.6. *If $x \in Y$, then $\bigoplus_{i=1}^{p(x)} H_i(x) = \mathbb{R}^n$.*

Proof of the lemma. Since

$$A(x)E_i^+(x) = E_i^+(f(x)) \quad \text{and} \quad A(x)E_i^-(x) = E_i^-(f(x)),$$

we have $A(x)H_i(x) = H_i(f(x))$ for each i. Take $v \in E_i^+(x) \cap E_{i+1}^-(x)$. Then $\chi^+(x,v) \leq \chi_i^+(x)$ and $\chi^-(x,v) \leq -\chi_{i+1}^+(x)$. Hence,

$$\chi^+(x,v) + \chi^-(x,v) \leq \chi_i^+(x) - \chi_{i+1}^+(x) < 0,$$

and by Lemma 6.5 we obtain $v = 0$. Therefore, $E_i^+(x) \cap E_{i+1}^-(x) = \{0\}$. Moreover, for each $i < j$ we have

$$H_i(x) \cap H_j(x) \subset E_i^+(x) \cap E_j^-(x) \subset E_i^+(x) \cap E_{i+1}^-(x) = \{0\}.$$

Observe that

$$\begin{aligned}\dim H_i(x) &\geq \dim E_i^+(x) + \dim E_i^-(x) - n \\ &= (k_1(x) + \cdots + k_i(x)) + (k_i(x) + \cdots + k_{p(x)}(x)) - n = k_i(x).\end{aligned}$$

Therefore,

$$\dim \bigoplus_{i=1}^{p(x)} H_i(x) \geq k_1(x) + \cdots + k_{p(x)}(x) = n$$

and hence $\bigoplus_{i=1}^{p(x)} H_i(x) = \mathbb{R}^n$. This completes the proof of the lemma. □

It remains to show that the convergence in (5.5) is uniform on $C_i(x) = \{v \in H_i(x) : \|v\| = 1\}$.

Lemma 6.7. *If $x \in Y$, then*

$$\lim_{m\to\pm\infty} \frac{1}{|m|} \log \inf_v \|\mathcal{A}(x,m)v\| = \lim_{m\to\pm\infty} \frac{1}{|m|} \log \sup_v \|\mathcal{A}(x,m)v\| = \chi_i(x),$$

with the infimum and supremum taken over all $v \in C_i(x)$.

Proof of the lemma. Let $x \in Y$. Consider an orthonormal basis $\{e_1, \ldots, e_{k_i(x)}\}$ of $H_i(x)$. Let now

$$u_m = \sum_{j=1}^{k_i(x)} c_{m,j} e_j$$

be a vector in $C_i(x)$ at which $v \mapsto \|\mathcal{A}(x,m)v\|$ attains its minimum. Choose an integer $j(m)$ such that $|c_{m,j(m)}| = \max_j |c_{m,j}|$. Since

$$\sum_{j=1}^{k_i(x)} {c_{m,j}}^2 = 1,$$

we have $|c_{m,j(m)}| \geq 1/\sqrt{k_i(x)}$. We denote by $\rho_{m,j}$ and $\varphi_{m,j}$, respectively, the distance and the angle between $\mathcal{A}(x,m)e_j$ and $\mathcal{A}(x,m)\operatorname{span}\{e_i : i \neq j\}$. Note that

$$\rho_{m,j} = \|\mathcal{A}(x,m)e_j\| \sin \varphi_{m,j}.$$

We have

$$\mathcal{A}(x,m)u_m = c_{m,j(m)} \left(\mathcal{A}(x,m)e_{j(m)} + \sum_{j\neq j(m)} \frac{c_{m,j}}{c_{m,j(m)}} \mathcal{A}(x,m)e_j \right)$$

and thus,

$$\|\mathcal{A}(x,m)u_m\| \geq |c_{m,j(m)}|\rho_{m,j(m)} \geq \frac{1}{\sqrt{k_i(x)}} \|\mathcal{A}(x,m)e_{j(m)}\| \sin \varphi_{m,j(m)}.$$

Since x is forward and backward regular, it follows from Theorem 2.24 (and the analogous statement in the case of backward regularity) that

$$\lim_{m\to\pm\infty} \frac{1}{m} \log|\sin \varphi_{m,j}| = 0.$$

Since $j(m)$ can only take on a finite number of values, we obtain

$$\begin{aligned}
&\liminf_{m\to\pm\infty}\frac{1}{|m|}\log\|\mathcal{A}(x,m)u_m\| \\
&\geq \liminf_{m\to\pm\infty}\frac{1}{|m|}\log\|\mathcal{A}(x,m)e_{j(m)}\| + \liminf_{m\to\pm\infty}\frac{1}{|m|}\log|\sin\varphi_{m,j(m)}| \\
&\geq \min_j \liminf_{m\to\pm\infty}\frac{1}{|m|}\log\|\mathcal{A}(x,m)e_j\| \\
&\quad + \min_j \liminf_{m\to\pm\infty}\frac{1}{|m|}\log|\sin\varphi_{m,j}| = \chi_i(x),
\end{aligned} \tag{6.16}$$

where the minima are taken over all $j \in \{1,\ldots,k_i(x)\}$. Choose a vector

$$v_m = \sum_{j=1}^{k_i(x)} d_{m,j}e_j$$

in $C_i(x)$ for which the function $v \mapsto \|\mathcal{A}(x,m)v\|$ attains its maximum. We have

$$\|\mathcal{A}(x,m)v_m\| \leq \sum_{j=1}^{k_i(x)}|d_{m,j}|\cdot\|\mathcal{A}(x,m)e_j\| \leq \sum_{j=1}^{k_i(x)}\|\mathcal{A}(x,m)e_j\|$$

and hence,

$$\limsup_{m\to\pm\infty}\frac{1}{|m|}\log\|\mathcal{A}(x,m)v_m\| \leq \chi_i(x). \tag{6.17}$$

The desired result now follows from (6.16) and (6.17). □

Lemmas 6.6 and 6.7 show that the filtrations $\mathcal{V}^+$ and $\mathcal{V}^-$ are coherent at every point $x \in Y$, and hence the set Y consists of LP-regular points for $\mathcal{A}$. This implies that the set of LP-regular points for $\mathcal{A}$ has full ν-measure.

6.3. Normal forms of measurable cocycles

The goal of this section is to describe a "normal form" of a general measurable cocycle associated with its Lyapunov exponent. A construction of such normal forms is a manifestation of the Multiplicative Ergodic Theorem 6.1. In the simple case of a rigid cocycle $\mathcal{A}$ whose generator is a constant map (i.e., $\mathcal{A}(x,m) = A^m$ for some matrix $A \in GL(n,\mathbb{R})$) it is easy to verify that the cocycle $\mathcal{A}$ is equivalent to the rigid cocycle $\mathcal{B}$ whose generator is the Jordan block form of the matrix A. We consider $\mathcal{B}$ as the "normal form" of $\mathcal{A}$ and say that $\mathcal{A}$ is reduced to $\mathcal{B}$.

A general measurable cocycle $\mathcal{A}$ satisfying the integrability condition (6.1) is, so to speak, "weakly" rigid; in other words, it can be reduced to a constant cocycle up to arbitrarily small error (see Theorem 6.10 below). We consider this constant cocycle as a "normal form" of $\mathcal{A}$.

6.3.1. Lyapunov inner products. Consider a measurable cocycle $\mathcal{A}$ over a measurable transformation $f\colon X \to X$ and a family of inner products $\langle \cdot, \cdot \rangle = \langle \cdot, \cdot \rangle_x$ on $\mathbb{R}^n$ for $x \in X$. For each LP-regular point $x \in X$ we will replace the standard inner product $\langle \cdot, \cdot \rangle = \langle \cdot, \cdot \rangle_x$ in $\mathbb{R}^n$ with a special one that is generated by the action of the cocycle along the trajectory $\{f^n(x)\}_{n \geq 0}$. It provides an important technical tool in studying hyperbolic properties of cocycles over dynamical systems (see also Section 7.1). We start with the following auxiliary result.

Exercise 6.8. For each $\varepsilon > 0$, each LP-regular point $x \in X$, and each $i = 1, \ldots, p(x)$, show that the formula

$$\langle u, v \rangle'_{x,i} = \sum_{m \in \mathbb{Z}} \langle \mathcal{A}(x, m)u, \mathcal{A}(x, m)v \rangle e^{-2\chi_i(x)m - 2\varepsilon|m|} \tag{6.18}$$

determines a scalar product on $H_i(x)$.

Let $x \in X$ be an LP-regular point. We fix $\varepsilon > 0$ and introduce a new inner product in $\mathbb{R}^n$ (which will thus depend on ε) by the formula

$$\langle u, v \rangle'_x = \sum_{i=1}^{p(x)} \langle u_i, v_i \rangle'_{x,i},$$

where u_i and v_i are the projections of u and v over $H_i(x)$ along $\bigoplus_{j \neq i} H_j(x)$. We call $\langle \cdot, \cdot \rangle'_x$ the *Lyapunov inner product at x*, and we call the corresponding norm $\|\cdot\|'_x$ the *Lyapunov norm at x*. The sequence of weights $\{e^{-2\chi_i(x)m - 2\varepsilon|m|}\}_{m \in \mathbb{Z}}$ in (6.18) is called the *Pesin tempering kernel* (this is a particular case of the tempering kernels that we will introduce in Section 6.4.1).

Exercise 6.9. Let $\langle \cdot, \cdot \rangle$ be a Lyapunov inner product at x. Show that:

(1) it depends Borel measurably on x on the set of LP-regular points in X;

(2) for every LP-regular point $x \in X$ and $i \neq j$, the spaces $H_i(x)$ and $H_j(x)$ are orthogonal with respect to the Lyapunov inner product.

We stress that, in general, the Lyapunov inner product does not determine a Riemannian metric on M. It will only be used to study the action of the cocycle along its LP-regular trajectories.

A coordinate change $C_\varepsilon\colon X \to GL(n, \mathbb{R})$ is called a *Lyapunov change of coordinates* if for each LP-regular point $x \in X$ and $u, v \in \mathbb{R}^n$ it satisfies

$$\langle u, v \rangle_x = \langle C_\varepsilon(x)u, C_\varepsilon(x)v \rangle'_x. \tag{6.19}$$

We note that the formula (6.19) does not determine the function $C_\varepsilon(x)$ uniquely.

6.3.2. The Oseledets–Pesin Reduction Theorem.

The following principal result describes "normal forms" of measurable cocycles (see [9] for more details).

Theorem 6.10 (Oseledets–Pesin Reduction Theorem). *Let $f\colon X \to X$ be an invertible measure-preserving transformation of the Lebesgue space (X, ν), and let $\mathcal{A}$ be a measurable cocycle over f. Given $\varepsilon > 0$, if x is an LP-regular point for $\mathcal{A}$ then:*

(1) *any Lyapunov change of coordinates C_ε sends the orthogonal decomposition $\bigoplus_{i=1}^{p(x)} \mathbb{R}^{k_i(x)}$ to the decomposition $\bigoplus_{i=1}^{p(x)} H_i(x)$ of $\mathbb{R}^n$;*

(2) *the cocycle $A_\varepsilon(x) = C_\varepsilon(f(x))^{-1}A(x)C_\varepsilon(x)$ has the block form*

$$A_\varepsilon(x) = \begin{pmatrix} A_\varepsilon^1(x) & & \\ & \ddots & \\ & & A_\varepsilon^{p(x)}(x) \end{pmatrix}, \tag{6.20}$$

where each block $A_\varepsilon^i(x)$ is a $k_i(x) \times k_i(x)$ matrix and the entries are zero above and below the matrices $A_\varepsilon^i(x)$;

(3) *each block $A_\varepsilon^i(x)$ satisfies*

$$e^{\chi_i(x)-\varepsilon} \le \|A_\varepsilon^i(x)^{-1}\|^{-1} \le \|A_\varepsilon^i(x)\| \le e^{\chi_i(x)+\varepsilon}; \tag{6.21}$$

(4) *if the integrability condition (6.1) holds, then the map C_ε is tempered ν-almost everywhere, and the spectra of $\mathcal{A}$ and $\mathcal{A}_\varepsilon$ coincide ν-almost everywhere.*

Proof. Statement (1) follows immediately from (6.19). For each i we have

$$\begin{aligned} A_\varepsilon(x)\mathbb{R}^{k_i(x)} &= C_\varepsilon(f(x))^{-1}A(x)H_i(x) \\ &= C_\varepsilon(f(x))^{-1}H_i(f(x)) = \mathbb{R}^{k_i(f(x))} = \mathbb{R}^{k_i(x)}. \end{aligned}$$

Thus, $A_\varepsilon(x)$ has the block form (6.20).

Since $\mathcal{A}(x, m+1) = \mathcal{A}(f(x), m)\mathcal{A}(x)$, for every $v \in H_i(x)$ we have that

$$\begin{aligned} \left(\|A(x)v\|'_{f(x)}\right)^2 &= \sum_{m\in\mathbb{Z}} \|\mathcal{A}(f(x), m)A(x)v\|^2 e^{-2m\chi_i(x)-2\varepsilon|m|} \\ &= \sum_{m\in\mathbb{Z}} \|\mathcal{A}(x, m+1)v\|^2 e^{-2m\chi_i(x)-2\varepsilon|m|} \\ &= \sum_{k\in\mathbb{Z}} \|\mathcal{A}(x, k)v\|^2 e^{-2k\chi_i(x)-2\varepsilon|k|+\eta(x,\varepsilon)}, \end{aligned}$$

where $\eta(x, \varepsilon) = 2\chi_i(x) + 2\varepsilon(|k| - |k-1|)$. Therefore,

$$e^{\chi_i(x)-\varepsilon}\|v\|'_x \le \|A(x)v\|'_{f(x)} \le e^{\chi_i(x)+\varepsilon}\|v\|'_x \tag{6.22}$$

for every $v \in H_i(x)$. By (6.19), for each $w \in \mathbb{R}^n$ we obtain $\|w\|_x = \|C_\varepsilon(x)w\|'_x$ and

$$\|A_\varepsilon(x)w\|_{f(x)} = \|C_\varepsilon(f(x))^{-1}A(x)C_\varepsilon(x)w\|_{f(x)} = \|A(x)C_\varepsilon(x)w\|'_{f(x)}.$$

Hence if $v = C_\varepsilon(x)w \in H_i(x)$, then

$$\|w\|_x = \|v\|'_x \quad \text{and} \quad \|A_\varepsilon(x)w\|_{f(x)} = \|A(x)v\|'_{f(x)}.$$

By (6.22), for $w \in \mathbb{R}^{k_i(x)}$,

$$e^{\chi_i(x)-\varepsilon}\|w\|_x \le \|A_\varepsilon(x)w\|_{f(x)} \le e^{\chi_i(x)+\varepsilon}\|w\|_x.$$

This implies the inequalities in (6.21).

We now proceed with the proof of statement (4). Since

$$\|w\|_x = \|C_\varepsilon(x)w\|'_x \ge \|C_\varepsilon(x)w\|_x,$$

we have

$$\|C_\varepsilon(x)\| \le 1 \quad \text{and} \quad \|C_\varepsilon(x)^{-1}\| \ge 1 \tag{6.23}$$

for every LP-regular point $x \in X$. Now observe that

$$\mathcal{A}_\varepsilon(x,m) = C_\varepsilon(f^m(x))^{-1}\mathcal{A}(x,m)C_\varepsilon(x).$$

For each $N > 0$, consider the set

$$X_N = \big\{x \in X : x \text{ is LP-regular},$$
$$\|C_\varepsilon(x)\| > N^{-1} \text{ and } \|C_\varepsilon(x)^{-1}\| < N\big\}.$$

By the Multiplicative Ergodic Theorem 6.1, $\nu(X_N) \to 1$ as $N \to \infty$. Let us fix $N > 0$ such that $\nu(X_N) > 0$. It follows from Poincaré's Recurrence Theorem that there exists a set $Y_N \subset X_N$ such that $\nu(X_N \setminus Y_N) = 0$ and the forward and backward orbits of each point $y \in Y_N$ return both infinitely many times to Y_N. Given $x \in Y_N$, let $m_k = m_k(x)$ be a sequence of positive integers such that $f^{m_k}(y) \in Y_N$ for all $k \ge 1$. For $i = 1, \dots, p(x)$ we have

$$\begin{aligned}\|\mathcal{A}^i_\varepsilon(x,m_k)\| &\le \|C_\varepsilon(f^{m_k}(x))^{-1}\| \cdot \|\mathcal{A}(x,m_k)|H_i(x)\| \cdot \|C_\varepsilon(x)\| \\ &\le N\|\mathcal{A}(x,m_k)|H_i(x)\|,\end{aligned}$$

where $\mathcal{A}^i_\varepsilon(x,m)$ is the cocycle generated by the block $A^i_\varepsilon(x)$. Hence,

$$\limsup_{m\to+\infty}\frac{1}{m}\log\|\mathcal{A}^i_\varepsilon(x,m)\| \le \chi_i(x). \tag{6.24}$$

In a similar way, one can show that

$$\begin{aligned}\|\mathcal{A}(x,m_k)|H_i(x)\| &\le \|C_\varepsilon(f^{m_k}(x))\| \cdot \|\mathcal{A}^i_\varepsilon(x,m_k)\| \cdot \|C_\varepsilon(x)^{-1}\| \\ &\le N\|\mathcal{A}^i_\varepsilon(x,m_k)\|\end{aligned}$$

and hence,

$$\chi_i(x) \le \liminf_{m\to+\infty}\frac{1}{m}\log\|\mathcal{A}^i_\varepsilon(x,m)\|. \tag{6.25}$$

We conclude from (6.24) and (6.25) that for each $x \in Y_N$ and each $v \in \mathbb{R}^{k_i(x)} = C_\varepsilon(x)^{-1}H_i(x)$,

$$\lim_{m\to+\infty}\frac{1}{m}\log\|\mathcal{A}_\varepsilon(x,m)v\| = \chi_i(x).$$

In a similar way, one can prove that

$$\lim_{m\to-\infty}\frac{1}{|m|}\log\|\mathcal{A}_\varepsilon(x,m)v\| = \chi_i(x).$$

Since $\nu(Y_N)\to 1$ as $N\to\infty$, the spectra of $\mathcal{A}$ and $\mathcal{A}_\varepsilon$ coincide ν-almost everywhere.

We now show that the function $C_\varepsilon\colon X\to GL(n,\mathbb{R})$ is tempered. Observe that

$$\begin{aligned}\|C_\varepsilon(f^m(x))^{-1}\| &= \sup_{v\neq 0}\frac{\|\mathcal{A}_\varepsilon(x,m)C_\varepsilon(x)^{-1}\mathcal{A}(x,m)^{-1}v\|}{\|v\|}\\ &= \max_i \sup_{w\in H_i(x)}\frac{\|\mathcal{A}_\varepsilon(x,m)C_\varepsilon(x)^{-1}w\|}{\|\mathcal{A}(x,m)w\|}.\end{aligned}$$

Since $\|C_\varepsilon(x)^{-1}\|\geq 1$, using (6.24), we obtain that for ν-almost all LP-regular points $x\in X$,

$$\begin{aligned}0 &\leq \liminf_{m\to\pm\infty}\frac{1}{|m|}\log\|C_\varepsilon(f^m(x))^{-1}\|\\ &\leq \limsup_{m\to\pm\infty}\frac{1}{|m|}\log\|C_\varepsilon(f^m(x))^{-1}\|\\ &= \max_i \sup_{w\in H_i(x)}\lim_{m\to\pm\infty}\frac{1}{|m|}\log\frac{\|\mathcal{A}_\varepsilon(x,m)C_\varepsilon(x)^{-1}w\|}{\|\mathcal{A}(x,m)w\|} = 0.\end{aligned}$$

This implies that for ν-almost every $x\in X$,

$$\lim_{m\to\pm\infty}\frac{1}{m}\log\|C_\varepsilon(f^m(x))^{-1}\| = 0.$$

In a similar way one can show that for ν-almost every $x\in X$,

$$\lim_{m\to\pm\infty}\frac{1}{m}\log\|C_\varepsilon(f^m(x))\| = 0.$$

This implies that C_ε is a tempered function ν-almost everywhere and completes the proof of Theorem 6.10. □

6.4. Lyapunov charts

We recall that for a diffeomorphism $f\colon M\to M$ of a compact smooth n-dimensional Riemannian manifold M, we can consider (by introducing local coordinates) the derivative df as a linear cocycle over f. We shall now use the theory of linear extensions of cocycles. Applying the Reduction

Theorem 6.10, given $\varepsilon > 0$ and an LP-regular point $x \in M$, there exists a linear transformation $C_\varepsilon(x)\colon \mathbb{R}^n \to T_xM$ such that:

(1) the matrix

$$A_\varepsilon(x) = C_\varepsilon(f(x))^{-1} \circ d_xf \circ C_\varepsilon(x)$$

has the Lyapunov block form (6.20) (see Theorem 6.10);

(2) $\{C_\varepsilon(f^m(x))\}_{m\in\mathbb{Z}}$ is a tempered sequence of linear transformations.

We wish to construct, for every LP-regular point $x \in M$, a neighborhood $N(x)$ of x such that f acts in $N(x)$ very much like the linear map $A_\varepsilon(x)$ in a neighborhood of the origin.

6.4.1. A tempering kernel. First we shall show that the function $x \mapsto \|v(x)\|'_x/\|v(x)\|_x$ is tempered on the set of LP-regular points for every measurable vector field $X \ni x \mapsto v(x) \in \mathbb{R}^n \setminus \{0\}$ (see (4.12) for the definition of tempered function). We need a technical but crucial statement known as the Tempering Kernel Lemma.

Lemma 6.11. *Let $f\colon X \to X$ be a measurable transformation. If $K\colon X \to \mathbb{R}$ is a positive measurable function tempered on some subset $Z \subset X$, then for any $\varepsilon > 0$ there exists a positive measurable function $K_\varepsilon\colon Z \to \mathbb{R}$ such that $K(x) \le K_\varepsilon(x)$ and for $x \in Z$,*

$$e^{-\varepsilon} \le \frac{K_\varepsilon(f(x))}{K_\varepsilon(x)} \le e^{\varepsilon}. \tag{6.26}$$

Exercise 6.12. Prove the lemma following the argument in the proof of Lemma 4.4.

The function K_ε satisfying (6.26) is called a *tempering kernel.* Note that if f preserves a Lebesgue measure ν on the space X, then by Exercise 5.6, any positive function $K\colon X \to \mathbb{R}$ with $\log K \in L^1(X,\nu)$ satisfies (4.12).

Proposition 6.13 (see [**9**]). *Given $\varepsilon > 0$, there is a positive measurable function $K_\varepsilon\colon X \to \mathbb{R}$ such that if $x \in X$ is an LP-regular point, then:*

(1) *$K_\varepsilon(x)e^{-\varepsilon|m|} \le K_\varepsilon(f^m(x)) \le K_\varepsilon(x)e^{\varepsilon|m|}$ for every $m \in \mathbb{Z}$;*

(2) *$n^{-1/2}\|v\|_x \le \|v\|'_x \le K_\varepsilon(x)\|v\|_x$ for every $v \in \mathbb{R}^n$;*

(3) *for every measurable vector field $X \ni x \mapsto v(x) \in \mathbb{R}^n \setminus \{0\}$, the function $x \mapsto \|v(x)\|'_x/\|v(x)\|_x$ is tempered on the set of LP-regular points.*

Proof. The first inequality in statement (2) follows from the well-known inequality

$$n\sum_{i=1}^n {a_i}^2 \ge \left(\sum_{i=1}^n a_i\right)^2,$$

with equality if and only if $a_1 = \cdots = a_n$. Given an LP-regular point $x \in X$ and $\varepsilon > 0$, let $C_\varepsilon(x)$ be the matrix defined by (6.19). Then $\|v\|'_x \le \|C_\varepsilon(x)^{-1}\| \cdot \|v\|_x$ for each $v \in \mathbb{R}^n$. By Theorem 6.10, the function C_ε is tempered. Thus, we can apply the Tempering Kernel Lemma 6.11 to the positive function $K(x) = \|C_\varepsilon(x)^{-1}\|$, and we find a function K_ε satisfying the desired properties. □

6.4.2. Construction of Lyapunov charts. We denote by $\mathcal{R}$ the set of LP-regular points for f and by $B(0,r)$ the standard Euclidean r-ball in $\mathbb{R}^n$ centered at the origin. The following result describes a particular class of local coordinate charts around LP-regular points that are very useful in applications of nonuniform hyperbolicity theory (see [**9**] for more details).

Theorem 6.14. *For every $\varepsilon > 0$ the following properties hold:*

(1) *there exists a tempered function $q\colon \mathcal{R} \to (0,1]$ and a collection of embeddings $\Psi_x\colon B(0,q(x)) \to M$ for each $x \in \mathcal{R}$ such that $\Psi_x(0) = x$ and $e^{-\varepsilon} < q(f(x))/q(x) < e^{\varepsilon}$; these embeddings satisfy $\Psi_x = \exp_x \circ C_\varepsilon(x)$, where $C_\varepsilon(x)$ is the Lyapunov change of coordinates;*
(2) *if $f_x := \Psi_{f(x)}^{-1} \circ f \circ \Psi_x\colon B(0,q(x)) \to \mathbb{R}^n$, then $d_0 f_x$ has the Lyapunov block form (6.20);*
(3) *the C^1 distance $d_{C^1}(f_x, d_0 f_x) < \varepsilon$ in $B(0,q(x))$;*
(4) *there exist a constant $K > 0$ and a measurable function $A\colon \mathcal{R} \to \mathbb{R}$ such that for every $y, z \in B(0,q(x))$,*
$$K^{-1}\rho(\Psi_x y, \Psi_x z) \le \|y - z\| \le A(x)\rho(\Psi_x y, \Psi_x z)$$
with $e^{-\varepsilon} < A(f(x))/A(x) < e^{\varepsilon}$.

Proof. For each $x \in X$ consider the Lyapunov change of coordinates $C_\varepsilon(x)$ from $T_x M$ onto $\mathbb{R}^n$ (see (6.19)). Thus for each $x \in \mathcal{R}$, the matrix
$$A_\varepsilon(x) = C_\varepsilon(f(x))^{-1} \circ d_x f \circ C_\varepsilon(x) \tag{6.27}$$
has the Lyapunov block form.

For each $x \in M$ and $r > 0$ set
$$T_x M(r) = \{w \in T_x M : \|w\| < r\}.$$
Now choose $r_0 > 0$ such that for every $x \in M$ the exponential map $\exp_x\colon T_x M(r_0) \to M$ is an embedding, $\|d_w \exp_x\|$, $\|d_{\exp_x w} {\exp_x}^{-1}\| \le 2$ for every $w \in T_x M(r_0)$, and $\exp_{f(x)}$ is injective on
$$\exp_{f(x)}^{-1} \circ f \circ \exp_x(T_x M(r_0)).$$
Define the map
$$f_x = C_\varepsilon(f(x))^{-1} \circ \exp_{f(x)}^{-1} \circ f \circ \exp_x \circ C_\varepsilon(x) \tag{6.28}$$

on $C_\varepsilon(x)T_xM(r_0) \subset \mathbb{R}^n$. Observe that the Euclidean ball

$$\{w \in T_xM : \|w\| < r_0\|C_\varepsilon(x)^{-1}\|^{-1}\}$$

is contained in $C_\varepsilon(x)T_xM(r_0)$ up to identification.

We wish to compare f_x with $A_\varepsilon(x)$. In order to do that, we first introduce the maps

$$r_x(u) = f_x(u) - A_\varepsilon(x)u,$$
$$g_x(w) = (\exp_{f(x)}^{-1} \circ f \circ \exp_x)w - dfw.$$

Since f is of class $C^{1+\alpha}$ and $d_0g_x = 0$, there exists a positive constant L, such that $\|d_wg_x\| \le L\|w\|^\alpha$, and thus

$$\begin{aligned}\|d_ur_x\| &= \|d_u(C_\varepsilon(f(x)) \circ g_x \circ C_\varepsilon(x)^{-1})\| \\ &\le L\|C_\varepsilon(f(x))\| \cdot \|C_\varepsilon(x)^{-1}\|^{1+\alpha}\|u\|^\alpha.\end{aligned}$$

Hence if $\|w\|$ is sufficiently small, the contribution of the nonlinear part of f_x is negligible. In particular, $\|d_ur_x\| < \varepsilon$ for

$$\|u\| < \delta(x) := \varepsilon^{1/\alpha}L^{-1/\alpha}\|C_\varepsilon(f(x))\|^{-1/\alpha}\|C_\varepsilon(x)^{-1}\|^{-1-1/\alpha}.$$

By the Mean Value Theorem we also have

$$\|r_x(u)\| < \varepsilon \text{ for } \|u\| \le \delta(x), \tag{6.29}$$

since $r_x(0) = 0$. It follows from the definition of $\delta(x)$ that

$$\begin{aligned}\lim_{m\to+\infty}\frac{1}{m}\log\delta(f^m(x)) = &-\lim_{m\to+\infty}\frac{1}{m\alpha}\log\|C_\varepsilon(f^{m+1}(x))\| \\ &-\lim_{m\to+\infty}\frac{1}{m(1+1/\alpha)}\log\|C_\varepsilon(f^m(x))^{-1}\| = 0,\end{aligned}$$

since C_ε is tempered. Applying the Tempering Kernel Lemma 6.11 to the function δ, we find a measurable function $q\colon X \to \mathbb{R}$ such that $q(x) \ge \delta(x)$ and

$$e^{-\varepsilon} \le q(f(x))/q(x) \le e^\varepsilon.$$

We define the map $\Psi_x\colon B(0,q(x)) \to M$ by $\Psi_x = \exp_x \circ C_\varepsilon(x)$. It is clear that Ψ_x is an embedding for each $x \in \mathcal{R}$, such that $\Psi_x(0) = x$. This proves statements (1) and (2). Statement (3) follows from the identity $d_0f_x = A_\varepsilon(x)$ (that in turn follows immediately from (6.27), (6.28), and the chain rule) and from (6.29).

Now we prove statement (4). It follows from (6.23) that

$$\|y - z\| = \|\Psi_x^{-1}(\Psi_xy) - \Psi_x^{-1}(\Psi_xz)\| \le 2\|C_\varepsilon(x)^{-1}\|\rho(\Psi_xy, \Psi_xz).$$

On the other hand, we have

$$\rho(\Psi_xy, \Psi_xz) \le 2\|C_\varepsilon(x)\| \cdot \|y - z\| \le 2\|y - z\|.$$

This means that if $K_\varepsilon\colon \mathcal{R} \to \mathbb{R}$ is a function as in the Tempering Kernel Lemma 6.11 such that

$$K_\varepsilon(x) \geq \|C_\varepsilon(x)^{-1}\| \quad \text{and} \quad e^{-\varepsilon} \leq K_\varepsilon(f(x))/K_\varepsilon(x) \leq e^\varepsilon,$$

then in statement (4) we can take $A(x) = 2K_\varepsilon(f(x))$. □

We note that for each $x \in \mathcal{R}$ there exists a constant $B(x) \geq 1$ such that for every y, $z \in B(0, q(x))$,

$$B(x)^{-1}\rho(\Psi_x y, \Psi_x z) \leq \rho'_x(\exp_x y, \exp_x z) \leq B(x)\rho(\Psi_x y, \Psi_x z),$$

where $\rho'_x(\cdot,\cdot)$ is the distance on $\exp_x B(0, q(x))$ with respect to the Lyapunov norm $\|\cdot\|'_x$. By Luzin's theorem, given $\delta > 0$, there exists a set of measure at least $1 - \delta$ where $x \mapsto B(x)$ as well as $x \mapsto A(x)$ in Theorem 6.14 are bounded.

For each LP-regular point x the set

$$R(x) := \Psi_x(B(0, q(x)))$$

is called a *regular neighborhood* of x or a *Lyapunov chart* at x.

We stress that the existence of regular neighborhoods uses the fact that f is of class $C^{1+\alpha}$ in an essential way.

Chapter 7

Local Manifold Theory

One of the goals of hyperbolicity theory is to describe the behavior of trajectories near a hyperbolic trajectory. In the case of uniform hyperbolicity such a description is provided by the classical Hadamard–Perron theorem, which effectively reduces the study of a nonlinear system in a small neighborhood of a hyperbolic trajectory to the study of the corresponding system of variational equations (in the continuous-time case) or the study of the action of the differential (in the discrete-time case) along the hyperbolic trajectory. A crucial manifestation of the Hadamard–Perron theorem is a construction of a local stable manifold which consists of all the points whose trajectories start near the hyperbolic one and approach it with an exponential rate.

In the case of the nonuniform hyperbolicity theory one can generalize the Hadamard–Perron theorem and construct local stable manifolds for trajectories which are LP-regular. However, local manifolds depend "wildly" on the base points: their sizes are measurable functions of the base point (that in general are not continuous) and may decrease along the trajectory with a subexponential rate. In this chapter we carry out a detailed study of local manifolds for nonuniformly hyperbolic systems.

We stress that constructing local stable and unstable manifolds requires a higher regularity of the system, i.e., that f is of class $C^{1+\alpha}$. This reflects a principal difference between uniform and nonuniform hyperbolicity—while the local manifold theory and the ergodic theory of Anosov diffeomorphisms can be developed under the assumption that the systems are of class C^1, the corresponding results for nonuniformly hyperbolic C^1 diffeomorphisms fail to be true. See Chapter 15 for some "pathological" phenomena in the C^1 topology.

7.1. Local stable manifolds

Let f be a $C^{1+\alpha}$ diffeomorphism of a compact smooth Riemannian manifold M. Recall that a *local smooth submanifold* of M is the graph of a smooth injective function $\psi\colon B \to M$, where B is the open unit ball in $\mathbb{R}^k$ for $k \leq \dim M$.

Consider a nonuniformly completely hyperbolic f-invariant set Y and let $\Lambda_{\lambda\mu\varepsilon j}$ be the corresponding collection of level sets (see (4.24)). In what follows, we fix numbers $\lambda, \mu, \varepsilon$, and $1 \leq j < p$ and assume that the level set $\Lambda = \Lambda_{\lambda\mu\varepsilon j}$ is not empty. We stress that we allow the case when this set consists of a single trajectory. In this section we prove the following theorem, which is one of the key results in hyperbolicity theory.

Theorem 7.1 (Stable Manifold Theorem). *For every $x \in \Lambda$ there exists a local stable manifold $V^s(x)$ such that $x \in V^s(x)$, $T_xV^s(x) = E^s(x)$, and for every $y \in V^s(x)$ and $n \geq 0$,*

$$\rho(f^n(x), f^n(y)) \leq T(x)\lambda^n \rho(x, y), \tag{7.1}$$

where ρ is the distance in M induced by the Riemannian metric, $0 < \lambda < 1$, and $T\colon \Lambda \to (0, \infty)$ is a Borel function satisfying

$$T(f^m(x)) \leq T(x)e^{10\varepsilon|m|}, \quad m \in \mathbb{Z}. \tag{7.2}$$

Furthermore, $f^m(V^s(x)) \subset V^s(f^m(x))$ for every $m \in \mathbb{Z}$.

Inequality (7.2) illustrates the crucial fact that while the estimate (7.1) may deteriorate along the trajectory $\{f^n(x)\}$, it only happens with "subexponential" rate. This fact is essentially due to condition (H5) in Section 4.2.

The proof of the Stable Manifold Theorem 7.1 presented below is due to Pesin [**68**] and is an elaboration of Perron's approach [**66**].

Remark 7.2. One can extend the Stable Manifold Theorem 7.1 to the case when the invariant set Y is nonuniformly *partially* hyperbolic in the broad sense.

We obtain the local stable manifold in the form

$$V^s(x) = \exp_x\{(v, \psi_x^s(v)) : v \in B^s(r)\}, \tag{7.3}$$

where $\psi_x^s\colon B^s(r) \to E^u(x)$ is a smooth map satisfying

$$\psi_x^s(0) = 0 \quad \text{and} \quad d\psi_x^s(0) = 0.$$

Here $B^s(r) \subset E^s(x)$ is the ball of radius r centered at the origin; $r = r(x)$ is called the *size* of the local stable manifold. We now describe how to construct the function ψ_x^s. Fix $x \in M$ and consider the map

$$\tilde{f}_x = \exp_{f(x)}^{-1} \circ f \circ \exp_x\colon B^s(r) \times B^u(r) \to T_{f(x)}M, \tag{7.4}$$

which is well-defined if r is sufficiently small.[1] Here $B^u(r)$ is the ball of radius r in $E^u(x)$ centered at the origin. Since the stable and unstable subspaces are invariant under the action of the differential df (see condition (H1) in Section 4.2), the map $\tilde{f}$ can be written in the following form:

$$\tilde{f}_x(v,w) = (A_x v + g_x(v,w), B_x w + h_x(v,w)), \tag{7.5}$$

where $v \in B^s(x)$ and $w \in B^u(x)$ and where

$$A_x \colon E^s(x) \to E^s(f(x)) \quad \text{and} \quad B_x \colon E^u(x) \to E^u(f(x))$$

are linear maps. In view of conditions (H2) and (H3) in Section 4.2 the map A_x acts along the trajectory $\{f^n(x)\}$ as a contraction while the map B_x acts as an expansion. Moreover,

$$g_x(0) = 0, \quad h_x(0) = 0, \quad dg_x(0) = 0, \quad \text{and} \quad h_x(0) = 0.$$

Since f is of class $C^{1+\alpha}$, we also have that for every $x \in \Lambda$,

$$\begin{aligned} \|dg_x(v_1,w_1) - dg_x(v_2,w_2)\| &\le C_1(\|v_1 - v_2\| + \|w_1 - w_2\|)^\alpha, \\ \|dh_x(v_1,w_1) - dh_x(v_2,w_2)\| &\le C_2(\|v_1 - v_2\| + \|w_1 - w_2\|)^\alpha, \end{aligned} \tag{7.6}$$

where $C_1 > 0$ and $C_2 > 0$ are constants independent of x.

In other words, the map $\tilde{f}_x$ can be viewed as a small perturbation of the linear map $(v,w) \mapsto (A_x v, B_x w)$ by the map $(g_x(v,w), h_x(v,w))$ satisfying condition (7.6), which is analogous to condition (3.10).

The local stable manifold must be invariant under the action of the map f. In other words, one should effectively construct a collection of local stable manifolds $V^s(f^m(x))$ along the trajectory of x in such a way that

$$f(V^s(f^m(x))) \subset V^s(f^{m+1}(x)).$$

Exercise 7.3. Using the relations (7.3) and (7.5), show that the above equation yields the following relation on the collection of functions $\psi^s_{f^m(x)}$:

$$\begin{aligned} &\psi^s_{f^{m+1}(x)}(A_{f^m(x)}v + g_{f^m(x)}(v, \psi^s_{f^m(x)}(v))) \\ &\qquad = B_{f^m(x)}\psi^s_{f^m(x)}(v) + h_{f^m(x)}(v, \psi^s_{f^m(x)}(v)). \end{aligned} \tag{7.7}$$

This means that constructing local stable manifolds $V^s(f^m(x))$ boils down to solving the functional equation (7.7) for the collection of locally defined functions $\psi^s_{f^m(x)}$ along the trajectory of x.

To proceed with finding a solution of equation (7.7) we use the Lyapunov inner product in the tangent bundle $T\Lambda$ introduced in Section 6.3. We present here a slightly modified version of this inner product that is better adapted to the construction of stable manifolds.

[1]The number $r = r(x)$ should be less than the radius of injectivity of the exponential map $\exp_x$ at the point x.

Choose numbers $0 < \lambda' < \mu' < \infty$ such that

$$\lambda e^{\varepsilon} < \lambda', \quad \mu' < \mu e^{-\varepsilon} \tag{7.8}$$

and define a new inner product $\langle \cdot, \cdot \rangle'_x$, called a *Lyapunov inner product*, as follows. Set

$$\langle v, w \rangle'_x = \sum_{k=0}^{\infty} \langle df^k v, df^k w \rangle_{f^k(x)} (\lambda')^{-2k}$$

for v, $w \in E^s(x)$, and set

$$\langle v, w \rangle'_x = \sum_{k=0}^{\infty} \langle df^{-k} v, df^{-k} w \rangle_{f^{-k}(x)} (\mu')^{2k}$$

for v, $w \in E^u(x)$.

Exercise 7.4. Using the Cauchy–Schwarz inequality, the conditions of nonuniform hyperbolicity (H2), (H3), and (H5) in Section 4.2, and (7.8), show that each of the above series converges; indeed show that if $v, w \in E^s(x)$, then

$$\begin{aligned}\langle v, w \rangle'_x &\leq \sum_{k=0}^{\infty} C(f^k(x))^2 \lambda^{2k} (\lambda')^{-2k} \|v\|_x \|w\|_x \\ &\leq C(x)^2 (1 - \lambda e^{\varepsilon}/\lambda')^{-1} \|v\|_x \|w\|_x < \infty,\end{aligned}$$

and if v, $w \in E^u(x)$, then

$$\begin{aligned}\langle v, w \rangle'_x &\leq \sum_{k=0}^{\infty} C(f^{-k}(x))^2 \mu^{-2k} (\mu')^{2k} \|v\|_x \|w\|_x \\ &\leq C(x)^2 (1 - \mu'/(\mu e^{-\varepsilon}))^{-1} \|v\|_x \|w\|_x < \infty.\end{aligned}$$

We extend $\langle \cdot, \cdot \rangle'_x$ to all vectors in $T_x M$ by declaring the subspaces $E^s(x)$ and $E^u(x)$ to be mutually orthogonal with respect to $\langle \cdot, \cdot \rangle'_x$; i.e., we set

$$\langle v, w \rangle'_x = \langle v^s, w^s \rangle'_x + \langle v^u, w^u \rangle'_x,$$

where $v = v^s + v^u$ and $w = w^s + w^u$ with v^s, $w^s \in E^s(x)$ and v^u, $w^u \in E^u(x)$.

The norm induced by the Lyapunov inner product is called a *Lyapunov norm* and is denoted by $\|\cdot\|'$. We emphasize that the Lyapunov inner product and, hence, the norm $\|\cdot\|'$ depend on the choice of numbers λ' and μ'.

It is worth mentioning that in general the Lyapunov inner product $\langle \cdot, \cdot \rangle'_x$ is *not* a Riemannian metric. While in the case of uniform hyperbolicity it is defined for all $x \in M$, it depends continuously in x (but in general not smoothly). In the case of nonuniform hyperbolicity $\langle , \rangle'_x$ is defined only for $x \in \Lambda$ and depends Borel measurably in x.

Exercise 7.5. For the Lyapunov inner product, show that for each $x \in \Lambda$:

(N1) the angle between the subspaces $E^s(x)$ and $E^u(x)$ in the inner product $\langle\cdot,\cdot\rangle'_x$ is $\pi/2$;

(N2) $\|A_x\|' \leq \lambda'$ and $\|B_x{}^{-1}\|' \leq (\mu')^{-1}$;

(N3) the relation between the Lyapunov inner product and the Riemannian inner product is given by

$$\frac{1}{\sqrt{2}}\|w\|_x \leq \|w\|'_x \leq D(x)\|w\|_x,$$

where $w \in T_xM$ and

$$D(x) = C(x)K(x)^{-1}\big[(1-\lambda e^{\varepsilon}/\lambda')^{-1} + (1-\mu'/(\mu e^{-\varepsilon}))^{-1}\big]^{1/2}$$

is a measurable function satisfying

$$D(f^m(x)) \leq D(x)e^{2\varepsilon|m|}, \quad m \in \mathbb{Z}. \tag{7.9}$$

Properties (N1) and (N2) show that the action of the differential *df* on $T\Lambda$ is *uniformly* hyperbolic with respect to the Lyapunov inner product.

To construct local stable manifolds, one can now use either the Hadamard method or the Perron method, which are well known in uniform hyperbolicity theory. Note that the "perturbation map" $\tilde{f}_x$ (see (7.4) and (7.5)) satisfies condition (7.6) in a neighborhood U_x of the point x whose size depends on x. Moreover, the size of U_x decays along the trajectory of x with "subexponential" rate (see (7.9)). This requires a substantial modification of the classical Hadamard–Perron approach.

7.2. An abstract version of the Stable Manifold Theorem

We present an approach based upon the Perron method. Fix $x \in \Lambda$ and consider the positive semitrajectory $\{f^m(x)\}_{m\geq 0}$ and the family of maps $\{\tilde{F}_m = \tilde{f}_{f^m(x)}\}_{m\geq 0}$. We identify the tangent spaces $T_{f^m(x)}M$ with $\mathbb{R}^p = \mathbb{R}^k \times \mathbb{R}^{p-k}$ (recall that $p = \dim M$ and $1 \leq k < p$) via an isomorphism τ_m such that

$$\tau_m(E^s(f^m(x))) = \mathbb{R}^k \quad \text{and} \quad \tau_m(E^u(f^m(x))) = \mathbb{R}^{p-k}.$$

In view of (7.5) the map $F_m = \tau_{m+1} \circ \tilde{F}_m \circ \tau_m{}^{-1}$ is of the form

$$F_m(v,w) = (A_m v + g_m(v,w), B_m w + h_m(v,w)), \tag{7.10}$$

where $A_m\colon \mathbb{R}^k \to \mathbb{R}^k$ and $B_m\colon \mathbb{R}^{p-k} \to \mathbb{R}^{p-k}$ are linear maps and g, h are nonlinear maps defined for each $v \in B^s(r_0) \subset \mathbb{R}^k$ and $w \in B^u(r_0) \subset \mathbb{R}^{p-k}$ (recall that $B^s(r_0)$ and $B^u(r_0)$ are balls centered at 0 of radius r_0). In view of (7.6) and conditions (N2) and (N3), with respect to the Lyapunov metric these maps satisfy

$$\|A_m\|' \le \lambda', \ (\|B_m{}^{-1}\|')^{-1} \ge \mu', \quad \text{where} \quad 0 < \lambda' < 1 < \mu' \tag{7.11}$$

and

$$g_m(0,0) = 0, \quad dg_m(0,0) = 0, \quad h_m(0,0) = 0, \quad dh_m(0,0) = 0, \tag{7.12}$$

$$\begin{aligned} \|dg_m(v_1,w_1) - dg_m(v_2,w_2)\|' &\le C\gamma^{-m}(\|v_1 - v_2\|' + \|w_1 - w_2\|')^\alpha, \\ \|dh_m(v_1,w_1) - dh_m(v_2,w_2)\|' &\le C\gamma^{-m}(\|v_1 - v_2\|' + \|w_1 - w_2\|')^\alpha, \end{aligned} \tag{7.13}$$

where

$$(\lambda')^\alpha < \gamma < 1, \quad 0 < \alpha \le 1, \quad C > 0. \tag{7.14}$$

Let

$$\mathcal{F}_m = \prod_{i=0}^{m-1} F_i = F_{m-1} \circ \cdots \circ F_0.$$

Exercise 7.6. Writing

$$\mathcal{F}_m(v,w) = \left(\mathcal{F}_m^{(1)}(v,w), \mathcal{F}_m^{(2)}(v,w)\right),$$

show that

$$\mathcal{F}_m^{(1)}(v,w) = \left(\prod_{i=0}^{m-1} A_i\right) v + \sum_{n=0}^{m-1} \left(\prod_{i=n+1}^{m-1} A_i\right) g_n(\mathcal{F}_n(v,w)) \tag{7.15}$$

for $m > 0$ and that

$$\mathcal{F}_m^{(1)}(v,w) = -\sum_{n=0}^{\infty} \left(\prod_{i=0}^{n} B_{i+m}\right)^{-1} h_{n+m}(\mathcal{F}_{m+n}(v,w)) \tag{7.16}$$

for $m < 0$. Hint: Observe that

$$\mathcal{F}_{m+1}^{(1)}(v,w) = A_m \mathcal{F}_m^{(1)}(v,w) + g_m(\mathcal{F}_m(v,w)),$$

$$\mathcal{F}_{m+1}^{(2)}(v,w) = B_m \mathcal{F}_m^{(2)}(v,w) + h_m(\mathcal{F}_m(v,w)).$$

To show the relation (7.15), iterate the first equality "forward". To obtain (7.16), rewrite the second equality in the form

$$\mathcal{F}_m^{(2)}(v,w) = B_m^{-1} \mathcal{F}_{m+1}^{(2)}(v,w) - B_m^{-1} h_m(\mathcal{F}_m(v,w))$$

and iterate it "backward".

We now state a general version of the Stable Manifold Theorem 7.1.

Theorem 7.7 (Abstract Stable Manifold Theorem). *Let κ be any number satisfying*

$$\lambda' < \kappa < \gamma^{1/\alpha}. \tag{7.17}$$

Then there exist constants $D > 0$ and $r_0 > r > 0$ and a map $\psi^s \colon B^s(r) \to \mathbb{R}^{p-k}$ such that:

(1) smoothness: *ψ^s is of class C^1 and for any $v_1, v_2 \in B^s(r)$,*

$$\|d\psi^s(v_1) - d\psi^s(v_2)\|' \le D(\|v_1 - v_2\|')^\alpha;$$

in addition, $\psi^s(0) = 0$ and $d\psi^s(0) = 0$;

(2) stability: *if $m \ge 0$ and $v \in B^s(r)$, then*

$$\mathcal{F}_m(v, \psi^s(v)) \in B^s(r) \times B^u(r),$$

$$\|\mathcal{F}_m(v, \psi^s(v))\|' \le D\kappa^m \|(v, \psi^s(v))\|';$$

(3) uniqueness: *given $v \in B^s(r)$ and $w \in B^u(r)$, if there is a number $K > 0$ such that*

$$\mathcal{F}_m(v, w) \in B^s(r) \times B^u(r) \quad \text{and} \quad \|\mathcal{F}_m(v, w)\|' \le K\kappa^m$$

for every $m \ge 0$, then $w = \psi^s(v)$;

(4) *the numbers D and r depend only on the numbers λ', μ', γ, α, κ, and C.*

Proof. Consider the linear space Γ_κ of sequences of vectors

$$z = \{z(m) \in \mathbb{R}^p\}_{m \in \mathbb{N}},$$

satisfying the following condition:

$$\|z\|_\kappa = \sup_{m \ge 0} (\kappa^{-m} \|z(m)\|') < \infty.$$

Γ_κ is a Banach space with the norm $\|z\|_\kappa$. Given $r > 0$, let

$$W = \{z \in \Gamma_\kappa \colon z(m) \in B^s(r) \times B^u(r) \text{ for every } m \in \mathbb{N}\}.$$

Since $0 < \kappa < 1$, the set W is open. Consider the map $\Phi_\kappa \colon B^s(r_0) \times W \to \Gamma_\kappa$ given by[2]

$$\Phi_\kappa(y, z)(0) = \left(y, -\sum_{k=0}^{\infty} \left(\prod_{i=0}^{k} B_i \right)^{-1} h_k(z(k)) \right),$$

and for $m > 0$ by

$$\Phi_\kappa(y, z)(m) = -z(m) + \left(\left(\prod_{i=0}^{m-1} A_i \right) y, 0 \right)$$

$$+ \left(\sum_{n=0}^{m-1} \left(\prod_{i=n+1}^{m-1} A_i \right) g_n(z(n)), -\sum_{n=0}^{\infty} \left(\prod_{i=0}^{n} B_{i+m} \right)^{-1} h_{n+m}(z(n+m)) \right).$$

[2]The formulae in the definition of the operator Φ_κ are motivated by Exercise 7.6 and will guarantee the invariance relations (7.7) (see (7.34) below).

Using conditions (7.11)–(7.17), we will show that the map Φ_κ is well-defined and is continuously differentiable over y and z. Indeed, by (7.12) and (7.13), for $z \in B^s(r_0) \times B^u(r_0)$ and $n \geq 0$,

$$\begin{aligned}\|g_n(z)\|' &= \|g_n(z) - g_n(0)\|' \leq \|dg_n(\xi)\|' \cdot \|z\|' \\ &= \|dg_n(\xi) - dg_n(0)\|' \cdot \|z\|' \\ &\leq C\gamma^{-n}(\|\xi\|')^\alpha \|z\|' \leq C\gamma^{-n}(\|z\|')^{1+\alpha},\end{aligned} \tag{7.18}$$

where ξ lies on the interval joining the points 0 and z. Similarly, we have that

$$\|h_n(z)\|' \leq C\gamma^{-n}(\|z\|')^{1+\alpha}. \tag{7.19}$$

Using (7.11), (7.18), and (7.19), we obtain

$$\begin{aligned}\|\Phi_\kappa(y,z)\|_\kappa &= \sup_{m\geq 0}(\kappa^{-m}\|\Phi_\kappa(y,z)(m)\|') \\ &\leq \sup_{m\geq 0}(\kappa^{-m}\|z(m)\|') \\ &\quad + \sup_{m\geq 0}\left\{\kappa^{-m}\left[\prod_{i=0}^{m-1}\|A_i\|'\cdot\|y\|' + \sum_{n=0}^{m-1}\left(\prod_{i=n+1}^{m-1}\|A_i\|'\right)\right.\right. \\ &\quad \times C\gamma^{-n}(\|z(n)\|')^{1+\alpha} \\ &\quad \left.\left. + \sum_{n=0}^{\infty}\left(\prod_{i=0}^{n}\|B_{i+m}^{-1}\|'\right)C\gamma^{-(m+n)}(\|z(n+m)\|')^{1+\alpha}\right]\right\} \\ &\leq \|z\|_\kappa + \sup_{m\geq 0}(\kappa^{-m}(\lambda')^m)\|y\|' \\ &\quad + \sup_{m\geq 0}\left(\kappa^{-m}C\|z\|_\kappa{}^{1+\alpha}\left[\sum_{n=0}^{m-1}(\lambda')^{m-n-1}\gamma^{-n}\kappa^{(1+\alpha)n}\right.\right. \\ &\quad \left.\left. + \sum_{n=0}^{\infty}(\mu')^{-(n+1)}\gamma^{-(m+n)}\kappa^{(1+\alpha)(m+n)}\right]\right).\end{aligned} \tag{7.20}$$

In view of (7.17), we have

$$\sup_{m\geq 0}(\kappa^{-m}(\lambda')^m) = 1.$$

Since the function $x \mapsto xa^x$ reaches its maximum $-1/e \log a$ at $x = -1/\log a$, we obtain

$$\sup_{m\geq 0}\left[\kappa^{-m}(\lambda')^{m-1}\sum_{n=0}^{m-1}\left((\lambda')^{-1}\gamma^{-1}\kappa^{(1+\alpha)}\right)^n\right]$$
$$\leq \begin{cases}(\lambda')^{-1}(\kappa^{-1}\lambda')^m m & \text{if } (\lambda')^{-1}\gamma^{-1}\kappa^{(1+\alpha)} \leq 1,\\ (\lambda')^{-1}(\gamma^{-1}\kappa^{\alpha})^m m & \text{if } (\lambda')^{-1}\gamma^{-1}\kappa^{(1+\alpha)} \geq 1\end{cases}$$
$$\leq (\lambda')^{-1}e^{-1}\left(\log\max\left\{\frac{\kappa^\alpha}{\gamma},\frac{\lambda'}{\kappa}\right\}\right)^{-1} =: M_1.$$

Furthermore, since $(\mu')^{-1}\gamma^{-1}\kappa^{1+\alpha} = ((\mu')^{-1}\kappa)(\kappa^\alpha\gamma^{-1}) < 1$, we have

$$\sup_{m\geq 0}\left(\kappa^{-m}\gamma^{-m}\kappa^{(1+\alpha)m}(\mu')^{-1}\sum_{n=0}^{m-1}\left((\mu')^{-1}\gamma^{-1}\kappa^{(1+\alpha)}\right)^n\right)$$
$$= \frac{1}{\mu' - \gamma^{-1}\kappa^{1+\alpha}} =: M_2. \tag{7.21}$$

Setting $M = M_1 + M_2$, we conclude that

$$\|\Phi_\kappa(y,z)\|_\kappa \leq \|z\|_\kappa + \|y\|' + CM\|z\|_\kappa^{1+\alpha}.$$

This implies that the map Φ_κ is well-defined. Moreover, $\Phi_\kappa(0,0) = (0,0)$.

We shall show that the map Φ_κ is of class C^1. Indeed, for any $y \in B^s(r_0)$ and $t \in E^s$ such that $y + t \in B^s(r_0)$, and any $z \in W$, and $m \geq 0$ we have

$$\Phi_\kappa(y+t,z)(m) - \Phi_\kappa(y,z)(m) = \left(\left(\prod_{i=0}^{m-1} A_i\right)t, 0\right).$$

It follows that

$$\partial_y\Phi_\kappa(y,z)(m) = \left(\prod_{i=0}^{m-1} A_i, 0\right). \tag{7.22}$$

For any $y \in B^s(r_0)$, $z \in W$, $t \in \Gamma_\kappa$ such that $z + t \in W$ we write

$$\Phi_\kappa(y, z+t) - \Phi_\kappa(y,z) = (\mathcal{A}_\kappa(z) - \mathrm{Id})t + o(z,t),$$

where Id is the identity map,

$$(\mathcal{A}_\kappa(z))t(m) = \left(\sum_{n=0}^{m-1}\left(\prod_{i=n+1}^{m-1} A_i\right) dg_n(z(n))t(n),\right.$$
$$\left. -\sum_{n=0}^{\infty}\left(\prod_{i=0}^{n} B_{i+m}\right)^{-1} dh_{n+m}(z(n+m))t(m+n)\right)$$

and

$$o(z,t)(m) =$$

$$\left(\sum_{n=0}^{m-1}\left(\prod_{i=n+1}^{m-1} A_i\right) o_1(z,t)(n), -\sum_{n=0}^{\infty}\left(\prod_{i=0}^{n} B_{i+m}\right)^{-1} o_2(z,t)(m+n)\right).$$

Here $o_i(z,t)(m)$ for $i = 1,2$ are defined by

$$o_1(z,t)(m) = g_m((z+t)(m)) - g_m(z(m)) - dg_m(z(m))t(m),$$
$$o_2(z,t)(m) = h_m((z+t)(m)) - h_m(z(m)) - dh_m(z(m))t(m).$$

If $z_1, z_2 \in W$ and $t \in \Gamma_\kappa$, then

$$\|(\mathcal{A}_\kappa(z_1) - \mathcal{A}_\kappa(z_2))t\|_\kappa$$
$$\le \sup_{m\ge 0}\left\{\kappa^{-m}\left[\sum_{n=0}^{m-1}\left(\prod_{i=n+1}^{m-1}\|A_i\|'\right) C\gamma^{-n}(\|z_1(n) - z_2(n)\|')^\alpha \|t(n)\|'\right.\right.$$
$$+\sum_{n=0}^{\infty}\left(\prod_{i=0}^{n}\|B_{i+m}^{-1}\|'\right) C\gamma^{-(n+m)}(\|z_1(n+m) - z_2(n+m)\|')^\alpha$$
$$\left.\left.\times \|t(n+m)\|'\right]\right\}$$
$$\le \sup_{m\ge 0}\left\{\kappa^{-m} C\left[\sum_{n=0}^{m-1}(\lambda')^{m-n-1}\gamma^{-n}\kappa^{(1+\alpha)n}\right.\right.$$
$$\left.\left.+\sum_{n=0}^{\infty}(\mu')^{-(n+1)}\gamma^{-(n+m)}\kappa^{(1+\alpha)(n+m)}\right]\right\} \times \|z_1 - z_2\|_\kappa{}^\alpha \|t\|_\kappa.$$

It follows that (see (7.20)–(7.21))

$$\|(\mathcal{A}_\kappa(z_1) - \mathcal{A}_\kappa(z_2))t\|_\kappa \le CM\|z_1 - z_2\|_\kappa{}^\alpha \|t\|_\kappa. \tag{7.23}$$

By (7.13) and the Mean Value Theorem, we find that

$$\|o_i(z,t)(m)\|' \le C\gamma^{-m}(\|t(m)\|')^{1+\alpha}.$$

This implies that

$$\|o(z,t)\|_\kappa \le CM\|t\|_\kappa{}^{1+\alpha}.$$

We conclude that $\partial_z\Phi_\kappa(y,z) = \mathcal{A}_\kappa(z) - \mathrm{Id}$. In particular, $\partial_z\Phi_\kappa(y,0) = -\mathrm{Id}$. By (7.23), the map $\partial_z\Phi_\kappa(y,z)$ is continuous. Therefore, the map Φ_κ satisfies the conditions of the Implicit Function Theorem, and hence, there exist a number $r \le r_0$ and a map $\varphi\colon B^s(r) \to W$ of class C^1 with

$$\varphi(0) = 0 \quad\text{and}\quad \Phi_\kappa(y, \varphi(y)) = 0. \tag{7.24}$$

Now we use a special version of the Implicit Function Theorem which enables us to obtain an explicit estimate of the number r and, thus, to show that this number depends only on the numbers λ', μ', γ, α, κ, and C. More precisely, the following statement holds.

Lemma 7.8 (Implicit Function Theorem). *Let E_1, E_2, and G be Banach spaces and let $g\colon A_1 \times A_2 \to G$ be a C^1 map, where $A_i \subset E_i$ is a ball centered at 0 of radius r_i, for $i = 1, 2$. Assume that $g(0,0) = 0$ and that the partial derivative (with respect to the second coordinate) $D_2 g(0,0)\colon E_2 \to G$ is a linear homeomorphism. Assume also that Dg is Hölder continuous in $A_1 \times A_2$ with Hölder constant a and Hölder exponent α. Let B be the ball in E_1 centered at 0 of radius*

$$\rho_1 = \min\left\{ r_1, r_2, \frac{r_2}{2bc}, \frac{1}{(1+2bc)(2ac)^{1/\alpha}} \right\}, \tag{7.25}$$

where

$$b = \sup_{x \in A_1} \|D_1 g(x,0)\|, \quad c = \|(D_2 g(0,0))^{-1}\|.$$

Then there exists a unique map $u\colon B \to A_2$ satisfying the following properties:

(1) *u is of class $C^{1+\alpha}$, $g(x, u(x)) = 0$ for every $x \in B$, and $u(0) = 0$;*

(2) *if $x \in B$, then*

$$\left\| \frac{du}{dx}(x) \right\| \le 1 + 2bc;$$

(3) *if $x_1, x_2 \in B$, then*

$$\left\| \frac{du}{dx}(x_1) - \frac{du}{dx}(x_2) \right\| \le 8ac(1+bc)^2 \|x_1 - x_2\|^\alpha.$$

Proof of the lemma. Let $\rho_2 = \max\{\rho_1, 2bc\rho_1\}$. By (7.25), we have that $\rho_1 \le r_1, \rho_2 \le r_2$. Also let $S_i \subset E_i$ be the open balls centered at zero of radii ρ_i, for $i = 1, 2$. We equip the space $E_1 \times E_2$ with the norm $\|(x,y)\| = \|x\| + \|y\|$, for $(x,y) \in E_1 \times E_2$. By (7.25), if $x \in S_1$ and $y_1, y_2 \in S_2$, then

$$\|g(x,y_1) - g(x,y_2) - D_2 g(0,0)(y_1 - y_2)\| \le a(\rho_1 + \rho_2)^\alpha \|y_1 - y_2\| \le \frac{1}{2c}\|y_1 - y_2\|.$$

Define a function $H\colon S_1 \times S_2 \to E_2$ by

$$H(x,y) = y - (D_2 g(0,0))^{-1} g(x,y).$$

We obtain

$$\begin{aligned} &\|H(x,y_1) - H(x,y_2)\| \\ &\quad = \|(D_2 g(0,0))^{-1}[D_2 g(0,0)(y_1 - y_2) - (g(x,y_1) - g(x,y_2))]\| \\ &\quad \le \frac{1}{2}\|y_1 - y_2\|. \end{aligned} \tag{7.26}$$

Furthermore,

$$\begin{aligned}\|H(x,0)\| &= \|-(D_2g(0,0))^{-1}g(x,0)\| \\ &= \|-(D_2g(0,0))^{-1}[g(x,0)-g(0,0)]\| \le cb\rho_1 \le \frac{\rho_2}{2}.\end{aligned} \tag{7.27}$$

We need the following statement.

Lemma 7.9. *Let E be a Banach space, let $S \subset E$ be the open ball of radius r centered at 0, and let $h\colon S \to E$ be a transformation. Assume that there exists $0 \le \beta < 1$ such that $\|h(y_1) - h(y_2)\| \le \beta\|y_1 - y_2\|$ for any $y_1, y_2 \in S$ and $\|h(0)\| \le r(1-\beta)$. Then there exists a unique $y \in S$ such that $h(y) = y$.*

Exercise 7.10. Prove Lemma 7.9.

By Lemma 7.9 with $r = \rho_2$ and $\beta = 1/2$ (see (7.26) and (7.27)), there exists a unique function $u\colon S_1 \to S_2$ such that $u(0) = 0$ and

$$g(x, u(x)) = 0$$

for every $x \in S_1$. Write $h_x(y) = H(x, y)$. Then

$$\|(h_x)^n(0) - (h_x)^{n-1}(0)\| \le \frac{\rho_2}{2^n}.$$

This implies that the series

$$\sum_{n=0}^{\infty}[{h_x}^n(0) - {h_x}^{n-1}(0)] = u(x)$$

converges uniformly and defines a continuous function.

We now show that u is differentiable. Choose $x, x+s \in S_1$ and set $t = u(x+s) - u(x)$. For any $\delta > 0$ there exists $r > 0$ such that

$$\begin{aligned}&\|D_1g(x,u(x))s + D_2g(x,u(x))t\| \\ &= \|g(x+s, u(x)+t) - g(x,u(x)) - D_1g(x,u(x))s - D_2g(x,u(x))t\| \\ &\le \delta\|(s,t)\|\end{aligned}$$

whenever $\|s\| < r$. Set

$$a = 2\|(D_2g(x,u(x)))^{-1}D_1g(x,u(x))\| + 1.$$

Choosing δ so small that $\delta\|(D_2g(x,u(x)))^{-1}\| \le 1/2$, we find that

$$\|t\| - \frac{a-1}{2}\|s\| \le \|t + (D_2g(x,u(x)))^{-1}D_1g(x,u(x))s\| \le \frac{\|s\| + \|t\|}{2}.$$

This implies that $\|t\| \le a\|s\|$ and thus,

$$\|t + (D_2g(x,u(x)))^{-1}D_1g(x,u(x))s\| \le \delta(a+1)\|(D_2g(x,u(x)))^{-1}\| \cdot \|s\|.$$

We conclude that u is differentiable and its derivative is given by

$$\frac{du}{dx}(x) = -(D_2g(x,u(x)))^{-1}D_1g(x,u(x)). \tag{7.28}$$

Therefore, u is of class C^1. Since Dg is Hölder continuous, it follows from (7.28) that u has a Hölder continuous derivative.

One can also verify that

$$\begin{aligned}\|D_1 g(x,y)\| &\le \|D_1 g(x,y) - D_1 g(0,0)\| + \|D_1 g(0,0)\| \\ &\le a(\rho_1+\rho_2)^\alpha + b \le \frac{1}{2c} + b.\end{aligned} \tag{7.29}$$

Furthermore,

$$\begin{aligned}\|D_2 g(x,y)\| &\ge \|D_2 g(0,0)\| - \|D_2 g(x,y) - D_2 g(0,0)\| \\ &\ge \frac{1}{c} - a(\rho_1+\rho_2)^\alpha \ge \frac{1}{2c}.\end{aligned}$$

This implies that $D_2 g(x,y)$ is invertible for every $(x,y)\in S_1\times S_2$ and that

$$\|(D_2 g(x,y))^{-1}\| \le 2c. \tag{7.30}$$

It follows from (7.29) and (7.30) that for every $x\in S_1$,

$$\begin{aligned}\left\|\frac{du}{dx}(x)\right\| &= \|-(D_2 g(x,u(x)))^{-1} D_1 g(x,u(x))\| \\ &\le 2c\left(\frac{1}{2c}+b\right) = 1+2bc.\end{aligned} \tag{7.31}$$

In view of (7.28), we obtain that

$$D_1 g(x,u(x)) + D_2 g(x,u(x))\frac{du}{dx}(x) = 0,$$

and thus,

$$\begin{aligned}D_1 g(x_1,u(x_1)) - D_1 g(x_2,u(x_2)) + [D_2 g(x_1,u(x_1)) - D_2 g(x_2,u(x_2))]\frac{du}{dx}(x_1) \\ + D_2 g(x_2,u(x_2))\left(\frac{du}{dx}(x_1) - \frac{du}{dx}(x_2)\right) = 0\end{aligned}$$

for every $x_1,x_2\in S_1$. Using (7.29) and (7.31), we conclude that

$$\begin{aligned}\left\|\frac{du}{dx}(x_1) - \frac{du}{dx}(x_2)\right\| &\le \|(D_2 g(x_2,u(x_2)))^{-1}\| \\ &\quad\times a\|(x_1,u(x_1)) - (x_2,u(x_2))\|^\alpha(2+2bc) \\ &\le 4ac(1+bc)(2+2bc)^\alpha\|x_1-x_2\|^\alpha \\ &\le 8ac(1+bc)^2\|x_1-x_2\|^\alpha.\end{aligned}$$

This completes the proof of Lemma 7.8. □

The map Φ_κ satisfies the conditions of Lemma 7.8 with

$$c=1, \quad b=1, \quad a=CM. \tag{7.32}$$

To confirm this, we need only to show that the derivatives $\partial_y \Phi_\kappa$ and $\partial_z \Phi_\kappa$ are Hölder continuous. This holds true for $\partial_y \Phi_\kappa$ in view of (7.22). Furthermore, by (7.23),

$$\begin{aligned}\|\partial_z \Phi_\kappa(y_1, z_1) - \partial_z \Phi_\kappa(y_2, z_2)\| &\le \|\partial_z \Phi_\kappa(y_1, z_1) - \partial_z \Phi_\kappa(y_1, z_2)\| \\ &\quad + \|\partial_z \Phi_\kappa(y_1, z_2) - \partial_z \Phi_\kappa(y_2, z_2)\| \\ &= 2\|\mathcal{A}_\kappa(z_1) - \mathcal{A}_\kappa(z_2)\| \le CM\|z_1 - z_2\|_\kappa{}^\alpha.\end{aligned}$$

We now describe some properties of the map φ. Differentiating the second equality in (7.24) with respect to y, we obtain

$$d\varphi(y) = -[\partial_z \Phi_\kappa(y, \varphi(y))]^{-1} \partial_y \Phi_\kappa(y, \varphi(y)).$$

Setting $y = 0$ in this equality yields

$$d\varphi(0)(m) = \left(\prod_{i=0}^{m-1} A_i, 0 \right). \tag{7.33}$$

One can write the vector $\varphi(y)(m)$ in the form

$$\varphi(y)(m) = (\varphi_1(y)(m), \varphi_2(y)(m)),$$

where $\varphi_1(y)(m) \in \mathbb{R}^k$ and $\varphi_2(y)(m) \in \mathbb{R}^{p-k}$. It follows from (7.24) that if $m \ge 0$, then

$$\varphi_1(y)(m) = \left(\prod_{i=0}^{m-1} A_i \right) y + \sum_{n=0}^{m-1} \left(\prod_{i=n+1}^{m-1} A_i \right) g_n(\varphi(y)(n)),$$

and

$$\varphi_2(y)(m) = -\sum_{n=0}^{\infty} \left(\prod_{i=0}^{n} B_{i+m} \right)^{-1} h_{n+m}(\varphi(y)(n+m)).$$

By Exercise 7.6, these equalities imply that

$$\begin{aligned}\varphi_1(y)(m+1) &= A_m \varphi_1(y)(m) + g_m(\varphi_1(y)(m), \varphi_2(y)(m)), \\ \varphi_2(y)(m+1) &= B_m \varphi_2(y)(m) + h_m(\varphi_1(y)(m), \varphi_2(y)(m)).\end{aligned}$$

Thus, we obtain that the function $\varphi(y)$ is invariant under the family of maps F_m, i.e.,

$$F_m(\varphi(y)(m)) = \varphi(y)(m+1) \tag{7.34}$$

(compare to (7.7)). The desired map ψ^s is now defined by

$$\psi^s(v) = \varphi_2(v)(0)$$

for each $v \in B^s(r)$. Note that $\varphi_1(v)(0) = v$. It follows from (7.34) that

$$\begin{aligned}\prod_{i=0}^{m-1} F_i(v, \psi^s(v)) &= \prod_{i=0}^{m-1} F_i(\varphi_1(v)(0), \varphi_2(v)(0)) \\ &= \prod_{i=0}^{m-1} F_i(\varphi(v)(0)) = \varphi(v)(m).\end{aligned}$$

Applying Lemma 7.8 and (7.32), we find that

$$\begin{aligned}\left\| \prod_{i=0}^{m-1} F_i(v, \psi^s(v)) \right\|' &\leq \kappa^m \|\varphi(v)\|_\kappa = \kappa^m \|\varphi(v) - \varphi(0)\|_\kappa \\ &\leq \kappa^m \sup_{\xi \in B^s(r)} \|d\varphi(\xi)\|_\kappa \|v\|' \\ &\leq 3\kappa^m \|v\|' \leq 3\kappa^m \|(v, \psi^s(v))\|'\end{aligned}$$

(we use here the fact that $\|v_1\|, \|v_2\| \leq \|v\|$ for every $v = (v_1, v_2) \in \mathbb{R}^2$). This proves statement (2) of Theorem 7.7. Furthermore, by (7.32) and Lemma 7.8, for any $u_1, u_2 \in B^s(r)$, we have

$$\begin{aligned}\|d\psi^s(v_1) - d\psi^s(v_2)\|' &= \|d\varphi_2(v_1)(0) - d\varphi_2(v_2)(0)\|' \\ &\leq \|d\varphi(v_1)(0) - d\varphi(v_2)(0)\|' \leq 32CM(\|v_1 - v_2\|')^\alpha.\end{aligned}$$

Statement (1) of the theorem now follows from (7.24) and (7.33) with the numbers D and r depending only on the numbers λ', μ', γ, α, κ, and C. This implies statement (4) of the theorem. Take $(v, w) \in B^s(r) \times B^u(r_0)$ satisfying the assumptions of statement (4) of the theorem. Set

$$z(m) = \left(\prod_{i=0}^{m-1} F_i(v, w) \right).$$

It follows that $z \in \Gamma_\kappa$ (with $\|z\|_\kappa \leq K$) and that $\Phi_k(v, z) = 0$. The uniqueness of the map φ implies that $z = \varphi(v)$ and hence,

$$w = \varphi_2(v)(0) = \psi^s(v).$$

This establishes statement (3) of the theorem and completes the proof. □

7.3. Basic properties of stable and unstable manifolds

In hyperbolicity theory there is a symmetry between the objects marked by the index "s" and those marked by the index "u". Namely, when the time direction is reversed, the statements concerning objects with index "s" become the statements about the corresponding objects with index "u". In particular, this allows one to define a *local unstable manifold* $V^u(x)$ at a point x in a nonuniformly completely hyperbolic set Y as a local stable manifold for f^{-1}. In this section we describe some basic properties of local stable and unstable manifolds.

Let Y be the set of nonuniformly completely hyperbolic trajectories for f. Consider the corresponding collection $\Lambda_{\lambda\mu\varepsilon j}$ of level sets and for each $\lambda, \mu, \varepsilon$, and $1 \le j < p$ consider the collection of regular sets $\{\Lambda^\ell_{\lambda\mu\varepsilon j} : \ell \ge 1\}$ (see (4.24) and (4.25)). In what follows, we fix numbers $\lambda, \mu, \varepsilon$, and j and set $\Lambda = \Lambda_{\lambda\mu\varepsilon j}$ and $\{\Lambda^\ell = \Lambda^\ell_{\lambda\mu\varepsilon j} : \ell \ge 1\}$. Note that the regular sets are compact.

7.3.1. Sizes of local manifolds. First, we discuss how the sizes of local manifolds vary. Recall that the size of the local stable manifold $V^s(x)$ at a point $x \in \Lambda$ is the number $r = r(x)$ that is determined by Theorem 7.7 and such that (7.3) holds.[3] Similarly, one defines the size of the local unstable manifold $V^u(x)$ at x.

It follows from statement (4) of Theorem 7.7 and condition (N3) in Section 7.1 that the sizes of the local stable and unstable manifolds at a point $x \in \Lambda$ and at any point $y = f^m(x)$ for $m \in \mathbb{Z}$ along the trajectory of x are related by

$$r(f^m(x)) \ge Ke^{-\varepsilon|m|}r(x), \tag{7.35}$$

where $K > 0$ is a constant.

Let μ be an ergodic invariant Borel measure for f such that $\mu(\Lambda) > 0$. Then for all sufficiently large ℓ the regular set Λ^ℓ has positive measure. Therefore, the trajectory of almost every point visits the set Λ^ℓ infinitely many times. It follows that for typical points x the function $r(f^m(x))$ is an oscillating function of m, which is of the same order as $r(x)$ for many values of m. Nevertheless, for some integers m the value $r(f^m(x))$ may become as small as is allowed by (7.35). Let us emphasize that the rate with which the sizes of the local stable manifolds $V^s(f^m(x))$ decrease as $m \to +\infty$ is smaller than the rate with which the trajectories $\{f^m(x)\}$ and $\{f^m(y)\}$, $y \in V^s(x)$, approach each other.

Consider the nonuniformly hyperbolic set Y which consists of a *single* nonuniformly hyperbolic trajectory $\{f^n(x)\}$. The Stable Manifold Theorem 7.1 applies. It characterizes the behavior of trajectories nearby an *individual* nonuniformly hyperbolic trajectory and does not need the presence of any other nonuniformly hyperbolic trajectories. Note also that one can replace the assumption that the manifold M is compact by the assumption that the diffeomorphism f satisfies (7.4)–(7.6) along a given nonuniformly hyperbolic trajectory.

Fix a regular set Λ^ℓ.

---

[3]We stress that the size of the local manifold is not uniquely defined as, for example, the number $r = \frac{1}{2}r(x)$ can be used as well.

Exercise 7.11. Using Theorem 7.7 and condition (N3) show that:

(1) the sizes of local manifolds are bounded from below on Λ^ℓ; i.e., there exists a number $r_\ell > 0$ that depends only on ℓ such that

$$r(x) \geq r_\ell \quad \text{for } x \in \Lambda^\ell; \tag{7.36}$$

(2) local manifolds depend uniformly continuously on $x \in \Lambda^\ell$ in the C^1 topology; i.e., if $x_n \in \Lambda^\ell$ is a sequence of points converging to a point $x \in \Lambda^\ell$, then

$$d_{C^1}(V^s(x_n), V^s(x)) \to 0, \quad d_{C^1}(V^u(x_n), V^u(x)) \to 0$$

as $n \to \infty$; here for two given local smooth submanifolds V_1 and V_2 in M of the same size r (i.e., V_i is of the form (7.3) for some r and some functions ψ_i, $i = 1, 2$), we have

$$d_{C^1}(V_1, V_2) = \sup_{x \in B^s(r)} \|\psi_1(x) - \psi_2(x)\| + \sup_{x \in B^s(r)} \|d_x\psi_1 - d_x\psi_2\|;$$

(3) there exists a number $\delta_\ell > 0$ such that for every $x \in \Lambda^\ell$ and $y \in \Lambda^\ell \cap B(x, \delta_\ell)$ the intersection $V^s(x) \cap V^u(y)$ is nonempty and consists of a single point which depends continuously on x and y;

(4) local stable and unstable distributions depend Hölder continuously on $x \in \Lambda^\ell$; more precisely, for every $\ell \geq 1$, $x \in \Lambda^\ell$, and points $z_1, z_2 \in V^s(x)$ or $z_1, z_2 \in V^u(x)$ we have

$$\begin{aligned} d(T_{z_1}V^s(x), T_{z_2}V^s(x)) &\leq C\rho(z_1, z_2)^\alpha, \\ d(T_{z_1}V^u(x), T_{z_2}V^u(x)) &\leq C\rho(z_1, z_2)^\alpha, \end{aligned} \tag{7.37}$$

where $C > 0$ is a constant and d is the distance in the Grassmannian bundle of TM generated by the Riemannian metric.

If f is an Anosov diffeomorphism (i.e., f is uniformly hyperbolic; see Section 1.1), then $\Lambda^\ell = M$ for some ℓ and hence,

$$r(x) \geq r_\ell \quad \text{for every} \quad x \in M.$$

7.3.2. Smoothness of local manifolds. One can obtain more refined information about smoothness of local stable manifolds. More precisely, assume that the diffeomorphism f is of class $C^{r+\alpha}$, with $r \geq 1$ and $0 < \alpha \leq 1$ (i.e., the differential $d^r f$ is Hölder continuous with Hölder exponent α). Then $V^s(x)$ is of class C^r; in particular, if f is of class C^r for some $r \geq 2$, then $V^s(x)$ is of class C^{r-1}. These results are immediate consequences of the following version of Theorem 7.7.

Theorem 7.12. *Assume that the conditions of Theorem 7.7 hold. In addition, assume that:*

(1) *g_m and h_m are of class C^r for some $r \geq 2$;*

(2) *there exists $K > 0$ such that for $\ell = 1, \dots, r$,*

$$\sup_{z\in B} \|d^\ell g_m(z)\|' \le K\gamma^{-m}, \quad \sup_{z\in B} \|d^\ell h_m(z)\|' \le K\gamma^{-m},$$

where $B = B^s(r_0) \times B^u(r_0)$ (see (7.10));

(3) *if $z_1, z_2 \in B$, then*

$$\|d^p g_m(z_1) - d^p g_m(z_2)\|' \le K\gamma^{-m}(\|z_1 - z_2\|')^\alpha,$$
$$\|d^p h_m(z_1) - d^p h_m(z_2)\|' \le K\gamma^{-m}(\|z_1 - z_2\|')^\alpha$$

for some $\alpha \in (0, 1)$.

If ψ^s is the map constructed in Theorem 7.7, then there exists a number $N > 0$, which depends only on the numbers λ', μ', γ, α, κ, and K such that:

(i) *ψ^s is of class $C^{r+\alpha}$;*

(ii) *$\sup_{u\in B^s(r)} \|d^\ell \psi^s(u)\|' \le N$ for $\ell = 1, \dots, r$.*

Sketch of the proof. It is sufficient to show that the map Φ_κ is of class C^r. Indeed, a simple modification of the argument in the proof of Theorem 7.7 allows one to prove that for $2 \le l \le r$,

$$d_y^l \Phi_\kappa(y, z) = (0, 0)$$

and for every $m \ge 1$,

$$d_z^l \Phi_\kappa(y, z)(m) =$$
$$\left(\sum_{n=0}^{m-1} \left(\prod_{i=n+1}^{m-1} A_i\right) d^l g_n(z(n)), -\sum_{n=0}^{\infty} \left(\prod_{i=0}^{n} B_{i+m}^{-1}\right) d^l h_{m+n}(z(m+n))\right).$$

□

Exercise 7.13. Complete the proof of Theorem 7.12.

It follows that if $f \in C^r$, then $V^s(x) \in C^{r-1+\alpha}$ for every $x \in \Lambda$ and $0 \le \alpha < 1$. One can in fact show that $V^s(x) \in C^r$; see [**75**].

7.3.3. Graph transform property. There is another proof of the Stable Manifold Theorem which is based on a version of the Graph Transform Property—a statement that is well known in the uniform hyperbolicity theory (and usually referred to as the Inclination Lemma or the λ-Lemma). This approach is essentially an elaboration of the Hadamard method.

Let $x \in \Lambda$. Choose numbers r_0, b_0, c_0, and d_0 and for every $m \ge 0$, set

$$r_m = r_0 e^{-\varepsilon m}, \quad b_m = b_0 e^{-\varepsilon m}, \quad c_m = c_0 e^{-\varepsilon m}, \quad d_m = d_0(\lambda')^m e^{\varepsilon m}.$$

Consider the set Ψ of $C^{1+\alpha}$ functions on $\{(m, v) : m \in \mathbb{N}, v \in B^s(r_m)\}$ such that

$$\psi(m, v) \in E^u(f^{-m}(x)) \quad \text{for every } m \ge 0 \text{ and } v \in B^s(r_m)$$

(where $B^s(r_m)$ is the ball in $E^s(f^{-m}(x))$ centered at 0 of radius r_m) and satisfying the following conditions:

$$\|\psi(m,0)\| \le b_m, \qquad \max_{v\in B^s(r_m)} \|d\psi(m,v)\| \le c_m,$$

and if $v_1, v_2 \in B^s(r_m)$, then

$$\|d\psi(m,v_1) - d\psi(m,v_2)\| \le d_m\|v_1 - v_2\|^\alpha. \tag{7.38}$$

Theorem 7.14 (Graph Transform Property). *For every $\ell \ge 1$ there are positive constants r_0, b_0, c_0, and d_0, which depend only on ℓ, such that for every $x \in \Lambda^\ell$ and every function $\psi \in \Psi$ there exists a function $\tilde\psi \in \Psi$ with the property that*

$$F_m^{-1}(\{(v,\psi(m,v)) : v \in B^s(r_m)\}) \supset \{(v,\tilde\psi(m+1,v)) : v \in B^s(r_{m+1})\} \tag{7.39}$$

for all $m \ge 0$.

Sketch of the proof. Let $(v^{(m+1)}, w^{(m+1)}) = F_m^{-1}(v,\psi(m,v))$. One can write

$$(v^{(m+1)}, w^{(m+1)}) = (A_m^{-1}v + \tilde g_m(v,\psi(m,v)), B_m^{-1}\psi(m,v) + \tilde h_m(v,\psi(m,v))),$$

where (compare with Section 7.1)

$$\tilde g_m(0,0) = 0, \quad d\tilde g_m(0,0) = 0, \quad \tilde h_m(0,0) = 0, \quad d\tilde h_m(0,0) = 0,$$

$$\|d\tilde g_m(v_1,w_1) - d\tilde g_m(v_2,w_2)\|' \le C\gamma^{-m}(\|v_1 - v_2\|' + \|w_1 - w_2\|')^\alpha,$$

$$\|d\tilde h_m(v_1,w_1) - d\tilde h_m(v_2,w_2)\|' \le C\gamma^{-m}(\|v_1 - v_2\|' + \|w_1 - w_2\|')^\alpha$$

for some constants satisfying (7.14). For some $c > 0$ we have that

$$\begin{aligned}\|v_1^{(m+1)} - v_2^{(m+1)}\|' &\ge (\lambda')^{-1}\|v_1 - v_2\|' \\ &\quad - \|\tilde g_m(v_1,\psi(m,v_1)) - \tilde g_m(v_2,\psi(m,v_2))\|' \\ &= [(\lambda')^{-1} - c(1+c_m)]\|v_1 - v_2\|'.\end{aligned}$$

Therefore, choosing r_0, b_0, $c_0 > 0$ sufficiently small, one can define a function $v \mapsto \tilde\psi(m+1,v)$ on $B^s(r_{m+1})$ by

$$\tilde\psi(m+1, v^{(m+1)}) = w^{(m+1)}. \tag{7.40}$$

This function satisfies (7.39). Furthermore, we have

$$\begin{aligned}\|w_1^{(m+1)} - w_2^{(m+1)}\|' &\le (\mu')^{-1}c_m\|v_1 - v_2\|' + c(1+c_m)\|v_1 - v_2\|' \\ &\le \frac{(\mu')^{-1}c_m + c(1+c_m)}{(\lambda')^{-1} - c(1+c_m)}\|v_1^{(m+1)} - v_2^{(m+1)}\|'.\end{aligned}$$

By eventually choosing a smaller $c_0 > 0$, this implies that

$$\|\tilde\psi(m+1,0)\|' \le b_{m+1}, \qquad \max_{v\in B^s(r_{m+1})} \|d\tilde\psi(m+1,v)\|' \le c_{m+1}.$$

Taking derivatives in (7.40) and using (7.38), one can show that if $v_1, v_2 \in B^s(r_{m+1})$, then

$$\|d\tilde{\psi}(m+1, v_1) - d\tilde{\psi}(m+1, v_2)\|' \le d_{m+1}(\|v_1 - v_2\|')^\alpha.$$

This completes the proof of the theorem. □

7.3.4. Global manifolds. For every $x \in \Lambda$ we define the *global stable* and *unstable manifolds* by

$$W^s(x) = \bigcup_{n=0}^{\infty} f^{-n}(V^s(f^n(x))), \quad W^u(x) = \bigcup_{n=0}^{\infty} f^n(V^u(f^{-n}(x))). \tag{7.41}$$

They are finite-dimensional immersed smooth submanifolds (of class $C^{r+\alpha}$ if f is of class $C^{r+\alpha}$) and are invariant under f. They have the following properties which are immediate consequences of the Stable Manifold Theorem 7.7.

Theorem 7.15. *If $x, y \in \Lambda$, then:*

(1) *$W^s(x) \cap W^s(y) = \varnothing$ if $y \notin W^s(x)$; $W^u(x) \cap W^u(y) = \varnothing$ if $y \notin W^u(x)$;*
(2) *$W^s(x) = W^s(y)$ if $y \in W^s(x)$; $W^u(x) = W^u(y)$ if $y \in W^u(x)$;*
(3) *for every $y \in W^s(x)$ (or $y \in W^u(x)$) we have $\rho(f^n(x), f^n(y)) \to 0$ as $n \to +\infty$ (respectively, $n \to -\infty$) with an exponential rate.*

Exercise 7.16. Prove Theorem 7.15.

Although global stable and unstable manifolds are smooth and pairwise disjoint, they may not form a foliation.[4] Moreover, these manifolds may be of *finite* size in some directions in the manifold; see Chapter 13 for more details.

7.3.5. Stable manifold theorem for flows. Let φ_t be a smooth flow on a compact smooth Riemannian manifold M and let Y be a nonuniformly hyperbolic set for φ_t. Also let $\Lambda_{\lambda\mu\varepsilon j}$ be the corresponding collection of level sets (see (4.24)). In what follows, we fix numbers $\lambda, \mu, \varepsilon$, and $1 \le j < p$ and assume that the level set $\Lambda = \Lambda_{\lambda\mu\varepsilon j}$ is not empty. The following statement is an analog of Theorem 7.1 for flows.

Theorem 7.17. *For every $x \in \Lambda$ there exists a local stable manifold $V^s(x)$ such that $x \in V^s(x)$, $T_xV^s(x) = E^s(x)$, and for every $y \in V^s(x)$ and $t > 0$,*

$$\rho(\varphi_t(x), \varphi_t(y)) \le T(x)\lambda^t \rho(x, y),$$

[4]Global manifolds form what can be called a *measurable lamination* of the set Λ.

where $0 < \lambda < 1$ *is a constant and* $T\colon \Lambda \to (0,\infty)$ *is a Borel function such that for every* $\tau \in \mathbb{R}$,

$$T(\varphi_\tau(x)) \le T(x)e^{10\varepsilon|\tau|}.$$

Furthermore, $\varphi_\tau(V^s(x)) \subset V^s(\varphi_\tau(x))$ *for every* $\tau \in \mathbb{R}$.

Proof. Consider the time-one map $f = \varphi_1$. It is nonuniformly partially hyperbolic in the broad sense (see Section 4.4) and the desired result follows from Remark 7.2. □

We call $V^s(x)$ a *local stable manifold* at x. In a similar fashion, by reversing the time, one can show that there exists a *local unstable manifold* $V^u(x)$ at x such that $T_xV^u(x) = E^u(x)$.

For every $x \in \Lambda$ we define the *global stable* and *unstable manifolds* at x by

$$W^s(x) = \bigcup_{t>0} \varphi_{-t}(V^s(\varphi_t(x))), \quad W^u(x) = \bigcup_{t>0} \varphi_t(V^u(\varphi_{-t}(x))). \tag{7.42}$$

These are finite-dimensional immersed smooth submanifolds (of class $C^{r+\alpha}$ if φ_t is of class $C^{r+\alpha}$). They satisfy properties (1) and (2) in Theorem 7.15. Furthermore, for every $y \in W^s(x)$ (or $y \in W^u(x)$) we have $\rho(\varphi_t(x), \varphi_t(y)) \to 0$ as $t \to +\infty$ (respectively, $t \to -\infty$) with an exponential rate.

We also define the *global weakly stable* and *unstable manifolds* at x by

$$W^{s0}(x) = \bigcup_{t\in\mathbb{R}} W^s(\varphi_t(x)), \quad W^{u0}(x) = \bigcup_{t\in\mathbb{R}} W^u(\varphi_t(x)).$$

It follows from (7.42) that

$$W^{s0}(x) = \bigcup_{t\in\mathbb{R}} \varphi_t(W^s(x)), \quad W^{u0}(x) = \bigcup_{t\in\mathbb{R}} \varphi_t(W^u(x)).$$

We remark that each of these two families of invariant immersed manifolds forms a partition of the set Λ.

Chapter 8

Absolute Continuity of Local Manifolds

Let f be a $C^{1+\alpha}$ diffeomorphism of a compact smooth Riemannian manifold M and let W be a continuous foliation of M with smooth leaves (see Section 1.1). Fix $x \in M$ and consider local manifolds $V(y)$ through points y in the ball $B(x,r)$ around x of a small radius r. Since $V(y)$ is a smooth submanifold, the restriction of the Riemannian metric to $V(y)$ generates a Riemannian volume on $V(y)$ that we call the *leaf-volume*. We denote by m the Riemannian volume on M and by $m_{V(y)}$ the leaf-volume on $V(y)$.

We describe one of the most crucial properties of foliations (and of their local manifolds) known as *absolute continuity*. It addresses the following question:

> *If $E \subset B(x,r)$ is a Borel set of positive volume, can the intersection $E \cap V(y)$ have zero leaf-volume for almost every $y \in E$?*

Absolute continuity is a property of the foliation with respect to volume and does not require the presence of the dynamics (nor of its invariant measure). This property can be understood in a variety of ways. The one that addresses the above question in the most straightforward way is the following: the foliation W is *absolutely continuous (in the weak sense)* if for almost every $x \in M$ (with respect to volume), any $q > 0$, any Borel subset $X \subset B(x,r)$ of positive volume, and almost every $y \in X$ the intersection $X \cap V(y)$ has positive leaf-volume.

If the foliation W is smooth, then by Fubini's theorem, the intersection $E \cap V(y)$ has positive leaf-volume for almost all $y \in B(x,r)$. If the foliation is only continuous, the absolute continuity property may not hold.

A simple example, which illustrates this paradoxical phenomenon, is presented in Section 8.4. In the example a set of full volume meets almost every leaf of the foliation at a single point—the phenomenon known as "Fubini's nightmare" or "Fubini foiled". Such pathological foliations are in a sense "typical" objects in the stable ergodicity theory (see Section 12.3) and the failure of absolute continuity is essentially due to the presence of zero Lyapunov exponents.

We describe a stronger version of the absolute continuity property. Fix $x \in M$ and consider the partition ξ of the ball $B(x,r)$ by local manifolds $V(y)$, $y \in B(x,r)$. Since the partition ξ is measurable, there exist the factor-measure $\tilde{m}$ in the factor-space $B(x,r)/\xi$ and the system of conditional measures m_y, $y \in B(x,r)/\xi$, such that for any Borel subset $E \subset B(x,r)$ of positive volume,

$$m(E) = \int_{B(x,r)/\xi} \int_{V(y)} \chi_E(y,z)\, dm_y(z)\, d\tilde{m}(y), \tag{8.1}$$

where χ_E is the characteristic function of the set E. Observe that the factor-space $B(x,r)/\xi$ can be naturally identified with a local transversal[1] T through x to the local manifolds $V(y)$, $y \in B(x,r)$. We say that the foliation W is *absolutely continuous (in the strong sense)* if for almost every $x \in M$, any $r > 0$, and any $y \in T$, $z \in V(y)$,

$$d\tilde{m}(y) = h(y)dm_T(y), \quad dm_y(z) = g(y,z)dm_{V(y)}(z),$$

where m_T is the leaf-volume on T and $h(y)$ and $g(y,z)$ are positive bounded Borel functions. In other words, the conditional measures on the elements $V(y)$ of the partition ξ are absolutely continuous with respect to the leaf-volume on $V(y)$ and the factor-measure is absolutely continuous with respect to the leaf-volume on the transversal T.[2] In particular, this implies that for any Borel subset $E \subset B(x,q)$ of positive volume, relation (8.1) can be rewritten as follows:

$$m(E) = \int_T h(y) \int_{V(y)} \chi_E(y,z)g(y,z)\, dm_{V(y)}(z)\, dm_T(y).$$

A celebrated result by Anosov claims that the stable and unstable invariant foliations for Anosov diffeomorphisms are absolutely continuous. We stress that generically (in the set of Anosov diffeomorphisms) these foliations are not smooth and therefore, the absolute continuity property is not at all trivial and requires a deep study of the structure of these foliations.

[1] Recall that a local transversal is a local smooth manifold that is uniformly transverse to local manifolds $V(y)$.

[2] Recall that if ν and μ are two measures on a measure space X, then ν is said to be absolutely continuous with respect to μ if for every $\varepsilon > 0$ there exists $\delta > 0$ such that $\nu(E) < \varepsilon$ for every measurable set E with $\mu(E) < \delta$.

An approach to establishing absolute continuity utilizes the holonomy maps associated with the foliation. To explain this, consider a foliation W. Given a point x, choose two transversals T^1 and T^2 to the family of local manifolds $V(y)$, $y \in B(x,r)$. The holonomy map associates to a point $z \in T^1$ the point $w = V(z) \cap T^2$. If it is absolutely continuous (see the definition below) for all points x and any choice of transversals T^1 and T^2, then the absolute continuity property follows.

For nonuniformly hyperbolic diffeomorphisms the study of absolute continuity is technically much more complicated due to the fact that the global stable and unstable manifolds may not form foliations (they may not even exist for some points in M) and the sizes of local manifolds may vary wildly from point to point. In order to overcome this difficulty, one should define and study the holonomy maps associated with local stable (or unstable) manifolds on regular sets.

8.1. Absolute continuity of the holonomy map

Let Y be the set of nonuniformly completely hyperbolic trajectories for f. Consider the corresponding collection $\Lambda_{\lambda\mu\varepsilon j}$ of level sets and for each $\lambda, \mu, \varepsilon$, and $1 \le j < p$ consider the collection of regular sets $\{\Lambda^\ell_{\lambda\mu\varepsilon j} : \ell \ge 1\}$ (see (4.24) and (4.25)). In what follows, we fix numbers λ, μ, ε, and j and set $\Lambda = \Lambda_{\lambda\mu\varepsilon j}$ and $\Lambda^\ell = \Lambda^\ell_{\lambda\mu\varepsilon j}$. Note that the regular sets are compact.

The stable and unstable subspaces $E^s(x)$ and $E^u(x)$ as well as the local stable manifolds $V^s(x)$ and local unstable manifolds $V^u(x)$ depend continuously on $x \in \Lambda^\ell$ and their sizes are bounded away from zero by a number r_ℓ (see (7.36)).

Fix $x \in \Lambda^\ell$, a number r, $0 < r \le r_\ell$, and let

$$Q^\ell(x,r) = \bigcup_{w \in \Lambda^\ell \cap B(x,r)} V^s(w), \tag{8.2}$$

where $B(x,r)$ is the ball centered at x of radius r.

Consider the family of local stable manifolds

$$\mathcal{L}^s(x) = \{V^s(w) : w \in \Lambda^\ell \cap B(x,r)\}. \tag{8.3}$$

Let T^1 and T^2 be two local open submanifolds which are *uniformly* transverse to the family $\mathcal{L}^s(x)$. We define the *holonomy map*

$$\pi^s : Q^\ell(x,r) \cap T^1 \to Q^\ell(x,r) \cap T^2$$

by setting

$$\pi^s(y) = T^2 \cap V^s(w) \quad \text{if } y = T^1 \cap V^s(w) \text{ and } w \in \Lambda^\ell \cap B(x,r)$$

(see Figure 8.1). The map π^s depends upon the choice of the point x, the number ℓ, and the transversals T^1 and T^2.

Exercise 8.1. Show that the holonomy map π^s is a homeomorphism between $Q^\ell(x,r)\cap T^1$ and $Q^\ell(x,r)\cap T^2$.

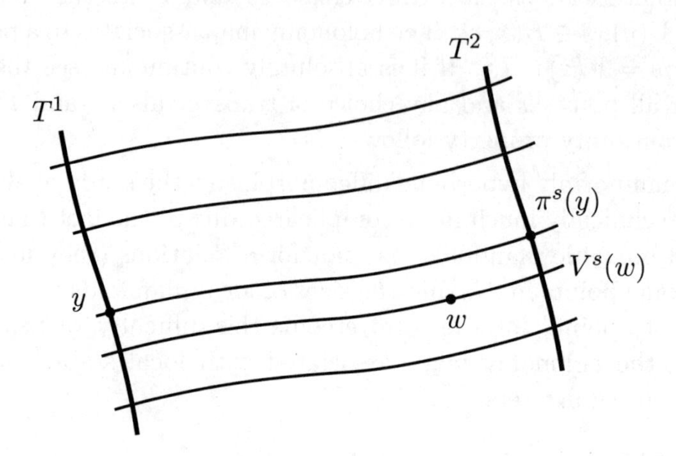

Figure 8.1. A family of stable leaves $V^s(w)$ and the holonomy map π^s

Recall that a measurable invertible transformation $S\colon X\to Y$ between measure spaces (X,ν) and (Y,μ) is said to be *absolutely continuous* if the measure μ is absolutely continuous with respect to the measure $S_*\nu$. In this case one defines the *Jacobian* $\mathrm{Jac}(S)(x)$ of S at a point $x\in X$ (specified by the measures ν and μ) to be the Radon–Nikodým derivative $d\mu/d(S_*\nu)$. If X is a metric space, then for μ-almost every $x\in X$ one has

$$\mathrm{Jac}(S)(x)=\lim_{r\to 0}\frac{\mu(S(B(x,r)))}{\nu(B(x,r))}. \tag{8.4}$$

We denote by

$$\mathcal{L}^u(x)=\{V^u(w): w\in\Lambda^\ell\cap B(x,r)\}$$

the family of local unstable manifolds.

Theorem 8.2 (Absolute Continuity). *Given $\ell\geq 1$, $x\in\Lambda^\ell$, and transversals T^1 and T^2 to the family $\mathcal{L}^s(x)$ such that*

$$m_{T^1}(Q^\ell(x,r)\cap T^1)>0, \tag{8.5}$$

the holonomy map π^s is absolutely continuous (with respect to the leaf-volumes m_{T^1} and m_{T^2}) and the Jacobian $\mathrm{Jac}(\pi^s)(y)$ is bounded from above and bounded away from zero for almost every point $y\in Q^\ell(x,r)\cap T^1$ (with respect to the leaf-volume m_{T^1}). The family $\mathcal{L}^u(x)$ of local unstable manifolds also satisfies the absolute continuity property.

The proof of Theorem 8.2 is given in Section 8.2. An important consequence of this theorem is the fact that local stable and unstable manifolds are absolutely continuous (in the strong sense). To see this, consider the set

Y of nonuniformly hyperbolic trajectories and assume that it has positive (not necessarily full) volume, i.e., $m(Y) > 0$. Let us stress that here the volume m need not be invariant under f. We can choose ℓ so large that the regular set Λ^ℓ has positive volume. Fix a Lebesgue point $x \in \Lambda^\ell$ and consider the family of local stable manifolds $\mathcal{L}^s(x)$ with $r = r(\ell)$ to be sufficiently small. For $u \in Q^\ell(x, r)$ there is a point $w \in \Lambda^\ell \cap B(x, r)$ such that $u \in V^s(w)$. Although u may not lie in $\Lambda^\ell \cap B(x, r)$, with some abuse of notation we denote by $V^s(u)$ the local stable manifold containing u and we denote by $m_{V^s(u)}$ the leaf-volume on $V^s(u)$.

There exists a family of local smooth submanifolds $T(y)$ for $y \in V^s(x)$ such that:

(1) every $T(y)$ is transverse to the family of local stable manifolds $\mathcal{L}^s(x)$ uniformly in y, i.e., the angle between $T(y)$ and $V^s(u)$ is uniformly away from zero for every $y \in V^s(x)$ and $u \in Q^\ell(x, r)$;

(2) $T(y_1) \cap T(y_2) = \varnothing$ for any $y_1 \neq y_2$;

(3) $\bigcup_{y \in V^s(u_0)} T(y) \supset B(x, r)$;

(4) $T(y)$ depends smoothly on $y \in V^s(x)$.

The collection of smooth submanifolds $T(y)$ generates a partition of the set $T = \bigcup_{y \in V^s(x)} T(y)$. We denote this partition by η. We also denote by $m_{T(y)}$ the leaf-volume on $T(y)$.

For $y \in V^s(x)$ let

$$A(y) = Q^\ell(x, r) \cap T(y).$$

Furthermore, for $u \in Q^\ell(x, r)$ we write $u = (y, z)$ where $y \in V^s(x)$ is such that $u \in T(y)$ and $z = V^s(u) \cap T(x)$.

Consider now the measurable partition ξ^s of the set $Q^\ell(x, r)$ into local stable manifolds $V^s(u)$ for $u \in Q^\ell(x, r)$. Denote by $m^s(u)$ the conditional measure on $V^s(u)$ generated by the partition ξ^s and volume m. For every $y \in V^s(x)$ the factor-space $Q^\ell(x, r)/\xi^s$ can be identified with the set $A(y)$. Denote by $\hat{m}^s$ the factor-measure generated by the partition ξ^s.

Theorem 8.3. *For m-almost every $u = (y, z) \in Q^\ell(x, r)$ the following statements hold:*

(1) *the measures $m^s(u)$ and $m_{V^s(u)}$ are equivalent;*

(2) *the factor-measure $\hat{m}^s$ is equivalent to the measure $m_{T(y)} | A(y)$.*

Proof. We follow the approach suggested in [4]. Consider the partition η. For $y \in V^s(x)$, denote by $\{\mu(y)\}$ the system of conditional measures on $T(y)$ and denote by $\hat{\mu}$ the factor-measure on T/η generated by the partition η and volume m. The factor-space T/η can be identified with the local stable manifold $V^s(x)$ and we can view the factor-measure $\hat{\mu}$ as a measure on $V^s(x)$. Since the partition η is smooth, there exist continuous functions $g(y,z)$ and $h(y)$ for $u=(y,z)\in Q^\ell(x,r)$ such that

$$d\mu(y)(z) = g(y,z)\,dm_{T(y)}(z), \quad d\hat{\mu}(y) = h(y)\,dm_{V^s(x)}(y).^3$$

Let $E \subset Q^\ell(x,r)$ be a Borel subset of positive volume. We have

$$\begin{aligned} m(E) &= \int_{T/\eta}\int_{T(y)} \chi_E(y,z)\,d\mu(y)(z)\,d\hat{\mu}(y) \\ &= \int_{T/\eta}\int_{T(y)} \chi_E(y,z)g(y,z)\,dm_{T(y)}(z)\,d\hat{\mu}(y) \\ &= \int_{T/\eta}\int_{T(x)} \chi_E(y,z)g(y,z)\,\mathrm{Jac}(\pi^s_{xy})(q)\,dm_{T(x)}(q)\,d\hat{\mu}(y), \end{aligned}$$

where χ_E is the characteristic function of the set E, π^s_{xy} is the holonomy map between $A(x)$ and $A(y)$, $z=\pi^s_{xy}(q)$, and $\mathrm{Jac}(\pi^s_{xy})(q)$ is the Jacobian of the map π^s_{xy} at q. Applying Fubini's theorem, we obtain

$$\begin{aligned} m(E) &= \int_{T(x)}\int_{T/\eta} \chi_E(y,z)g(y,z)\,\mathrm{Jac}(\pi^s_{xy})(q)\,d\hat{\mu}(y)\,dm_{T(x)}(q) \\ &= \int_{T(x)}\int_{V^s(x)} \chi_E(y,z)g(y,z)\,\mathrm{Jac}(\pi^s_{xy})(q)h(y)\,dm_{V^s(x)}(y)\,dm_{T(x)}(q). \end{aligned}$$

This implies that the conditional measure $m^s(z)$ on $V^s(z)$ is equivalent to the leaf-volume $m_{V^s(z)}$ with the density function

$$\kappa^s(y,z) = \mathrm{Jac}(\pi^s_{xy})((\pi^s_{xy})^{-1}(z))g(y,z)h(y) > 0$$

and the desired result follows. □

As an immediate consequence of Theorem 8.3 we obtain the following results.

Theorem 8.4. *The following statements hold:*

(1) $m_{V^s(x)}(V^s(x)\setminus\Lambda)=0$ *for m-almost every $x\in\Lambda$;*

(2) *for m-almost every $x\in\Lambda^\ell$ and $y\in V^s(x)$,*

$$dm^s(x)(y) = \rho(x,y)\,\mathrm{Jac}(\pi^s(y))dm_{V^s(x)}(y)$$

where $\rho(x,y)$ is a continuous function in x and y and the Jacobian $\mathrm{Jac}(\pi^s(y))$ is given by (8.10).

[3] Indeed, there is a local diffeomorphism $G\colon T\to\mathbb{R}^p=\mathbb{R}^k\times\mathbb{R}^{p-k}$ (where $p=\dim M$ and $k=\dim T(y)$, $y\in V^s(x)$) that *rectifies* the partition η, i.e., it maps it into a partition of the open set $B^k\times B^{p-k}$ (where $B^k\subset\mathbb{R}^k$ and $B^{p-k}\subset\mathbb{R}^{p-k}$ are open unit balls) by subspaces parallel to $\mathbb{R}^k$ and it maps $V^s(x)$ onto B^{p-k}. It remains to apply Fubini's theorem to the measure G_*m, which is equivalent to the volume in $\mathbb{R}^p$.

Let $x \in \Lambda^\ell$ and let T be a smooth submanifold which is transverse to the family of local smooth manifolds $\mathcal{L}^s(x)$. Let N be a set of zero leaf-volume in T.

Theorem 8.5. *We have $m(\bigcup V^s(w)) = 0$ where the union is taken over all points $w \in \Lambda^\ell \cap B(x,r)$ for which $V^s(w) \cap T \in N$.*

Exercise 8.6. Prove Theorems 8.4 and 8.5.

Let ν be a measure on M that is absolutely continuous with respect to volume m. Consider the measurable partition ξ^s of the set $Q^\ell(x,r)$ (defined by (8.2)) into local stable manifolds $V^s(w)$ for $w \in \Lambda^\ell \cap B(x,r)$. Denote by $\nu^s(w)$ the conditional measure on $V^s(w)$ generated by the partition ξ^s and the measure ν, and denote by $\hat{\nu}^s$ the factor-measure generated by the partition ξ^s.

Exercise 8.7. Prove a more general versions of Theorems 8.3 and 8.4. Namely:

Theorem 8.8. *The following statements hold:*

(1) *the measure $\nu^s(w)$ is absolutely continuous with respect to $m_{V^s(w)}$ for ν-almost every $w \in \Lambda^\ell \cap B(x,r)$; moreover,*

$$d\nu^s(x)(y) = \rho(x,y)\,\mathrm{Jac}(\pi^s(y))dm_{V^s(x)}(y)$$

for ν-almost every $x \in \Lambda^\ell$ and $y \in V^s(x)$, where $\rho(x,y)$ is a measurable bounded function and the Jacobian $\mathrm{Jac}(\pi^s(y))$ is given by (8.10);

(2) *the factor-measure $\hat{\nu}^s$ is equivalent to the measure $m_{T(w)}|A(w)$ for ν-almost every $w \in \Lambda^\ell \cap B(x,r)$.*

The proof of the absolute continuity property for Anosov diffeomorphisms was given by Anosov in [**3**] and by Anosov and Sinai in [**4**] where they used the approach based on the holonomy map. For nonuniformly hyperbolic systems the absolute continuity property was established by Pesin in [**68**]. The proof presented in the next section is an elaboration of his proof.

8.2. A proof of the absolute continuity theorem

Fix $x \in \Lambda$ and a number $0 < r \le r_\ell$. Consider the set $Q^\ell(x,r)$ and the family of local stable manifolds $\mathcal{L}^s(x)$ (see (8.2) and (8.3)). Let T^1 and T^2 be local transversals to the family $\mathcal{L}^s(x)$. Fix $w \in \Lambda^\ell \cup B(x,r)$ and let $y \in T^1 \cap V^s(w)$. Given a number $\tau > 0$, we let

$$B^1(\tau) = Q^\ell(x,r) \cap B^1(y,\tau), \quad B^2(\tau) = \pi^s(Q^\ell(x,r) \cap B^1(y,\tau)),$$

where $B^1(y,\tau) \subset T^1$ is the ball of radius τ centered at y. In view of (8.5) we may assume that y is a Lebesgue density point of the set $B^1(\tau)$. It suffices to show that there is a constant $K > 0$ that is independent of the choice of the point x, the transversals T^1, T^2, and the point y such that for every $\tau > 0$,

$$\frac{1}{K} \leq \frac{m_{T^2}(B^2(\tau))}{m_{T^1}(B^1(\tau))} \leq K.$$

Indeed, letting $\tau \to 0$, we conclude from (8.4) that

$$\frac{1}{K} \leq \operatorname{Jac}(\pi^s)(y) \leq K.$$

In what follows, we fix the number τ. We split the proof into several steps.

Step 1. Let $y^i = V^s(w) \cap T^i$ for $i = 1, 2$. Choose $n \geq 0$, $q > 0$ and for $i = 1, 2$ set

$$T_n^i = f^n(T^i), \quad w_n = f^n(w), \quad y_n^i = f^n(y^i), \quad q_n = qe^{-\varepsilon n}. \tag{8.6}$$

Note that $T_0^i = T^i$, $w_0 = w$, $y_0^i = y^i$, and $q_0 = q$. Consider the open smooth submanifolds T_n^1 and T_n^2 and the point w_n. In view of Theorem 7.14, for $i = 1, 2$, there exists an open neighborhood $T_n^i(w, q)$ of the point y_n^i such that

$$T_n^i(w,q) = \exp_{w_n}\{(\psi_n^i(v), v) \colon v \in B^u(q_n)\} \subset T_n^i, \tag{8.7}$$

where $B^u(q_n) \subset E^u(w_n)$ is the open ball centered at zero of radius q_n and $\psi_n^i \colon B^u(q_n) \to E^s(w_n)$ is a smooth map. If $q = q(\ell)$ is sufficiently small, then by Theorem 7.7, for any $w \in \Lambda^\ell \cap B(x,r)$, $n > 0$, and $k = 1, \dots, n$ we have that

$$f^{-1}(T_k^i(w,q)) \subset T_{k-1}^i(w,q), \quad i = 1, 2. \tag{8.8}$$

We wish to compare the measures $m_{T_n^1}|T_n^1(w,q)$ and $m_{T_n^2}|T_n^2(w,q)$ for sufficiently large n.

Lemma 8.9. *There exist $K_1 > 0$ and $q = q(\ell)$ such that for each $n > 0$,*

$$\frac{m_{T_n^1}(T_n^1(w,q))}{m_{T_n^2}(T_n^2(w,q))} \leq K_1.$$

Proof of the lemma. The result follows from Theorem 7.14. □

Step 2. We continue with the following covering lemma.

Lemma 8.10. *For any sufficiently large $n > 0$ there are points $w_j \in \Lambda^\ell \cap B(x,r)$, $j = 1, \dots, t = t(n)$, such that for $i = 1, 2$ the sets $T_n^i(w_j, q)$ form an open cover of the set $f^n(B^i(\tau))$. These covers have finite multiplicity which depends only on the dimension of T^1.*

Proof of the lemma. We recall the Besicovich Covering Lemma. It states that for each $Z \subset \mathbb{R}^k$, if $r\colon Z \to \mathbb{R}^+$ is a bounded function, then the cover $\{B(x, r(x)) : x \in Z\}$ of Z contains a countable (Besicovich) subcover of finite multiplicity depending only on k.

It follows from the definition of the sets $T_n^i(w, q)$ (see (8.7)) that there is a number $R = R(\ell, n) > 0$ such that for every $w \in \Lambda^\ell \cap B(x, r)$,

$$B_n^i(w, \frac{1}{2}R) \subset T_n^i(w, q) \subset B_n^i(w, R) \subset T_n^i(w, 10q),$$

where $B_n^i(w, R)$ is the ball in T_n^i centered at w of radius R. Moreover, for sufficiently large n,

$$\|d\psi_n^i v\| \le \varepsilon, \quad v \in B^u(10q_n), \quad i = 1, 2.$$

This allows one to apply the Besicovich Covering Lemma to the cover of the set $Z^i = f^n(B^i(\tau))$ by balls

$$\left\{ B_n^i(w, \frac{1}{2}R) : w \in \Lambda^\ell \cap B(x, r) \right\}.$$

Thus, we obtain a Besicovich subcover $\{B_n^i(w_j, \frac{1}{2}R) : j = 1, \dots, t\}$, $t = t(n)$, of the set Z^i of finite multiplicity M_i (which does not depend on n). The sets $T_n^i(w_j, q)$ also cover Z^i and the multiplicity L_i of this cover does not depend on n. Indeed, L_i does not exceed the multiplicity K_i of the cover of Z^i by balls $B_n^i(w_j, R)$. Note that every ball $B_n^i(w_j, R)$ can be covered by at most CM_i balls $B_n^1(w, \frac{1}{2}R)$ where C depends only on the dimension of T^1. Furthermore, given w, we have $w_j \in B_n^i(w, \frac{1}{2}R)$ for at most M_i points w_j (otherwise at least $M_i + 1$ balls $B_n^i(w_j, \frac{1}{2}R)$ would contain w). Therefore, $w_j \in B_n^i(w, R)$ for at most CM_i points w_j. This implies that $K_i \le CM_i$. □

Step 3. We now compute the measures that are the pullbacks under f^{-n} of the measures $m_{T_n^i} | T_n^i(w, q)$ for $i = 1, 2$. Note that for every $w \in \Lambda^\ell \cap B(x, r)$ the points y_n^1 and y_n^2 (see (8.6)) lie on the stable manifold $V^s(w_n)$. Choose $z \in f^{-n}(T_n^i(w, q))$ (with $i = 1$ or $i = 2$). Set $z_n = f^n(z)$ and

$$H^i(z, n) = \operatorname{Jac}(d_{z_n} f^{-n} | T_{z_n} T_n^i(w, q))$$

(recall that $\operatorname{Jac}(S)$ denotes the Jacobian of the transformation S). We need the following lemma.

Lemma 8.11. *There exist $K_2 > 0$ and $n_1(\ell) > 0$ such that for every $w \in \Lambda^\ell \cap B(x, r)$ and $n \ge n_1(\ell)$ one can find $q = q(n)$ such that*

$$\left| \frac{H^2(y_n^2, n)}{H^1(y_n^1, n)} \right| \le K_2,$$

and if $z \in T_0^1(w,q)$, *then*

$$\left|\frac{H^1(z_n,n)}{H^1(y_n^1,n)}\right| \le K_2.$$

Proof of the lemma. For any $0 < k \le n$ and $z \in T_0^1(w,q)$ we denote by $V(k,z)$ the tangent space $T_{z_k}T_k^1(w,q)$ to the submanifold $T_k^1(w,q)$ at the point z_k. We transport in a parallel manner the subspace $V(k,z) \subset T_{z_k}M$ along the geodesic connecting the points z_k and y_k^1 (this geodesic is uniquely defined since these points are sufficiently close to each other) to obtain a new subspace $\tilde{V}(k,z) \subset T_{y_k^1}M$. We have

$$\begin{aligned}
&\big|\operatorname{Jac}(d_{z_k}f^{-1}|V(k,z)) - \operatorname{Jac}(d_{y_k^1}f^{-1}|V(k,y_1))\big| \\
&\quad\le \big|\operatorname{Jac}(d_{z_k}f^{-1}|V(k,z)) - \operatorname{Jac}(d_{y_k^1}f^{-1}|\tilde{V}(k,z))\big| \\
&\quad\quad + \big|\operatorname{Jac}(d_{y_k^1}f^{-1}|\tilde{V}(k,z)) - \operatorname{Jac}(d_{y_k^1}f^{-1}|V(k,y_1))\big|.
\end{aligned}$$

Since the diffeomorphism f is of class $C^{1+\alpha}$, in view of (8.8) and Exercise 4.15, we obtain

$$\begin{aligned}
&\big|\operatorname{Jac}(d_{z_k}f^{-1}|V(k,z)) - \operatorname{Jac}(d_{y_k^1}f^{-1}|V(k,y_1))\big| \\
&\quad\le C_1\rho(z_k,y_k^1)^\alpha + C_2 d(\tilde{V}(k,z),V(k,y_1)),
\end{aligned}$$

where $C_1 > 0$ and $C_2 > 0$ are constants (recall that ρ is the distance in M and d is the distance in the Grassmannian bundle of TM generated by the Riemannian metric). It follows from (7.37) that

$$d(\tilde{V}(k,z),V(k,y_1)) \le C_3\rho(z_k,y_k^1)^\alpha,$$

where $C_3 > 0$ is a constant. Together with (7.1) this implies that

$$\big|\operatorname{Jac}(d_{z_k}f^{-1}|V(k,z)) - \operatorname{Jac}(d_{y_k^1}f^{-1}|V(k,y_1))\big| \le C_4(e^{-\varepsilon(n-k)}\rho(z,y_1))^\alpha,$$

where $C_4 > 0$ is a constant. Note that for any $0 < k \le n$ and $z \in T_0^1(w,q)$ we have

$$C_5{}^{-1} \le |\operatorname{Jac}(d_{z_k}f^{-1}|V(k,z))| \le C_5,$$

where $C_5 > 0$ is a constant. We obtain

$$\begin{aligned}
\frac{H^1(z_n,n)}{H^1(y_n^1,n)} &= \prod_{k=1}^{n} \frac{\operatorname{Jac}(d_{z_k}f^{-1}|V(k,z))}{\operatorname{Jac}(d_{y_k^1}f^{-1}|V(k,y_1))} \\
&= \exp\sum_{k=1}^{n} \log\frac{\operatorname{Jac}(d_{z_k}f^{-1}|V(k,z))}{\operatorname{Jac}(d_{y_k^1}f^{-1}|V(k,y_1))}
\end{aligned}$$

$$
\begin{aligned}
&\leq \exp \sum_{k=1}^{n} \left(\frac{\operatorname{Jac}(d_{z_k} f^{-1} | V(k,z))}{\operatorname{Jac}(d_{y_k^1} f^{-1} | V(k,y_1))} - 1 \right) \\
&\leq \exp \left(\sum_{k=1}^{n} C_4 C_5 (e^{-\varepsilon(n-k)} \rho(z,y_1))^{\alpha} \right) \\
&\leq \exp \left(\frac{C_4 C_5 q^{\alpha}}{1 - e^{-\varepsilon \alpha}} \right).
\end{aligned}
$$

The last expression can be made arbitrarily close to 1 by choosing q sufficiently small. This proves the second inequality. The first inequality can be proven in a similar fashion. □

Lemma 8.11 allows one to compare the measures of the preimages under f^{-n} of $T_n^1(w,q)$ and $T_n^2(w,q)$. More precisely, the following statement holds.

Lemma 8.12. *There exist $K_3 > 0$ and $n_2(\ell) > 0$ such that for any $w \in \Lambda^{\ell} \cap B(x,r)$ and $n \geq n_2(\ell)$ one can find $q = q(n)$ such that*

$$
\frac{m_{T^1}(f^{-n}(T_n^1(w,q)))}{m_{T^2}(f^{-n}(T_n^2(w,q)))} \leq K_3.
$$

Proof of the lemma. For $i = 1, 2$ we have

$$
\begin{aligned}
m_{T^i}(f^{-n}(T_n^i(w,q))) &= \int_{T_n^i(w,q)} H^i(z,n)\, dm_{T_n^i}(z) \\
&= H^i(z_n^i, n) m_{T_n^i}(T_n^i(w,q)),
\end{aligned}
$$

where $z_n^i \in T_n^i(w,q)$ are some points. It follows from the assumptions of the lemma, (8.8), and Lemmas 8.9 and 8.11 that for sufficiently large n and sufficiently small q,

$$
\begin{aligned}
&\frac{m_{T^1}(f^{-n}(T_n^1(w,q)))}{m_{T^2}(f^{-n}(T_n^2(w,q)))} \\
&\qquad \leq \left| \frac{H^1(z_n^1, n)}{H^2(z_n^2, n)} \right| \times \frac{\nu_{T_n^1}(T_n^1(w,q))}{m_{T_n^2}(T_n^1(w,q))} \\
&\qquad \leq \left| \frac{H^1(z_n^1, n)}{H^1(y_n^1, n)} \cdot \frac{H^2(y_n^2, n)}{H^2(z_n^2, n)} \cdot \frac{H^1(y_n^1, n)}{H^2(y_n^2, n)} \right| \\
&\qquad \leq K_2^3 + K_1 \leq K_3
\end{aligned}
$$

for some $K_3 > 0$. The lemma follows. □

Step 4. Given $n > 0$, choose points $w_j \in \Lambda^{\ell} \cap B(x,r)$ and a number $q = q(n)$ as in Lemma 8.10. Consider the sets

$$
\hat{T}_n^1 = \bigcup_{j=1}^{t} T_n^1(w_j, q), \quad \hat{T}_n^2 = \bigcup_{j=1}^{t} T_n^2(w_j, q).
$$

Note that $\hat{T}_n^1 \supset f^n(B^1(\tau))$ and $\hat{T}_n^2 \supset f^n(B^1(\tau))$ (see Figure 8.2).

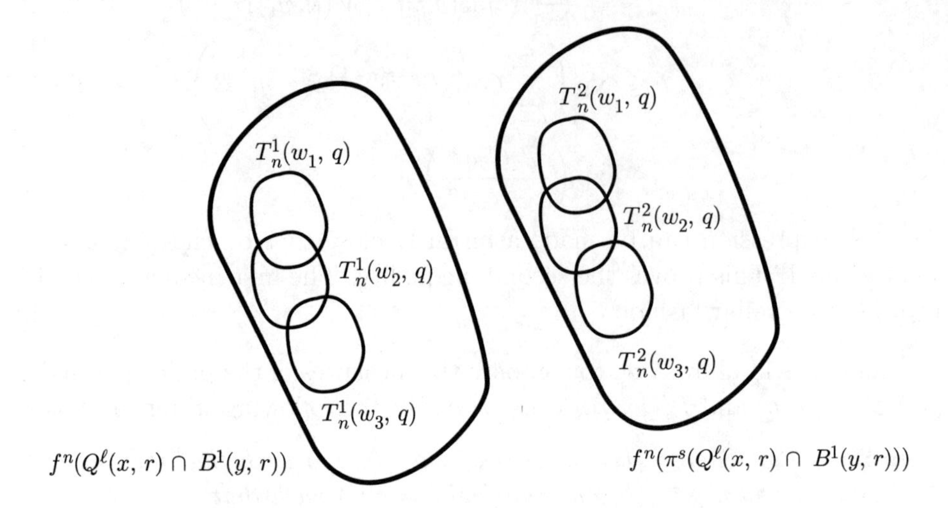

Figure 8.2. Sets $f^n(Q^\ell(x,r) \cap T^i)$ and their covers by $T_n^i(w_j, q)$

We wish to compare the measures $m_{T^1}|f^{-n}(\hat{T}_n^1)$ and $m_{T^2}|f^{-n}(\hat{T}_n^2)$. For $i = 1, 2$ let L_i be the multiplicity of the cover of the set $B^i(\tau)$ by the sets $\{f^{-n}(T_n^i(w_j, q))\}$ (see Lemma 8.10) and let $L = \max\{L_1, L_2\}$. We have

$$\frac{1}{L}\sum_{j=1}^{t} m_{T^i}(f^{-n}(T_n^i(w_j,q))) \le m_{T^i}(f^{-n}(\hat{T}_n^i))$$
$$\le \sum_{j=1}^{t} m_{T^i}(f^{-n}(T_n^i(w_j,q))).$$

It follows from Lemma 8.12 that

$$\begin{aligned}\frac{m_{T^1}(f^{-n}(\hat{T}_n^1))}{m_{T^2}(f^{-n}(\hat{T}_n^2))} &\le L\frac{\sum_{j=1}^t m_{T^1}(f^{-n}(T_n^1(w_j,q)))}{\sum_{j=1}^t m_{T^2}(f^{-n}(T_n^2(w_j,q)))}\\ &\le L\max\left\{\frac{m_{T^1}(f^{-n}(T_n^1(w_j,q)))}{m_{T^2}(f^{-n}(T_n^2(w_j,q)))} : j = 1,\dots,t\right\}\\ &\le LK_3,\end{aligned}$$

with a similar bound for the inverse ratio. We conclude that

$$K_4^{-1} \le \frac{m_{T^1}(f^{-n}(\hat{T}_n^1))}{m_{T^2}(f^{-n}(\hat{T}_n^2))} \le K_4, \tag{8.9}$$

where $K_4 > 0$ is a constant independent of n.

Step 5. Given a number $\beta > 0$, denote by U^i_β the β-neighborhood of the set $B^i(\tau)$, $i = 1, 2$. By (8.5), there exists $\beta_0 > 0$ such that for every $0 < \beta \le \beta_0$,

$$\frac{m_{T^1}(B^1(\tau))}{m_{T^1}(U^1_\beta)} \ge \frac{1}{2}.$$

For any $\beta > 0$, any sufficiently large $n > 0$, and any sufficiently small $q = q(n)$, for $i = 1, 2$ we have that $B^i(\tau) \subset f^{-n}(\hat{T}^i_n) \subset U^i_\beta$. It follows from (8.5) and (8.9) that $m_{T^2}(f^{-n}(\hat{T}^2_n)) > C$ where $C > 0$ is a constant independent of n and q. This implies that $m_{T^2}(B^2(\tau)) > 0$. Therefore, reducing the number β_0 if necessary, we obtain that for every $0 < \beta \le \beta_0$,

$$\frac{m_{T^2}(B^2(\tau))}{m_{T^2}(U^2_\beta)} \ge \frac{1}{2}.$$

It follows from (8.9) that

$$\frac{1}{2K_4} \le \frac{m_{T^1}(B^1(\tau))}{m_{T^2}(B^2(\tau))} \le 2K_4.$$

This completes the proof of Theorem 8.2.

Remark 8.13. One can extend the above result to the case when the invariant set Y is nonuniformly *partially* hyperbolic in the broad sense and show that the family of local stable manifolds satisfies the absolute continuity property (see Remark 7.2).

Remark 8.14. One can obtain a sharper estimate on the Jacobian of the holonomy map. Consider the family of local stable manifolds $\mathcal{L}^s(x)$ (see (8.3)). Let T^1 and T^2 be two transversals to this family. We can choose them such that for $i = 1, 2$ the set $\exp_x^{-1} T^i$ is the graph of a smooth map $\psi^i \colon B^u(q) \subset E^u(x) \to E^s(x)$ (for some $q > 0$) with sufficiently small C^1 norm. Set

$$\Delta = \Delta(T^1, T^2) = \|\psi^1 - \psi^2\|_{C^1}.$$

Theorem 8.15. *Under the assumption of Theorem* 8.2 *there exists a constant* $K = K(\ell) > 0$ *such that for almost every* $y \in Q^\ell(x, r) \cap T^1$,

$$|\mathrm{Jac}(\pi^s)(y) - 1| \le K\Delta.$$

8.3. Computing the Jacobian of the holonomy map

In this section we strengthen the Absolute Continuity Theorem 8.2 and obtain an explicit formula for the Jacobian of the holonomy map.

Theorem 8.16. *Under the assumption of Theorem* 8.2, *for almost every* $y \in Q^\ell(x, r) \cap T^1$,

$$\mathrm{Jac}(\pi^s)(y) = \prod_{k=0}^{\infty} \frac{\mathrm{Jac}(d_{f^k(\pi^s(y))} f^{-1} | T_{f^k(\pi^s(y))} f^k(T^2))}{\mathrm{Jac}(d_{f^k(y)} f^{-1} | T_{f^k(y)} f^k(T^1))} \tag{8.10}$$

(in particular, the infinite product on the right-hand side converges).

Proof. Fix a point $y \in Q^\ell(x,r) \cap T^1$. Repeating arguments in the proof of Lemma 8.11, it is straightforward to verify that the infinite product

$$\prod_{k=0}^{\infty} \frac{\operatorname{Jac}(d_{f^k(\pi^s(y))} f^{-1} | T_{f^k(\pi^s(y))} f^k(T^2))}{\operatorname{Jac}(d_{f^k(y)} f^{-1} | T_{f^k(y)} f^k(T^1))} =: J$$

converges. Therefore, given $\varepsilon > 0$, there exists $n_1 > 0$ such that for every $n \geq n_1$ we have

$$\left| J - \prod_{k=0}^{n-1} \frac{\operatorname{Jac}(d_{f^k(\pi^s(y))} f^{-1} | T_{f^k(\pi^s(y))} f^k(T^2))}{\operatorname{Jac}(d_{f^k(y)} f^{-1} | T_{f^k(y)} f^k(T^1))} \right| \leq \varepsilon. \tag{8.11}$$

Fix $n > 0$. One can choose small neighborhoods $U_n^1 \subset T^1$ and $U_n^2 \subset T^2$ of the points y and $\pi^s(y)$, respectively, and define the holonomy map

$$\pi_n^s \colon f^n(U_n^1) \to f^n(U_n^2) \quad \text{by} \quad \pi_n^s = f^n \circ \pi^s \circ f^{-n}.$$

We have

$$\operatorname{Jac}(\pi^s)(y) = \prod_{k=0}^{n-1} \frac{\operatorname{Jac}(d_{f^k(\pi^s(y))} f^{-1} | T_{f^k(\pi^s(y))} f^k(T^2))}{\operatorname{Jac}(d_{f^k(y)} f^{-1} | T_{f^k(y)} f^k(T^1))} \operatorname{Jac}(\pi_n^s)(f^n(y)). \tag{8.12}$$

We now choose $n > n_1$ so large and neighborhoods U_n^1 and U_n^2 so small that

$$\Delta(f^n(U_n^1), f^n(U_n^2)) \leq \varepsilon$$

(see Remark 8.14). Theorem 8.15, applied to the holonomy map π_n^s, yields that for almost every $y \in Q^\ell(x,r) \cap T^1$,

$$|\operatorname{Jac}(\pi_n^s)(f^n(y)) - 1| \leq K\varepsilon. \tag{8.13}$$

It follows from (8.11), (8.12), and (8.13) that

$$|\operatorname{Jac}(\pi^s)(y) - J| \leq \varepsilon + (J + \varepsilon)K\varepsilon.$$

Since ε is arbitrary, this completes the proof of the theorem. □

Exercise 8.17. Show that the function $\operatorname{Jac}(\pi^s)(y)$ depends Hölder continuously on $y \in Q^\ell(x,r) \cap T^1$.

8.4. An invariant foliation that is not absolutely continuous

We describe an example due to Katok of a continuous foliation with smooth leaves that is not absolutely continuous (another version of this example can be found in [**63**]). We will see below that the Lyapunov exponent along this foliation is zero at every point in the manifold. In Section 12.3 we present an elaborate discussion of foliations with smooth leaves that are not absolutely continuous such that the Lyapunov exponents along the foliation are all negative.

Consider a hyperbolic automorphism T of the torus $\mathbb{T}^2$ (see Section 1.1). Given $\varepsilon > 0$, let $\{\psi_t : t \in [0,1]\}$ be a one-parameter family of real-valued C^∞ functions on $[0,1]$ satisfying:

(a) $0 < 1 - \psi_t(u) \le \varepsilon$ for $u \in [0,1]$ and $\psi_t(u) = 1$ for $u \ge r_0$, for some $0 < r_0 < 1$;

(b) $\psi_t'(u) \ge 0$ for every $0 \le u < r_0$;

(c) ψ_t depends smoothly on t and $\psi_0(u) = \psi_1(u) = 1$ for $u \in [0,1]$.

For $t \in [0,1]$ let $f_t = G_{\mathbb{T}^2}(\psi_t)$ be the Katok map constructed via the function ψ_t (see the end of Section 1.3). The map f_t preserves the area m and $f_0 = f_1 = T$. Hence, f_t is a loop through T in the space of C^1 diffeomorphisms of $\mathbb{T}^2$.

Exercise 8.18. Show that for a sufficiently small $\varepsilon > 0$ and any family of functions $\{\psi_t : t \in [0,1]\}$ satisfying conditions (a)–(c) above, the family $\{f_t : t \in S^1\}$ has the following properties:

(1) f_t is a small perturbation of T for every $t \in S^1$; in particular, f_t is an Anosov diffeomorphism (see also Exercise 1.20);

(2) f_t depends smoothly on t.

Exercise 8.19. Show that for a sufficiently small $\varepsilon > 0$ one can choose a family of functions $\{\psi_t : t \in [0,1]\}$ satisfying conditions (a)–(c) such that the function

$$h(t) = h_m(f_t) = \int_{\mathbb{T}^2} \log \|d_x f_t | E_t^u(x)\| \, dm(x) \tag{8.14}$$

is strictly monotone in a small neighborhood of $t = 0$ (here $E_t^u(x)$ is the unstable subspace for f_t at the point x).

We introduce a diffeomorphism $F \colon \mathbb{T}^2 \times S^1 \to \mathbb{T}^2 \times S^1$ by $F(x,t) = (f_t(x), t)$. Since f_t is sufficiently close to T, they are conjugate via a Hölder homeomorphism g_t, i.e., $f_t = g_t \circ T \circ g_t^{-1}$ (see Section 1.1). Given $x \in \mathbb{T}^2$, consider the set

$$H(x) = \{(g_t(x), t) : t \in S^1\}.$$

It is diffeomorphic to the circle S^1 and the collection of these sets forms an F-invariant foliation H of $\mathbb{T}^2 \times S^1 = \mathbb{T}^3$ with $F(H(x)) = H(T(x))$. Note that $H(x)$ depends Hölder continuously on x. However, the holonomy maps associated with the foliation H are not absolutely continuous. To see this, consider the holonomy map

$$\pi_{t_1,t_2} \colon \mathbb{T}^2 \times \{t_1\} \to \mathbb{T}^2 \times \{t_2\}.$$

We have that

$$\pi_{0,t}(x,0) = (g_t(x), t)$$

and $F(\pi_{0,t}(x,0)) = \pi_{0,t}(T(x),0)$. If the map $\pi_{0,t}$ (with t being fixed) were absolutely continuous, the measure $(\pi_{0,t})_*m$ would be absolutely continuous with respect to m. Since f_t is Anosov, it is ergodic and hence, m is the only absolutely continuous f_t-invariant probability measure. This implies that $(\pi_{0,t})_*m = m$.

The function $h(t)$ given by (8.14) is the metric entropy of the map f_t (see Section 9.3.1 for the definition of the metric entropy and relevant results) and is preserved by conjugacy. Hence, $h(t) = h(0)$ for all sufficiently small t. This is a contradiction and hence, the holonomy maps associated with the foliation H are not absolutely continuous.

Chapter 9

Ergodic Properties of Smooth Hyperbolic Measures

In this chapter we move to the core of smooth ergodic theory and consider smooth dynamical systems preserving smooth hyperbolic invariant measures (i.e., invariant measures which are equivalent to the Riemannian volume and have nonzero Lyapunov exponents almost everywhere). A sufficiently complete description of their ergodic properties is one of the main manifestations of the above results on local instability (see Sections 7.1 and 7.3) and absolute continuity (see Chapter 8). It turns out that smooth hyperbolic invariant measures have an abundance of ergodic properties. This makes smooth ergodic theory a deep and well-developed part of the general theory of smooth dynamical systems.

9.1. Ergodicity of smooth hyperbolic measures

One of the main manifestations of absolute continuity is a description of the ergodic properties of diffeomorphisms preserving smooth hyperbolic measures.

Let f be a $C^{1+\alpha}$ diffeomorphism of a smooth compact Riemannian manifold M without boundary. Throughout this chapter we assume that f preserves a *smooth measure* ν, i.e., a measure which is equivalent to the Riemannian volume and whose density is bounded from above and bounded away from zero. Furthermore, we assume that this measure is hyperbolic, that is, the set $\mathcal{E}$ of points with nonzero Lyapunov exponents (defined by (4.5)) has positive ν-measure.

We begin by describing the decomposition of a smooth invariant measure into its ergodic components. Recall that a measurable partition χ is a *decomposition into ergodic components* if for almost every element C of χ the map $S|C$ is ergodic with respect to the conditional measure ν_C induced by ν.

The following statement is one of the main results of smooth ergodic theory. It describes the decomposition of the measure ν into ergodic components and was established in [**69**]. The proof exploits a simple yet deep argument due to Hopf [**40**] (see the proof of Lemma 9.3 below).

Theorem 9.1. *There exist invariant sets $\mathcal{E}_0, \mathcal{E}_1, \dots$ such that:*

(1) $\bigcup_{i\geq 0} \mathcal{E}_i = \mathcal{E}$, *and* $\mathcal{E}_i \cap \mathcal{E}_j = \varnothing$ *whenever* $i \neq j$;

(2) $\nu(\mathcal{E}_0) = 0$, *and* $\nu(\mathcal{E}_i) > 0$ *for each* $i \geq 1$;

(3) $f|\mathcal{E}_i$ *is ergodic for each* $i \geq 1$.

Proof. We begin with the following statement.

Lemma 9.2. *Given an f-invariant Borel function φ, there exists a set $N \subset M$ of measure zero such that if $y \in \mathcal{E} \cap B(x,r)$ and z, $w \in V^s(y) \setminus N$ or z, $w \in V^u(y) \setminus N$, then $\varphi(z) = \varphi(w)$.*

Proof of the lemma. Let $z, w \in V^s(y)$ and let ψ be a continuous function. By Birkhoff's Ergodic Theorem, the functions

$$\overline{\psi}(x) = \lim_{n\to\infty} \frac{1}{2n+1} \sum_{k=-n}^{n} \psi(f^k(x)),$$

$$\psi^+(x) = \lim_{n\to\infty} \frac{1}{n} \sum_{k=1}^{n} \psi(f^k(x)), \quad \text{and} \quad \psi^-(x) = \lim_{n\to\infty} \frac{1}{n} \sum_{k=1}^{n} \psi(f^{-k}(x))$$

are defined for ν-almost every point x. We also have that $\overline{\psi}(x) = \psi^+(x) = \psi^-(x)$ outside a subset $N \subset M$ of measure zero.

Since $\rho(f^n(z), f^n(w)) \to 0$ as $n \to \infty$ (see Theorem 7.7) and ψ is continuous, we obtain

$$\overline{\psi}(z) = \psi^+(z) = \psi^+(w) = \overline{\psi}(w).$$

Notice that continuous functions are dense in $L^1(M, \nu)$ and hence, the functions of the form $\overline{\psi}$ are dense in the set of f-invariant Borel functions. The lemma follows. □

Observe that the set $\mathcal{E}$ is nonuniformly completely hyperbolic and we can consider the corresponding collection $\Lambda_{\lambda\mu\varepsilon j}$ of level sets and for each $\lambda, \mu, \varepsilon$, and $1 \leq j < p$ we can consider the collection of regular sets $\{\Lambda^{\ell}_{\lambda\mu\varepsilon j} : \ell \geq 1\}$ (see (4.24) and (4.25)). Note that

$$\mathcal{E} \subset \bigcup \Lambda_{\lambda\mu\varepsilon j}, \tag{9.1}$$

where the union is taken over all *rational* numbers λ, μ, ε and all $1 \leq j < p$. Therefore it suffices to prove the theorem for the restriction $f|\Lambda_{\lambda\mu\varepsilon j}$ to each set $\Lambda_{\lambda\mu\varepsilon j}$ of positive measure. We therefore fix numbers $\lambda, \mu, \varepsilon$, and j and set $\Lambda = \Lambda_{\lambda\mu\varepsilon j}$ and $\Lambda^{\ell} = \Lambda^{\ell}_{\lambda\mu\varepsilon j}$.

Let $x \in \Lambda^{\ell}$ be a Lebesgue density point. For each $r > 0$ set (see Figure 9.1)

$$P^{\ell}(x,r) = \bigcup_{y \in \Lambda^{\ell} \cap B(x,r)} (V^s(y) \cup V^u(y)). \tag{9.2}$$

Lemma 9.3. *There exists $r = r(\ell) > 0$ such that the map f is ergodic on the set*

$$P(x) = \bigcup_{n \in \mathbb{Z}} f^n(P^{\ell}(x,r)). \tag{9.3}$$

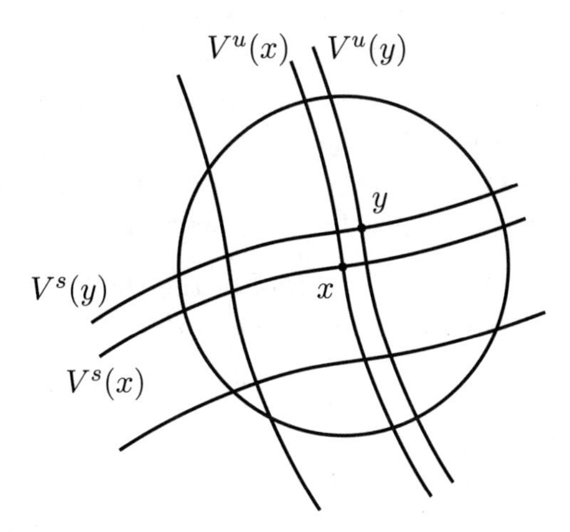

Figure 9.1. Stable and unstable manifolds forming the set $P^{\ell}(x,r)$

Proof of the lemma. Let φ be an f-invariant function, and let N be the set of measure zero constructed in Lemma 9.2. Choose $0 < r < \min\{r_{\ell}, \delta_{\ell}\}$ (see (7.36)). Consider the family $\mathcal{L}^u(x)$ of local unstable manifolds $V^u(x)$, $y \in \Lambda^{\ell} \cap B(x,r)$. By Theorem 8.4 (applied to this family), there exists $y \in \Lambda^{\ell} \cap B(x,r)$ such that $m_{V^u}(y)(V^u(y) \cap N) = 0$ (indeed, almost every $y \in \Lambda^{\ell} \cap B(x,r)$ has this property), where $m_{V^u}(y)$ is the leaf-volume on $V^u(y)$. Let

$$R^s = \bigcup V^s(w),$$

where the union is taken over all points $w \in \Lambda^\ell \cap B(x, r_\ell)$ for which $V^s(w) \cap V^u(y) \in N$. By Theorem 8.5, we have $\nu(R^s) = 0$.

Let $z_1, z_2 \in P^\ell(x, r) \setminus (R^s \cup N)$. There are points $w_i \in \Lambda^\ell \cap B(x, r)$ such that $z_i \in V^s(w_i)$ for $i = 1, 2$. Note that the intersection $V^s(w_i) \cap V^u(y)$ is nonempty and consists of a single point y_i for $i = 1, 2$ (see Figure 9.1). By Lemma 9.2,

$$\varphi(z_1) = \varphi(y_1) = \varphi(y_2) = \varphi(z_2).$$

This completes the proof of the lemma. □

Lemma 9.4. *We have $\nu(P(x)) > 0$.*

Proof of the lemma. Since $P(x) \supset P^\ell(x, r) \supset \Lambda^\ell \cap B(x, r)$, we have

$$\nu(P(x)) \geq \nu(\Lambda_n^\ell \cap B(x, r)) > 0$$

and the lemma follows. □

Since almost every point $x \in \mathcal{E}$ is a Lebesgue density point of Λ^ℓ for some ℓ, the invariant sets $P(x)$ (defined by (9.3) for different values of ℓ) cover the set $\mathcal{E}$ (mod 0). By Lemma 9.4, there are at most countably many such sets. We denote them by $P_1, P_2, \dots$. We have $\nu(P_i) > 0$ for each $i \geq 1$, and the set $\mathcal{E}_0 = \mathcal{E} \setminus \bigcup_{i \geq 1} P_i$ has measure zero. By Lemma 9.3, for each $i \geq 1$ the map $f|P_i$ is ergodic and hence $P_i \cap P_j = \varnothing$ (mod 0) whenever $i \neq j$. Set $\mathcal{E}_n = \mathcal{E} \cap (P_n \setminus \bigcup_{i=1}^{n-1} P_i)$. We have that $\mathcal{E}_n \cap \mathcal{E}_n = \varnothing$ and $f|\mathcal{E}_n$ is ergodic. This completes the proof of the theorem. □

Remark 9.5. It follows from the proof of the theorem that every ergodic component $\mathcal{E}_n$ of positive measure is of the form

$$\mathcal{E}_n = \bigcup_{n \in \mathbb{Z}} f^n(P^\ell(x, r)),$$

where $P^\ell(x, r)$ is given by (9.2) and x is a Lebesgue density point of $\mathcal{E}$.

We describe an example of a diffeomorphism with nonzero Lyapunov exponents that has more than one ergodic component. Consider Katok's diffeomorphism $G_{\mathbb{T}^2}$ of the torus $\mathbb{T}^2$ constructed in Section 1.4. The map $G_{\mathbb{T}^2}$ preserves area and is ergodic. The punched torus $\mathbb{T}^2 \setminus \{0\}$ is C^∞-diffeomorphic to the manifold $\mathbb{T}^2 \setminus \overline{U}$, where U is a small open disk around 0 and $\overline{U}$ denotes its closure. Therefore, we obtain a C^∞ diffeomorphism $F_{\mathbb{T}^2}$ of the manifold $\mathbb{T}^2 \setminus U$ with $F_{\mathbb{T}^2}|\partial U = \mathrm{Id}$. The map $F_{\mathbb{T}^2}$ preserves a smooth measure, has nonzero Lyapunov exponents, and is ergodic.

Let $(\tilde{M}, \tilde{F}_{\mathbb{T}^2})$ be a copy of $(M, F_{\mathbb{T}^2})$. By gluing the manifolds M and $\tilde{M}$ along ∂U, we obtain a smooth compact manifold $\mathcal{M}$ without boundary and a diffeomorphism $\mathcal{F}$ of $\mathcal{M}$.

Exercise 9.6. Show that $\mathcal{F}$ preserves a smooth measure and has nonzero Lyapunov exponents almost everywhere.

However, the map $\mathcal{F}$ is not ergodic and has two ergodic components of positive measure (M and $\tilde{M}$). Similarly, one can obtain a diffeomorphism with nonzero Lyapunov exponents with n ergodic components of positive measure for an arbitrary n. However, this construction cannot be used to obtain a diffeomorphism with nonzero Lyapunov exponents which has countably many ergodic components of positive measure. Such an example is constructed in Section 13.1 and it illustrates that Theorem 9.1 cannot be improved.

As an immediate consequence of Theorem 9.1 we obtain the following result. It is a generalization of Theorem 8.8 to the case when the set $\mathcal{E}$ has positive but not necessarily full measure. Let ν_x^s and ν_x^u be, respectively, the conditional measure on $V^s(x)$ and $V^u(x)$ generated by the measure ν.

Theorem 9.7. *For ν-almost every $x \in \mathcal{E}$ we have*

$$\nu_x^s(V^s(x) \setminus \mathcal{E}) = 0, \quad \nu_x^u(V^u(x) \setminus \mathcal{E}) = 0.$$

In ergodic theory there is a hierarchy of ergodic properties of which ergodicity (or the description of ergodic components) is the weakest one. Among the stronger properties let us mention, without going into detail, mixing and K-property (including the description of the π-partition). The strongest property that implies all others is the Bernoulli property or the description of Bernoulli components. In connection with smooth ergodic measures the latter is known as the Spectral Decomposition Theorem . It was established in [**69**].

Theorem 9.8. *For each $i \geq 1$ the following properties hold:*

(1) *the set $\mathcal{E}_i$ is a disjoint union of sets $\mathcal{E}_i^j$, for $j = 1, \ldots, n_i$, which are cyclically permuted by f, i.e., $f(\mathcal{E}_i^j) = \mathcal{E}_i^{j+1}$ for $j = 1, \ldots, n_i - 1$ and $f(\mathcal{E}_i^{n_i}) = \mathcal{E}_i^1$;*

(2) *$f^{n_i}|\mathcal{E}_i^j$ is a Bernoulli automorphism for each j.*

We illustrate the statement of this theorem by considering the diffeomorphism $\mathcal{F}\circ I$, where $\mathcal{F}$ is the diffeomorphism constructed above and $I\colon \mathcal{M} \to \mathcal{M}$ is the radial symmetry along the boundary of U. This diffeomorphism preserves a smooth measure, has nonzero Lyapunov exponents, and is ergodic. However, the map $\mathcal{G}^2$, where $\mathcal{G} = \mathcal{F} \circ I$, is not ergodic and has two ergodic components of positive measure (M and $\tilde{M}$) which are cyclically permuted by $\mathcal{G}$.

We now consider the case of a smooth flow φ_t on a compact manifold M. We assume that ν is a smooth measure which is φ_t-invariant. This means that $\nu(\varphi_t A) = \nu(A)$ for any Borel set $A \subset M$ and $t \in \mathbb{R}$.

Note that the Lyapunov exponent along the flow direction is zero. We assume that all other values of the Lyapunov exponent are nonzero for ν-almost every point, and hence, φ_t is a nonuniformly hyperbolic flow (see Section 4.2). We also assume that ν vanishes on the set of fixed points of φ_t.

Since the time-one map of the flow is nonuniformly partially hyperbolic, we conclude that the families of stable and unstable local manifolds possess the absolute continuity property (see Remark 8.13). This allows one to study ergodic properties of nonuniformly hyperbolic flows. The following result from [**71**] describes the decomposition into ergodic components for flows with nonzero Lyapunov exponents with respect to smooth invariant measures.

Theorem 9.9. *There exist invariant sets $\mathcal{E}_0, \mathcal{E}_1, \dots$ such that:*

(1) $\bigcup_{i \geq 0} \mathcal{E}_i = \mathcal{E}$ *and* $\mathcal{E}_i \cap \mathcal{E}_j = \varnothing$ *whenever* $i \neq j$*;*

(2) $\nu(\mathcal{E}_0) = 0$ *and* $\nu(\mathcal{E}_i) > 0$ *for each* $i \geq 1$*;*

(3) $\varphi_t | \mathcal{E}_i$ *is ergodic for each* $i \geq 1$.

Using the flow described in Section 1.5, one can construct an example of a flow with nonzero Lyapunov exponents having an arbitrary finite number of ergodic components.

9.2. Local ergodicity

For a general $C^{1+\alpha}$ diffeomorphism of a compact manifold M preserving a smooth hyperbolic measure, the number of ergodic components is countable (but *not* necessarily finite; see Section 13.1). One can wonder whether the ergodic components are open (up to a set of measure zero) and try to find additional conditions which would guarantee that. This problem is often referred to as the *local ergodicity problem.*

The main obstacles for local ergodicity are the following:

(1) the stable and unstable distributions are measurable but not necessarily continuous;

(2) the global stable (or unstable) leaves may not form a foliation;

(3) the unstable leaves may not expand under the action of f^n (we remind the reader that they were defined as being exponentially *contracting* under f^{-n}); the same is true for stable leaves with respect to the action of f^{-n}.

We describe the approach to local ergodicity developed in [**69**]. Roughly speaking, it requires that the stable (or unstable) leaves form a foliation of a measurable subset in M of positive measure. First, we extend the notion of continuous foliation of M with smooth leaves introduced in Section 1.1 to the foliation of a measurable subset.

Given a subset $X \subset M$, we call a partition ξ of X a *(δ, q)-foliation of X with smooth leaves* or simply a *(δ, q)-foliation of X* if there exist continuous functions $\delta\colon X \to (0, \infty)$ and $q\colon X \to (0, \infty)$ and an integer $k > 0$ such that for each $x \in X$:

(1) there exists a smooth immersed k-dimensional submanifold $W(x)$ containing x for which $\xi(x) = W(x) \cap X$ where $\xi(x)$ is the element of the partition ξ containing x; the manifold $W(x)$ is called the *(global) leaf* of the (δ, q)-foliation at x; the connected component of the intersection $W(x) \cap B(x, \delta(x))$ containing x is called the *local leaf* at x and is denoted by $V(x)$;

(2) there exists a continuous map $\varphi_x\colon B(x, q(x)) \to C^1(D, M)$ (where $D \subset \mathbb{R}^k$ is the open unit ball) such that for every $y \in X \cap B(x, q(x))$ the manifold $V(y)$ is the image of the map $\varphi_x(y)\colon D \to M$.

For every $x \in X$ and $y \in B(x, q(x))$, the set $U(y) = \varphi(y)(D)$ is called the *local leaf* of the (δ, q)-foliation at y. Note that $U(y) = V(y)$ for $y \in X$.

Theorem 9.10. *Let f be a $C^{1+\alpha}$ diffeomorphism of a compact smooth Riemannian manifold M preserving a smooth measure ν with nonzero Lyapunov exponents on a set $\mathcal{E}$ of positive measure (see* (4.5)*). Assume that there exists a (δ, q)-foliation W of $\mathcal{E}$ such that $W(x) = W^s(x)$ for every $x \in \mathcal{E}$ (where $W^s(x)$ is the global stable manifold at x; see Section* 7.3*). Then every ergodic component of f of positive measure is open* (mod 0) *in $\mathcal{E}$ (with respect to the induced topology).*

Proof. Observe that $\mathcal{E}$ is nonuniformly completely hyperbolic for f and that we can consider the corresponding collection $\Lambda_{\lambda\mu\varepsilon j}$ of level sets and for each $\lambda, \mu, \varepsilon$, and $1 \le j < p$ we can consider the collection of regular sets $\{\Lambda^\ell_{\lambda\mu\varepsilon j} : \ell \ge 1\}$ (see (4.24) and (4.25)). In view of (9.1) it suffices to prove the theorem for each level set of positive measure and, therefore, in what follows, we fix numbers $\lambda, \mu, \varepsilon$, and j and set $\Lambda = \Lambda_{\lambda\mu\varepsilon j}$ and $\Lambda^\ell = \Lambda^\ell_{\lambda\mu\varepsilon j}$.

We need the following statement, which is an immediate consequence of Theorem 8.8.

Lemma 9.11. *There exists a set $N \subset M$ of measure zero such that for every $x \in \Lambda \setminus N$,*

$$\nu^s_x(V^s(x) \setminus \Lambda) = \nu^u_x(V^u(x) \setminus \Lambda) = 0.$$

Let $P \subset \Lambda$ be an f-invariant set with $\nu(P) > 0$. We assume that $f|P$ is ergodic, and we will show that P is open (mod 0). By Lemma 9.3, there exist a number $\ell > 0$ and a Lebesgue density point x of the set Λ^ℓ such that $P = P(x)$ (mod 0) (see (9.3)). By Lemma 9.11, ν_x^u-almost every point $y \in V^u(x)$ belongs to Λ. Let $B_U(y, r)$ be the ball in $U(y)$ centered at y of radius r (with respect to the induced metric). For a ν_x^u-measurable set $Y \subset V^u(x)$, let

$$R(x, r, Y) = \bigcup_{y \in Y} B_U(y, r)$$

be a "fence" through the set Y. We also introduce the "fences" of local stable manifolds passing through $V^u(x)$, i.e., the sets (see Figure 9.2)

$$R(r) = R(x, r, V^u(x)), \quad \tilde{R}(r) = R(x, r, V^u(x) \cap \Lambda),$$

and

$$R^m(r) = R(x, r, V^u(x) \cap \Lambda^m), \quad m > 0.$$

Clearly,

$$R^m(r) \subset \tilde{R}(r) \subset R(r).$$

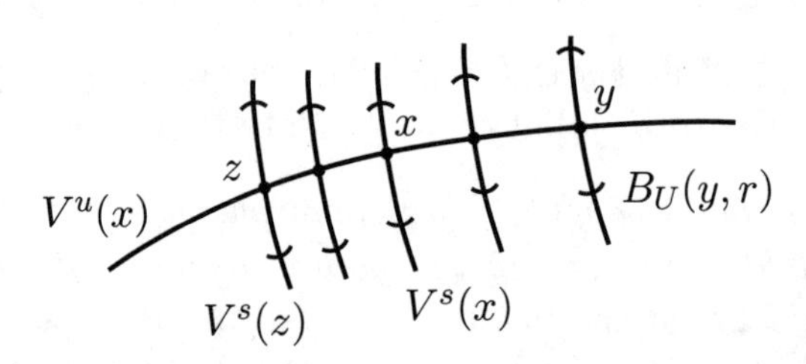

Figure 9.2. The "fences" of local stable manifolds

Since W is a (δ, q)-foliation of Λ, one can find $\delta_0 > 0$ such that $B_U(y, \delta_0) \subset U(y)$ for any $y \in V^u(x)$. By Theorem 7.7 and (7.36), there exists $r_m > 0$ such that $B_U(y, r_m) \subset V^s(y)$ for any $y \in V^u(x) \cap \Lambda^m$.

Fix $r \in (0, r_m]$. Given $y \in R^m(r/2)$, we denote by $n_i(y)$ the successive return times of the positive semitrajectory of y to the set $R^m(r/2)$. We also denote by $z_i \in V^u(x) \cap \Lambda^m$ a point for which $f^{n_i(y)}(y) \in B_U(z_i, r/2)$. In view of the Poincaré recurrence theorem, one can find a subset $N \subset M$ of zero ν-measure for which the sequence $n_i(y)$ is well-defined provided that $y \in R^m(r/2) \setminus N$.

Lemma 9.12. *For ν-almost every $y \in R^m(r/2)$ we have*

$$W^s(y) = \bigcup_{i=1}^{\infty} f^{-n_i(y)}(B_U(z_i, r)). \tag{9.4}$$

Proof of the lemma. Let $y \in R^m(r/2) \setminus N$ and $z \in W^s(y)$. By Theorem 7.15, we have $\rho(f^{n_i(y)}(y), f^{n_i(y)}(z)) \le r/2$ for all sufficiently large i. Therefore,

$$f^{n_i(y)}(z) \in B_U(f^{n_i(y)}(y), r/2) \subset B_U(z_i, r)$$

and the lemma follows. □

Denote by $\xi^m(\delta_0)$ the partition of the set $R^m(\delta_0)$ into the sets $B_U(y, \delta_0)$.

Lemma 9.13. *The partition $\xi^m(\delta_0/2)$ is measurable and has the following properties:*

(1) *the conditional measure on the element $B_U(y, \delta_0/2)$ of this partition is absolutely continuous with respect to the measure ν^s_y;*
(2) *the factor measure on the factor space $R^m(\delta_0/2)/\xi^m(\delta_0/2)$ is absolutely continuous with respect to the measure $\nu^u_x | V^u(x) \cap \Lambda^m$.*

Proof of the lemma. Choose $r = \min\{\delta_0/100, r_m\}$. By Lemma 9.12, for almost every point $w \in V^u(x) \cap \Lambda^m$ one can find a point $y(w) \in B_U(w, r/2) \subset R^m(r/2)$ for which (9.4) holds. Moreover, the point $y(w)$ can be chosen in such a way that the map

$$w \in V^u(x) \cap \Lambda^m \mapsto y(w) \in R^m(r/2)$$

is measurable. For each $n > 0$, set

$$R_n = \bigcup_{w \in V^u(x) \cap \Lambda^m} \bigcup_{n_i(y) \le n} \left(f^{-n_i(y(w))}(B_U(z_i, r)) \cap R^m(3\delta_0/4) \right).$$

Observe that

$$R^m\left(\frac{3}{4}\delta_0\right) = \bigcup_{n \in \mathbb{N}} R_n.$$

Given $\varepsilon > 0$, there exists $p > 0$ and a set $Y \subset V^u(x) \cap \Lambda^m$ such that

$$\nu^u_x((V^u(x) \cap \Lambda^m) \setminus Y) \le \varepsilon \quad \text{and} \quad R(\delta_0/2, Y) \subset \bigcup_{n \le p} R_n. \tag{9.5}$$

It follows from Theorem 8.3 that the partition $\xi^m(\delta_0/2) | R_n$ satisfies statements (1) and (2) of the lemma for each $n > 0$. Since ε is arbitrary, the desired result follows. □

We proceed with the proof of the theorem. Denote by $\xi(\delta_0)$ the partition of the set $R(\delta_0)$ into the sets $B_U(y, \delta_0)$ and denote by $\tilde{\xi}(\delta_0)$ the partition of the set $\tilde{R}(\delta_0)$ into the sets $B_U(y, \delta_0)$. Since W is a (δ, q)-foliation of Λ, the factor space $R(\delta_0)/\xi(\delta_0)$ can be identified with $V^u(x)$ and the factor space $\tilde{R}(\delta_0)/\xi(\delta_0)$ can be identified with $V^u(x) \cap \Lambda$.

Letting $m \to \infty$ in Lemma 9.13, we conclude that the partition $\tilde{\xi}(\delta_0/2)$ also satisfies statements (1) and (2) of the lemma. By Lemma 9.3, we have $P \supset \tilde{R}(\delta_0/2)$. This implies that $P = \bigcup_{n\in\mathbb{Z}} f^n(\tilde{R}(\delta_0/2))$.

Note that the set $R(\delta_0/2)$ is open. We will show that the set

$$A = \Lambda \cap (R(\delta_0/2) \setminus \tilde{R}(\delta_0/2)) \tag{9.6}$$

has measure zero. Assuming the contrary, we have that the set

$$A^m = \Lambda^m \cap (R(\delta_0/2) \setminus \tilde{R}(\delta_0/2)) \tag{9.7}$$

has positive measure for all sufficiently large m. Therefore, for ν-almost every $z \in A^m$ we obtain that $\nu_z^u(V^u(z) \cap A^m) > 0$. Consider the set $R(z, \delta_0/2, A^m)$. Clearly,

$$R(z, \delta_0/2, A^m) \subset R(\delta_0/2) \setminus \tilde{R}(\delta_0/2).$$

Let $B^m = R(z, \delta_0/2, A^m) \cap V^u(x)$. This set is nonempty. It follows from (9.7) that $B^m \subset V^u(x) \setminus \Lambda$ and thus, $\nu_x^u(B^m) = 0$ (see Lemma 9.11). Moreover, repeating arguments in the proof of Lemma 9.13 (see (9.5)), we find that the holonomy map π, which moves $V^u(z) \cap A^m$ onto B^m, is absolutely continuous. Hence, $\nu_x^u(B^m) > 0$. This contradiction implies that the set A in (9.6) has measure zero and completes the proof of Theorem 9.10. □

Theorem 9.10 provides a way to establish ergodicity of the map $f|\mathcal{E}$. Recall that a map f is said to be *topologically transitive* if given two nonempty open sets A and B, there exists a positive integer n such that $f^n(A) \cap B \neq \varnothing$.

Theorem 9.14. *Assume that the conditions of Theorem* 9.10 *hold and that* $\mathcal{E}$ *is an open* (mod 0) *set. Then:*

(1) *every ergodic component of f of positive measure lying in $\mathcal{E}$ is open* (mod 0)*;*
(2) *if $f|\mathcal{E}$ is topologically transitive, then $f|\mathcal{E}$ is ergodic.*

Proof. The first statement follows from Theorem 9.10. Assume that the map $f|\mathcal{E}$ is topologically transitive. Let $C, D \subset \mathcal{E}$ (mod 0) be two distinct ergodic components of positive measure. We have that $\nu(f^n(C) \cap D) = 0$ for any integer n. By the first statement, the sets C and D are open (mod 0). Since f is topologically transitive, we have $f^m(C) \cap D \neq \varnothing$ for some m. Furthermore, the set $f^m(C) \cap D$ is also open (mod 0). Since ν is equivalent to the Riemannian volume, we conclude that $\nu(f^m(C) \cap D) > 0$. This contradiction implies that $f|\mathcal{E}$ is ergodic. □

For a general diffeomorphism preserving a smooth hyperbolic measure, one should not expect the unstable (and stable) leaves to form a (δ, q)-foliation for some functions $\delta(x)$ and $q(x)$. In order to explain why this is so, consider a local unstable manifold $V^u(x)$ passing through a point $x \in \Lambda$. For a typical x and sufficiently large ℓ, the set $V^u(x) \cap \Lambda^\ell$ has positive Riemannian volume (as a subset of the smooth manifold $V^u(x)$) but, in general, is a Cantor-like set. When the local manifold is moved forward a given time n, one should expect a sufficiently small neighborhood of the set $V^u(x) \cap \Lambda^\ell$ to expand. Other pieces of the local manifold (corresponding to bigger values of ℓ) will also expand but with smaller rates. This implies that the global leaf $W^u(x)$ (defined by (7.41)) may bend "uncontrollably". As a result, the map $x \mapsto \varphi_x$ in the definition of (δ, q)-foliations may not be, indeed, continuous.

Furthermore, the global manifold $W^u(x)$ may turn out to be "bounded"; i.e., it may not admit an embedding of an arbitrarily large ball in $\mathbb{R}^k$ (where $k = \dim W^u(x)$) (see Chapter 13).

The local continuity of the global unstable leaves often comes up in the following setting. Using some additional information on the system, one can build an invariant foliation whose leaves contain local unstable leaves. This alone may not yet guarantee that global unstable leaves form a foliation. However, one often may find that the local unstable leaves expand in a "controllable" and somewhat uniform way when they are moved forward. As we see below, this guarantees the desired properties of unstable leaves. Such a situation occurs, for example, for geodesic flows on compact Riemannian manifolds of nonpositive curvature (see Section 10.4).

We now state a formal criterion for local ergodicity. Let f be a $C^{1+\alpha}$ diffeomorphism of a compact smooth Riemannian manifold, preserving a smooth hyperbolic measure ν, and let $\mathcal{E}$ be the set of points with nonzero Lyapunov exponents (which has full measure). As before, in the proofs of statements below we can replace the set $\mathcal{E}$ by an appropriate level set $\Lambda = \Lambda_{\lambda\mu\varepsilon j}$ and we consider the corresponding collection of regular sets $\{\Lambda^\ell = \Lambda^\ell_{\lambda\mu\varepsilon j} : \ell \geq 1\}$ (see (4.24) and (4.25)).

Theorem 9.15. *Let W be a (δ, q)-foliation of $\mathcal{E}$ satisfying:*

(1) *$W(x) \supset V^s(x)$ for every $x \in \mathcal{E}$;*

(2) *there exists a number $\delta_0 > 0$ and a measurable function $n(x)$ on $\mathcal{E}$ such that for almost every $x \in \mathcal{E}$ and any $n \geq n(x)$,*

$$f^{-n}(V^s(x)) \supset B_U(f^{-n}(x), \delta_0).$$

Then every ergodic component of f of positive measure is open (mod 0).

Proof. Let x be a Lebesgue density point of Λ^ℓ for some sufficiently large $\ell > 0$. Set $A(r) = \Lambda^\ell \cap B(x, r)$ where $B(x, r)$ is the ball in M centered at x of radius r. Applying Lemma 9.12 to the set $A(r)$ for sufficiently small r and using the conditions of the theorem, we find that $W^s(y) \supset B_U(y, \delta_0)$

for almost every $y \in A(r)$. One obtains the desired result by repeating arguments in the proof of Theorem 9.10. □

In the case of one-dimensional (δ, q)-foliations the second condition of Theorem 9.15 holds automatically and, hence, can be omitted.

Theorem 9.16. *Let W be a one-dimensional (δ, q)-foliation of $\mathcal{E}$, satisfying the following property: $W(x) \supset V^s(x)$ for every $x \in \mathcal{E}$. Then every ergodic component of f of positive measure is open* (mod 0). *Moreover, $W^s(x) = W(x)$ for almost every $x \in \mathcal{E}$.*

Proof. Fix $\ell > 1$ sufficiently large. For almost every point $x \in \Lambda^\ell$, the intersection $A(x) = V^s(x) \cap \Lambda^\ell$ has positive Lebesgue measure in $V^s(x)$. For every $y \in A(x)$, let $s(y)$ be the distance between x and y measured along $V^s(x)$. Then there exists a differentiable curve $\gamma\colon [0, s(y)] \to V^s(x)$ with $\gamma(0) = x$ and $\gamma(s(y)) = y$, satisfying

$$\begin{aligned}\rho_{W(f^{-n}(x))}(f^{-n}(x), f^{-n}(y)) &= \int_0^{s(y)} \|d_{\gamma(t)} f^{-n} \gamma'(t)\|\, dt \\ &\geq \int_0^{s(y)} \|d_{f^n\gamma(t)} f^n \gamma'(t)\|^{-1}\, dt \\ &\geq \ell^{-1} e^{-\varepsilon n} \lambda^{-n} \geq \delta_0\end{aligned}$$

for sufficiently large n (see Section 4.2), where δ_0 is a positive constant. Therefore, the second condition of Theorem 9.15 holds and the desired result follows. □

One can readily extend Theorems 9.15 and 9.16 to the case when the set $\mathcal{E}$ is open (mod 0) and has positive (but not necessarily full) measure as well as to dynamical systems with continuous time.

Theorem 9.17. *Let f be a $C^{1+\alpha}$ diffeomorphism of a compact smooth Riemannian manifold preserving a smooth measure ν and let $\mathcal{E}$ be the set of points with nonzero Lyapunov exponents. Assume that $\mathcal{E}$ is open* (mod 0) *and has positive measure. Also let W be a (δ, q)-foliation of $\mathcal{E}$ which satisfies properties* (1) *and* (2) *in Theorem* 9.15. *Then every ergodic component of $f|\mathcal{E}$ of positive measure is open* (mod 0).

Theorem 9.18. *Let φ_t be a smooth flow of a compact smooth Riemannian manifold preserving a smooth measure ν and let $\mathcal{E}$ be the set of points with nonzero Lyapunov exponents. Assume that the set $\mathcal{E}$ is open* (mod 0) *and has positive measure. Also let W be a (δ, q)-foliation of $\mathcal{E}$ with the following properties:*

(1) *$W(x) \supset V^s(x)$ for every $x \in \mathcal{E}$;*

(2) *there exists a number* $\delta_0 > 0$ *and a measurable function* $t(x)$ *on* $\mathcal{E}$ *such that for almost every* $x \in \mathcal{E}$ *and any* $t \geq t(x)$,

$$\varphi_{-t}(V^s(x)) \supset B_U(\varphi_{-t}(x), \delta_0).$$

Then every ergodic component of $\varphi_t|\mathcal{E}$ *of positive measure is open* (mod 0). *Furthermore, if the foliation* $W(x)$ *is one-dimensional, then the second requirement can be dropped.*

9.3. The entropy formula

One of the main concepts of smooth ergodic theory is that sufficient instability of trajectories yields rich ergodic properties of the system. The entropy formula is in a sense a "quantitative manifestation" of this idea and is yet another pearl of the theory. It expresses the metric entropy $h_\nu(f)$ of a diffeomorphism with respect to a smooth hyperbolic measure, in terms of the values of the Lyapunov exponent.[1]

9.3.1. The metric entropy of a diffeomorphism. We briefly describe some relevant notions from the theory of measurable partitions and entropy theory.

Let $(X, \mathcal{B}, \mu)$ be a Lebesgue measure space. A finite or countable family $\xi \subset \mathcal{B}$ is a *measurable partition* of X if $\mu(\bigcup_{C\in\xi} C) = \mu(X)$ and $\mu(C \cap D) = 0$ for every $C, D \in \xi$ with $C \neq D$. Given two partitions ξ and η, we say that η is a *refinement* of ξ and we write $\xi \subset \eta$ if every element of η is contained (mod 0) in an element of ξ. We say that η is *equivalent* to ξ and we write $\xi = \eta$ if these partitions are refinements of each other. The *common refinement* is the partition $\xi \vee \eta$ with elements $C \cap C'$ where $C \in \xi$ and $C' \in \eta$. Finally, we say that ξ is *independent* of η if $\mu(C \cap C') = \mu(C)\mu(C')$ for all $C \in \xi$ and $C' \in \eta$.

The *entropy* of the measurable partition ξ (with respect to μ) is given by

$$H_\mu(\xi) = -\sum_{C\in\xi} \mu(C) \log \mu(C),$$

with the convention that $0 \log 0 = 0$.

Exercise 9.19. Let ξ and η be two finite or countable partitions of X. Show that:

(1) $H_\mu(\xi) \geq 0$ and $H_\mu(\xi) = 0$ if and only if ξ is the trivial partition;

(2) if $\xi \subset \eta$, then $H_\mu(\xi) \leq H_\mu(\eta)$ and equality holds if and only if $\xi = \eta$;

[1] Other commonly used terms are *measure-theoretic entropy* and *Kolmogorov–Sinai* entropy.

(3) if ξ has n elements, then $H_\mu(\xi) \le \log n$ and equality holds if and only if each element has measure $1/n$;

(4) $H_\mu(\xi \vee \eta) \le H_\mu(\xi) + H_\mu(\eta)$ and equality holds if and only if ξ is independent of η.

Hint: Use the fact that the function $x \log x$ is continuous and strictly concave.

Given two measurable partitions ξ and ζ, we also define the *conditional entropy of ξ with respect to ζ* by

$$H_\mu(\xi|\zeta) = -\sum_{C\in\xi}\sum_{D\in\zeta} \mu(C\cap D)\log\frac{\mu(C\cap D)}{\mu(D)}.$$

One can show that the conditional entropy has the following properties. Let ξ, ζ, and η be three finite or countable partitions of X. Then:

(1) $H_\mu(\xi|\zeta) \ge 0$ and $H_\mu(\xi|\zeta) = 0$ if and only if $\xi \subset \zeta$;

(2) if $\zeta \subset \eta$, then $H_\mu(\xi|\zeta) \ge H_\mu(\xi|\eta)$ and equality holds if and only if $\zeta = \eta$;

(3) if $\zeta \subset \eta$, then $H_\mu(\xi \vee \zeta|\eta) = H_\mu(\xi|\eta)$;

(4) if $\xi \subset \eta$, then $H_\mu(\xi|\zeta) \le H_\mu(\eta|\zeta)$ and equality holds if and only if $\xi \vee \zeta = \eta \vee \zeta$;

(5) $H_\mu(\xi \vee \zeta|\eta) = H_\mu(\xi|\eta) + H_\mu(\zeta|\chi \vee \eta)$ and $H_\mu(\xi \vee \zeta) = H_\mu(\xi) + H_\mu(\zeta|\xi)$;

(6) $H_\mu(\xi|\zeta \vee \eta) \le H_\mu(\xi|\eta)$;

(7) $H_\mu(\xi|\zeta) = \sum_{D\in\zeta} \nu(D)H(\xi|D)$ where $H(\xi|D)$ is the entropy of ξ with respect to the conditional measure on D induced by μ;

(8) $H_\mu(\xi|\zeta) \le H_\mu(\xi)$ and equality holds if and only if ξ is independent of ζ.

Now consider a measurable transformation $T\colon X \to X$. Given a partition ξ and a number $k \ge 0$, denote by $T^{-k}\xi$ the partition of X by the sets $T^{-k}(C)$ with $C \in \xi$.

We define the *entropy* of T with respect to the partition ξ by the formula

$$h_\mu(T,\xi) = \inf_n \frac{1}{n} H_\mu\left(\bigvee_{k=0}^{n-1} T^{-k}\xi\right). \tag{9.8}$$

Using the properties of the conditional entropy described above, one can show that

$$\begin{aligned} h_\mu(T,\xi) &= \lim_{n\to\infty} \frac{1}{n} H_\mu\left(\bigvee_{k=0}^{n-1} T^{-k}\xi\right) \\ &= \lim_{n\to\infty} H_\mu\left(\xi \Big| \bigvee_{k=1}^{n} T^{-k}\xi\right) = H_\mu(\xi|\xi^-), \end{aligned} \tag{9.9}$$

where $\xi^- = \bigvee_{k=0}^{\infty} T^{-k}\xi$. One can show that the entropy of T with respect to the partition ξ has the following properties:

(1) $h_\mu(T,\xi) \le H_\mu(\xi)$;

(2) $h_\mu(T, T^{-1}\xi) = h_\mu(T,\xi)$;

(3) $h_\mu(T,\xi) = h_\mu(T, \bigvee_{i=0}^{n} T^{-i}\xi)$.

We define the *metric entropy* of T with respect to μ by

$$h_\mu(T) = \sup_\xi h_\mu(T,\xi),$$

where the supremum is taken over all measurable partitions ξ with finite entropy (which actually coincides with the supremum over all finite measurable partitions). Using (9.9), one can show that:

(1) $h_\mu(T^m) = m h_\mu(T)$ for every $m \in \mathbb{N}$;

(2) if T is invertible with measurable inverse, then $h_\mu(T^m) = |m| h_\mu(T)$ for every $m \in \mathbb{Z}$;[2]

(3) let $(X, \mathcal{B}, \mu)$ and $(Y, \mathcal{A}, \nu)$ be two Lebesgue measure spaces, let $T\colon X \to X$ and $S\colon Y \to Y$ be two measurable transformations, and let $P\colon X \to Y$ be a measurable transformation with measurable inverse such that $P \circ T = S \circ P$ and $P_*\mu = \nu$ (in other words, P is an isomorphism of measure spaces that conjugates T and S); then $h_\mu(T) = h_\nu(S)$.

9.3.2. Upper bound for the metric entropy. Let f be a C^1 diffeomorphism of a smooth compact Riemannian manifold M and let ν be an invariant Borel probability measure on M. Consider the Lyapunov exponent $\chi(x,\cdot)$ and its Lyapunov spectrum

$$\operatorname{Sp}\chi(x) = \{(\chi_i(x), k_i(x)) : 1 \le i \le s(x)\};$$

see Section 4.1.

[2] To see this, observe that one can replace $T^{-k}\xi$ by $T^k\xi$ in (9.8) and (9.9) and conclude that $h_\mu(T^{-1},\xi) = h_\mu(T,\xi)$ and hence that $h_\mu(T^{-1}) = h_\mu(T)$.

Theorem 9.20. *The entropy of f admits the upper bound*

$$h_\nu(f) \leq \int_M \sum_{i:\chi_i(x)>0} k_i(x)\chi_i(x)\, d\nu(x). \tag{9.10}$$

This upper bound was established by Ruelle in [**77**]. Independently, Margulis obtained this estimate in the case of volume-preserving diffeomorphisms (unpublished). The inequality (9.10) is often referred to as the Margulis–Ruelle Inequality. The proof presented below is due to Katok and Mendoza [**48**].

Proof of the theorem. By decomposing ν into its ergodic components, we may assume without loss of generality that ν is ergodic. Then $s(x) = s$, $k_i(x) = k_i$, and $\chi_i(x) = \chi_i$ are constant ν-almost everywhere for each $1 \leq i \leq s$. Fix $m > 0$. Since the manifold M is compact, there exists a number $r_m > 0$ such that for every $0 < r \leq r_m$, $y \in M$, and $x \in B(y,r)$ we have

$$\begin{aligned}\frac{1}{2} d_x f^m \left(\exp_x^{-1} B(y,r)\right) &\subset \exp_{f^m(x)}^{-1} f^m(B(y,r)) \\ &\subset 2 d_x f^m \left(\exp_x^{-1} B(y,r)\right),\end{aligned} \tag{9.11}$$

where for a set $A \subset T_z M$ and $z \in M$, we write $\alpha A = \{\alpha v : v \in A\}$.

We begin by constructing a special partition ξ of the manifold M with some "nice" properties that will be used to estimate the metric entropy of f from above. First, each element C of the partition is roughly a ball of a small radius (which is independent of the element of the partition) such that the image of the element C under f^m is also approximated by the image of the ball. Second, the metric entropy of f^m with respect to ξ is close to the metric entropy of f^m. More precisely, the following statement holds.

Lemma 9.21. *Given $\varepsilon > 0$, there exists a partition ξ of M satisfying the following conditions:*

(1) $h_\nu(f^m, \xi) \geq h_\nu(f^m) - \varepsilon$;

(2) *there are numbers $r < 2r' \leq r_m/20$ such that $B(x,r') \subset C \subset B(x,r)$ for every element $C \in \xi$ and every $x \in C$;*

(3) *there exists $0 < r < r_m/20$ such that*
   (a) *if $C \in \xi$, then $C \subset B(y,r)$ for some $y \in M$ and*
   (b) *for every $x \in C$,*

$$\frac{1}{2} d_x f^m \left(\exp_x^{-1} B(y,r)\right) \subset \exp_{f^m(x)}^{-1} f^m(C) \subset 2 d_x f^m \left(\exp_x^{-1} B(y,r)\right).$$

Proof of the lemma. Given $\alpha > 0$, consider a maximal α-separated set Γ, i.e., a finite set of points such that $d(x,y) > \alpha$ whenever x, $y \in \Gamma$. For each

$x \in \Gamma$ define

$$\mathcal{D}_\Gamma(x) = \{y \in M : d(y,x) \le d(y,z) \text{ for all } z \in \Gamma \setminus \{x\}\}.$$

Obviously, $B(x,\alpha/2) \subset \mathcal{D}_\Gamma(x) \subset B(x,\alpha)$. Note that the sets $\mathcal{D}_\Gamma(x)$ corresponding to different points $x \in \Gamma$ intersect only along their boundaries, i.e., at a finite number of submanifolds of codimension greater than zero. Since ν is a Borel measure, if necessary, we can move the boundaries slightly so that they have measure zero. Thus, we obtain a partition ξ with $\operatorname{diam}\xi \le \alpha$. Moreover, we can choose a partition ξ such that

$$h_\nu(f^m,\xi) > h_\nu(f^m) - \varepsilon \quad \text{and} \quad \operatorname{diam}\xi < r_m/10.$$

This implies statements (1) and (2). Statement (3) follows from (9.11). □

We proceed with the proof of the theorem. Using the properties of conditional entropy, we obtain that

$$\begin{aligned} h_\nu(f^m,\xi) &= \lim_{k\to\infty} H_\nu(\xi|f^m\xi \vee \cdots \vee f^{km}\xi) \\ &\le H_\nu(\xi|f^m\xi) = \sum_{D\in f^m\xi} \nu(D) H(\xi|D) \\ &\le \sum_{D\in f^m\xi} \nu(D)\log\left(\operatorname{card}\{C\in\xi : C\cap D \ne \varnothing\}\right), \end{aligned} \tag{9.12}$$

where card denotes the cardinality of the set. In view of this estimate our goal now is first to obtain a uniform exponential estimate for the number of those elements $C \in \xi$ which have nonempty intersection with a given element $D \in f^m\xi$ and then to establish an exponential bound on the number of those sets $D \in f^m\xi$ that contain LP-regular orbits $\{f^k(x)\}$ along which for $v \in T_xM$ the length $\|d_xf^kv\|$ admits effective exponential estimates in terms of the Lyapunov exponents $\chi(x,v)$.

Lemma 9.22. *There exists a constant $K_1 > 0$ such that for $D \in f^m\xi$,*

$$\operatorname{card}\{C\in\xi : D\cap C = \varnothing\} \le K_1 \sup\{\|d_xf\|^{mp} : x \in M\},$$

where $p = \dim M$.

Proof. By the Mean Value Theorem,

$$\operatorname{diam}(f^m(C)) \le \sup\{\|d_xf\|^m : x\in M\}\operatorname{diam} C.$$

Thus, if $C \cap D \ne \varnothing$, then C is contained in the $4r'$-neighborhood of D. Therefore,

$$\operatorname{diam} \bigcup_{C\cap D\ne\varnothing} C \le (\sup\{\|d_xf\|^m : x\in M\} + 2)4r'$$

and hence, the volume of $\bigcup_{C\cap D\ne\varnothing} C$ is bounded from above by

$$Kr^p \sup\{\|d_xf\|^{mp} : x\in M\},$$

where $K > 0$ is a constant. On the other hand, by property (2) of the partition ξ, the set C contains a ball $B(x, r')$, and hence, the volume of C is at least $K'(r')^p$, where K' is a positive constant. This implies the desired result. □

Fix $\varepsilon > 0$ and let $R_m = R_m(\varepsilon)$ be the set of LP-regular points $x \in M$ satisfying the following condition: for $k > m$ and $v \in T_x M$,

$$e^{k(\chi(x,v)-\varepsilon)}\|v\| \leq \|d_x f^k v\| \leq e^{k(\chi(x,v)+\varepsilon)}\|v\|.$$

Lemma 9.23. *If $D \in f^m\xi$ has nonempty intersection with R_m, then there exists a constant $K_2 > 0$ such that*

$$\operatorname{card}\{C \in \xi : D \cap C \neq \varnothing\} \leq K_2 e^{\varepsilon m} \prod_{i:\chi_i > 0} e^{m(\chi_i+\varepsilon)k_i}.$$

Proof. Let $C' \in \xi$ be such that $C' \cap R_m \neq \varnothing$ and $f^m(C') = D$. Pick a point $x \in C' \cap f^{-m}(R_m)$ and let $B = B(x, 2\operatorname{diam} C')$. The set

$$\tilde{B}_0 = d_x f^m(\exp_x^{-1} B) \subset T_{f^m x} M$$

is an ellipsoid and $D \subset B_0 = \exp_{f^m(x)}(\tilde{B}_0)$. If a set $C \in \xi$ has nonempty intersection with B_0, then it lies in the set $B_1 = \{y \in M : d(y, B_0) < \operatorname{diam}\xi\}$. Therefore,

$$\operatorname{card}\{C \in \xi : D \cap C \neq \varnothing\} \leq \operatorname{vol}(B_1)(\operatorname{diam}\xi)^{-p},$$

where $\operatorname{vol}(B_1)$ denotes the volume of B_1. Up to a bounded factor, $\operatorname{vol}(B_1)$ is bounded by the product of the lengths of the axes of the ellipsoid $\tilde{B}_0$. Those of them corresponding to nonpositive exponents are at most subexponentially large. The remaining ones are of size at most $e^{m(\chi_i+\varepsilon)}$, up to a bounded factor, for all sufficiently large m. Thus,

$$\begin{aligned}\operatorname{vol}(B_1) &\leq K e^{m\varepsilon}(\operatorname{diam} B)^p \prod_{i:\chi_i>0} e^{m(\chi_i+\varepsilon)k_i} \\ &\leq K e^{m\varepsilon}(2\operatorname{diam}\xi)^p \prod_{i:\chi_i>0} e^{m(\chi_i+\varepsilon)k_i},\end{aligned}$$

for some constant $K > 0$ and the lemma follows. □

We proceed with the proof of the theorem and estimate the metric entropy of f^m with respect to the partition ξ. Namely, by (9.12), we obtain that

$$\begin{aligned} h_\nu(f^m,\xi) &\le \sum_{D\cap R_m\neq\varnothing} \nu(D)\log\operatorname{card}\{C\in\xi : C\cap D\neq\varnothing\} \\ &\quad+ \sum_{D\cap R_m=\varnothing} \nu(D)\log\operatorname{card}\{C\in\xi : C\cap D\neq\varnothing\} \\ &= \sum_1 + \sum_2 . \end{aligned}$$

By Lemma 9.23, the first sum can be estimated as

$$\sum_1 \le \Big(\log K_2 + \varepsilon m + m\sum_{i:\chi_i>0}(\chi_i+\varepsilon)k_i\Big)\nu(R_m)$$

and by Lemma 9.22, the second sum can be estimated as

$$\sum_2 \le \big(\log K_1 + pm\log\sup\{\|d_xf\| : x\in M\}\big)\nu(M\setminus R_m).$$

Note that by the Multiplicative Ergodic Theorem 4.2, for every sufficiently small ε,

$$\bigcup_{m\ge 0} R_m(\varepsilon) = M \pmod 0$$

and hence, $\nu(M\setminus R_m)\to 0$ as $m\to\infty$. We are now ready to obtain the desired upper bound. By Lemma 9.21, we have that

$$\begin{aligned} mh_\nu(f)-\varepsilon = h_\nu(f^m)-\varepsilon &\le h_\nu(f^m,\xi) \\ &\le \log K_2 + \varepsilon m + m\sum_{i:\chi_i>0}(\chi_i+\varepsilon)k_i \\ &\quad+ \big(\log K_1 + pm\log\sup\{\|d_xf\| : x\in M\}\big)\nu(M\setminus R_m). \end{aligned}$$

Dividing by m and letting $m\to\infty$, we find that

$$h_\nu(f) \le 2\varepsilon + \sum_{i:\chi_i>0}(\chi_i+\varepsilon)k_i.$$

Letting $\varepsilon\to 0$, we obtain the desired upper bound. □

An important consequence of Theorem 9.20 is that any C^1 diffeomorphism with positive topological entropy has an invariant measure with at least one positive and one negative Lyapunov exponent. In particular, a surface diffeomorphism with positive topological entropy always has a hyperbolic invariant measure.

Exercise 9.24. Show that for an arbitrary invariant measure the inequality (9.10) may be strict. Hint: Examine a diffeomorphism with a hyperbolic fixed point and an atomic measure concentrated at this point.

9.3.3. Lower bound for the metric entropy. We shall now prove the lower bound and, hence, the entropy formula. We stress that while the upper bound for the metric entropy holds for diffeomorphisms of class C^1, the lower bound requires that f be of class $C^{1+\alpha}$.

Theorem 9.25 (Entropy Formula). *If f is of class $C^{1+\alpha}$ and ν is a smooth invariant measure on M, then*

$$h_\nu(f) = \int_M \sum_{i:\chi_i(x)>0} k_i(x)\chi_i(x)\, d\nu(x). \tag{9.13}$$

In the case of Anosov diffeomorphisms, a statement equivalent to the entropy formula was proved by Sinai in [**82**] and for a general diffeomorphism preserving a smooth measure, it was established by Pesin in [**69**] (see also [**70**]). The proof presented below follows the original argument in [**69**] and exploits in an essential way the machinery of stable manifolds and their absolute continuity. There is another proof of the lower bound due to Mañé [**62**] that is more straightforward and does not use this machinery. While the requirement that the map f be of class $C^{1+\alpha}$ for some $\alpha > 0$ is crucial for both proofs, it is worth mentioning that by a result of Tahzibi [**84**] for a C^1 *generic* surface diffeomorphism the lower bound still holds.

Proof of the theorem. We only need to show that

$$h_\nu(f) \geq \int_M \sum_{i:\chi_i(x)>0} k_i(x)\chi_i(x)\, d\nu(x),$$

or equivalently (by replacing f by f^{-1} and using Theorem 4.2) that

$$h_\nu(f) \geq -\int_M \sum_{i:\chi_i(x)<0} k_i(x)\chi_i(x)\, d\nu(x). \tag{9.14}$$

Consider the set

$$\Gamma = \big\{x \in M : \chi^+(x,v) < 0 \text{ for some } v \in T_xM\big\}.$$

If $\nu(\Gamma) = 0$, the desired result follows immediately from (9.14). We, therefore, assume that $\nu(\Gamma) > 0$. Let $\tilde{\Gamma} = \Gamma \cap \mathcal{R}$, where $\mathcal{R}$ is the set of LP-regular points. Observe that the set $\tilde{\Gamma}$ is nonuniformly partially hyperbolic and that we can consider the corresponding collection of level sets $\Lambda_{\lambda\mu\varepsilon j}$ and for each $\lambda, \mu, \varepsilon$, and $1 \leq j < p$ we can consider the collection of regular sets $\{\Lambda^\ell_{\lambda\mu\varepsilon j} : \ell \geq 1\}$ (see (4.24) and (4.25)). In view of (9.1) it suffices to prove the theorem for each level set of positive measure and therefore in what follows we fix numbers $\lambda, \mu, \varepsilon$, and j and set $\Lambda = \Lambda_{\lambda\mu\varepsilon j}$ and $\Lambda^\ell = \Lambda^\ell_{\lambda\mu\varepsilon j}$. Observe that for every $x \in \Lambda$ the largest negative Lyapunov exponent does not exceed λ uniformly in x.

For each $x \in \tilde{\Gamma} \cap \Lambda$ set

$$T_n(x) = \operatorname{Jac}(d_x f^n | E^s(x)) \quad \text{and} \quad g(x) = \prod_{i:\chi_i(x)<0} e^{k_i(x)\chi_i(x)}.$$

Fix $\varepsilon > 0$ and consider the invariant sets

$$\tilde{\Gamma}_n = \{x \in \tilde{\Gamma} : (1+\varepsilon)^{-n} < g(x) \leq (1+\varepsilon)^{-n+1}\}, \quad n > 0.$$

We shall evaluate the metric entropy of the restriction $f|\tilde{\Gamma}_n$ with respect to the measure $\nu_n = \nu|\tilde{\Gamma}_n$.

Lemma 9.26. *Given $\varepsilon > 0$, there exists a positive Borel function $L(x)$ such that for any $x \in \tilde{\Gamma}$ and $n > 0$,*

$$T_n(x) \leq L(x) g(x)^n e^{\varepsilon n}.$$

Proof of the lemma. This is an immediate consequence of Theorems 4.1 and 2.21. □

Fix $n > 0$ and $\ell > 0$, and define the collection of measurable sets

$$\tilde{\Gamma}_n^\ell = \{x \in \tilde{\Gamma}_n \cap \Lambda^\ell : L(x) \leq \ell\}, \quad \Gamma_n^\ell = \bigcup_{x \in \tilde{\Gamma}_{n,\ell}} V^s(x), \quad \hat{\Gamma}_n = \bigcup_{j \in \mathbb{Z}} f^j(\Gamma_n^\ell).$$

Note that $\hat{\Gamma}_n$ is f-invariant. Fix $\beta > 0$. By choosing a sufficiently large ℓ, we obtain $\nu_n(\tilde{\Gamma}_n^\ell) > 1 - \beta$. Since

$$\tilde{\Gamma}_n^\ell \subset \Gamma_n^\ell \subset \hat{\Gamma}_n \subset \tilde{\Gamma}_n \pmod 0,$$

we also have

$$\nu_n(\hat{\Gamma} \setminus \Gamma_n^\ell) \leq \beta \quad \text{and} \quad \nu_n(\tilde{\Gamma}_n \setminus \hat{\Gamma}_n) \leq \beta. \tag{9.15}$$

Let us now choose any finite measurable partition ξ of M such that each element $\xi(x)$ of ξ is homeomorphic to a ball, has piecewise smooth boundary, and has diameter at most r_ℓ (such a partition is constructed in Lemma 9.21).

We define a partition η of $\hat{\Gamma}_n$ composed of $\hat{\Gamma}_n \setminus \Gamma_n^\ell$ and of the elements $V^s(y) \cap \xi(x)$ for each $y \in \xi(x) \cap \tilde{\Gamma}_n^\ell$.

Given $x \in \Gamma_n^\ell$ and a sufficiently small $r \leq r_\ell$ (see Chapter 8), we set

$$B_\eta(x, r) = \{y \in \eta(x) : \rho(y, x) < r\}.$$

Lemma 9.27. *There exists $q_\ell > 0$ and a set $A^\ell \subset \Gamma_n^\ell$ with $\nu_n(\Gamma_n^\ell \setminus A^\ell) \leq \beta$ such that $\eta^-(x) \supset B_\eta(x, q_\ell)$ for every $x \in A^\ell$.*

Proof of the lemma. For each $\delta > 0$ set

$$\partial\xi = \bigcup_{y \in M} \partial(\xi(y)) \quad \text{and} \quad \partial\xi_\delta = \{y : \rho(y, \partial\xi) \leq \delta\}.$$

One can easily show that there exists $C_1 > 0$ such that

$$\nu_n(\partial\xi_\delta) \leq C_1 \delta. \tag{9.16}$$

Let

$$D_q = \{x \in \Gamma_n^\ell : B_\eta(x,q) \setminus \eta^-(x) \neq \varnothing\}.$$

If $x \in D_q$, then there exist $m \in \mathbb{N}$ and $y \in B_\eta(x,q)$ such that $y \notin (f^{-m}\eta)(x)$. Hence, $\partial\xi \cap f^m(B_\eta(x,q)) \neq \varnothing$. Therefore, by Theorem 7.1, if $x \in D_q$, then $f^m(x) \in \partial\xi_{C_2\lambda^m e^{\varepsilon m} q}$ for some $C_2 = C_2(\ell) > 0$. Thus, in view of (9.16),

$$\nu_n(D_q) \leq C_1 \sum_{m=0}^{\infty} C_2 \lambda^m e^{\varepsilon m} q \leq C_3 q$$

for some $C_3 = C_3(\ell) > 0$. The lemma follows by setting $q_\ell = \beta C_3{}^{-1}$ and $A^\ell = \Gamma_n^\ell \setminus D_{q_\ell}$. □

The following statement is crucial in our proof of the lower bound. It exploits the absolute continuity property of the measure ν in an essential way.

Lemma 9.28. *There exists $C_5 = C_5(\ell) \geq 1$ such that for $x \in A^\ell \cap V^s(y)$ and $y \in \tilde{\Gamma}_n^\ell$,*

$$C_5{}^{-1} \leq \frac{d\nu_x^-}{dm_y} \leq C_5,$$

where ν_x^- is the conditional measure on the element $\eta^-(x)$ of the partition η^- and m_y is the leaf-volume on $V^s(y)$.

Proof of the lemma. The statement is an immediate consequence of Theorem 8.3 and Lemma 9.27. □

We also need the following statement.

Lemma 9.29. *There exists $C_4 = C_4(\ell) > 0$ such that for every $x \in \Gamma_n^\ell \cap V^s(y)$ with $y \in \tilde{\Gamma}_n^\ell$ and for every $k > 0$ we have*

$$m_{f^k(y)}(f^k(\eta(x))) \leq C_4 T_k(y)$$

Proof of the lemma. We have

$$m_{f^k(y)}(f^k(\eta(x))) = \int_{\eta(x)} T_k(z)\, dm_y(z).$$

A similar argument to that in the proof of Lemma 8.11 shows that there exists $C' = C'(\ell) > 0$ such that for every $z \in \eta(x)$ and all sufficiently large $k \in \mathbb{N}$,

$$\left| \frac{T_k(z)}{T_k(y)} - 1 \right| \leq C'.$$

The desired result follows. □

We now complete the proof of the theorem. Write $\hat{f} = f|\hat{\Gamma}$. For every $k > 0$ we have

$$h_{\nu_n}(f|\tilde{\Gamma}_n) \geq h_{\nu_n}(\hat{f}) = \frac{1}{k}h_{\nu_n}(\hat{f}^k) \geq \frac{1}{k}H_{\nu_n}(\hat{f}^k\eta|\eta^-), \tag{9.17}$$

where $\eta^- = \bigvee_{k=0}^{\infty} \hat{f}^{-k}\eta$. To estimate the last expression, we find a lower bound for

$$H_{\nu_n}(\hat{f}^k\eta|\eta^-(x)) = \int_{\eta^-(x)} -\log \nu_x^-(\eta^-(x) \cap (f^k\eta)(y))\, d\nu_x^-(y).$$

Note that for $x \in A^\ell$,

$$\nu_x(B_\eta(x, q_\ell)) \geq C_6 {q_\ell}^{\dim M}$$

for some $C_6 = C_6(\ell) > 0$. It follows from Lemmas 9.26, 9.28, and 9.29 that for $x \in A^\ell$ and $n \in \mathbb{N}$,

$$\begin{aligned} H_{\nu_n}(\hat{f}^k\eta|\eta^-(x)) &\geq -C_5 \log\Big(C_5\nu_x(\eta^-(x) \cap (f^k\eta)(x))\Big)\, \nu_x(B_\eta(x, q_\ell)) \\ &\geq -C_5 \log(C_5 C_4 T_k(x))\nu_x(B_\eta(x, q_\ell)) \\ &\geq -C_5 \log(C_5 C_4 L(x) g(x)^k e^{\varepsilon k}) C_6 {q_\ell}^{\dim M} \\ &\geq -C_5 \log(C_5 C_4 \ell (1+\varepsilon)^{(-n+1)k} e^{\varepsilon k}) C_6 {q_\ell}^{\dim M}. \end{aligned}$$

Choosing q_ℓ sufficiently small, we may assume that $C_5 C_6 {q_\ell}^{\dim M} < 1$. Therefore,

$$\begin{aligned} H_{\nu_n}(\hat{f}^k\eta|\eta^-(x)) &\geq -k(-n+1)\log(1+\varepsilon) - k\varepsilon - C_7 \\ &\geq -k\log g(x) - k(\log(1+\varepsilon) + \varepsilon) - C_7 \end{aligned}$$

for some $C_7 = C_7(\ell) > 0$. Integrating this inequality over the elements of η^-, we obtain

$$\frac{1}{k}H_{\nu_n}(\hat{f}^k\eta|\eta^-) \geq \int_{A^\ell} \left(-\log g(x) - (\log(1+\varepsilon) + \varepsilon) - \frac{C_7}{k}\right) d\nu_n(x).$$

Given $\delta > 0$, we can choose numbers k and ℓ sufficiently large and ε sufficiently small such that (in view of (9.15), (9.17), and Lemma 9.27)

$$h_{\nu_n}(f|\tilde{\Gamma}_n) \geq -\frac{1}{\nu(\tilde{\Gamma}_n)} \int_{\tilde{\Gamma}_n} \sum_{i:\chi_i(x)<0} k_i(x)\chi_i(x)\, d\nu(x) - \delta.$$

Summing up these inequalities over n (note that the sets $\tilde{\Gamma}_n$ are disjoint and invariant), we obtain

$$\begin{aligned} h_\nu(f) &\geq \sum_{n=1}^{\infty} \nu(\tilde{\Gamma}_n) h_{\nu_n}(f|\tilde{\Gamma}_n) \\ &\geq \sum_{n=1}^{\infty} \left(-\int_{\tilde{\Gamma}_n} \sum_{i:\chi_i(x)<0} k_i(x)\chi_i(x)\,d\nu(x) - \delta\nu(\tilde{\Gamma}_n) \right) \\ &= -\int_M \sum_{i:\chi_i(x)<0} k_i(x)\chi_i(x)\,d\nu(x) - \delta. \end{aligned}$$

Since δ is arbitrary, the desired result follows. □

A slight modification of the above proof allows one to establish the entropy formula (9.13) for measures which are absolutely continuous with respect to the Riemannian volume (not necessarily equivalent to it).

Chapter 10

Geodesic Flows on Surfaces of Nonpositive Curvature

For a long time geodesic flows have played an important stimulating role in the development of hyperbolicity theory. In the beginning of the 20th century Hadamard and Morse, while studying the statistics of geodesics on surfaces of negative curvature, made the important observation that the local instability of trajectories gives rise to some global properties of dynamical systems such as ergodicity and topological transitivity. The results obtained during this period were summarized in the survey by Hedlund [**37**] (see also the article [**40**] by Hopf).

The subsequent study of geodesic flows has revealed the true nature and importance of instability and has marked a remarkable shift from differential-geometric methods to dynamical techniques (see the seminal book by Anosov [**3**] and the article [**4**]). Geodesic flows were a source of inspiration (and an excellent model) for introducing concepts of both uniform and nonuniform hyperbolicity.

On the other hand, they always were a touchstone for applying new advanced methods of the general theory of dynamical systems. This, in particular, has led to some new interesting results in differential and Riemannian geometry (such as lower and upper estimates of the number of closed geodesics and their distribution in the phase space). In this chapter we will present some of these results. While describing ergodic properties of geodesic flows in the spirit of the book, we consider the Liouville measure that is invariant under the flow (we allow other invariant measures when discussing an upper bound for the metric entropy), and we refer the reader to the excellent survey [**54**] where measures of maximal entropy are studied. We only consider the case of geodesic flows on surfaces and we refer the reader to [**9**] for the general case.

10.1. Preliminary information from Riemannian geometry

For the reader's convenience we collect here some basic notions and results from Riemannian geometry that will be used throughout this chapter. Consider a compact p-dimensional smooth manifold (of class C^r, $r \geq 3$) without boundary, endowed with a Riemannian metric of class C^r. For $x \in M$ we denote by $\langle \cdot, \cdot \rangle_x$ the inner product in T_xM.

10.1.1. The canonical Riemannian metric. We endow the second tangent bundle $T(TM)$ with a special Riemannian metric. Let $\pi\colon TM \to M$ be the natural projection, that is, $\pi(x, v) = x$ for each $x \in M$ and each $v \in T_xM$. The map $d_v\pi\colon T(TM) \to TM$ is linear and its p-dimensional kernel $H(v) \subset T_vTM$ is the space of horizontal vectors.

Consider the connection operator $K\colon T(TM) \to TM$ defined as follows. For $v \in TM$ and $\xi \in T_vTM$ let $Z(t)$ be any curve in TM such that $Z(0) = d\pi\xi$ and $\frac{d}{dt}Z(t)|_{t=0} = \xi$. Set $K\xi = (\nabla Z)(t)|_{t=0}$ where ∇ is the covariant derivative associated with the Riemannian (Levi-Civita) connection.[1] The map K is linear and its kernel is the p-dimensional space of vertical vectors $V(v) \subset T_vTM$. For every $v \in TM$ we have that $T_vTM = H(v) \oplus V(v)$. This allows one to introduce the special *canonical metric* on $T(TM)$ by

$$\langle \xi, \eta \rangle_v = \langle d_v\pi\xi, d_v\pi\eta \rangle_{\pi(v)} + \langle K\xi, K\eta \rangle_{\pi(v)}$$

in which the spaces $H(v)$ and $V(v)$ are orthogonal.

10.1.2. Geodesics. These are curves along which the tangential vector field is parallel. For any $x \in M$ there is a local coordinate chart in which for any $v \in T_xM$ the geodesic $\gamma_v(t) = (\gamma_1(t), \dots, \gamma_p(t))$ with $\gamma_v(0) = x$ and $\dot\gamma_v(0) = v$ satisfies a certain system of second-order differential equations on functions $\gamma_i(t)$. We shall always assume that the parameter t along the geodesic $\gamma(t)$ is the arc length; this ensures that $\|\dot\gamma(t)\| = 1$ and $(\nabla\dot\gamma)(t) = 0$.

On complete (in particular, compact) manifolds, geodesics are infinitely extendible and for any $x, y \in M$ there is a (perhaps nonunique) geodesic joining x and y. Among such geodesics there is always one whose length is the distance $\rho(x, y)$ between the points x and y. Furthermore, for any $v \in TM$ there is a unique geodesic $\gamma_v(t)$ such that $\gamma_v(0) = \pi(v)$ and $\dot\gamma_v(0) = v$.

[1] Recall that $\nabla Z(t)|_{t=0} = \frac{d}{dt}\tilde Z(t)|_{t=0}$ where the vector $\tilde Z(t)$ is the parallel translation of the vector $Z(t)$ along the curve $\gamma(t) = \pi(Z(t))$.

10.1.3. The universal Riemannian covering. This is a complete manifold H such that $M = H/\Gamma$ where Γ is the fundamental group $\pi(M)$ (and is a proper discrete subgroup of the group of isometries of H). The covering map $H \to M$ allows one to lift objects from M to H and to project objects from H to M.

10.1.4. Curvature. For $x \in M$ and a plane $P \subset T_xM$ given by vectors $v, w \in T_xM$, the sectional curvature $K_x(P)$ of M at x in the direction of P is given by

$$K_x(P) = \frac{\langle R(v,w)w, v\rangle}{\|v\|^2\|w\|^2 - \langle v, w\rangle^2},$$

where $R(v, w)$ is the curvature tensor associated with v and w.[2] It can be shown that $K_x(P)$ does not depend on the choice of v and w specifying the plane P. The Riemannian metric is said to be of negative (respectively, nonpositive) curvature if the sectional curvature at each point and in the direction of every plane P is negative (respectively, nonpositive). If M is a surface, then the curvature at a point $x \in M$ is a function of x only.

10.1.5. Fermi coordinates. Given a geodesic $\gamma(t)$, chose an orthonormal basis $\{e_i\}$, $i = 1, \dots, p$, in the space $T_{\gamma(0)}M$ such that $e_n = \dot\gamma(0)$. Let $\{e_i(t)\}$ be the vector field along $\gamma(t)$ obtained by the parallel translation of e_i along $\gamma(t)$ (so that $e_i(0) = e_i$). The vectors $\{e_i(t)\}$ form an orthonormal basis in $T_{\gamma(t)}M$ and the corresponding coordinates in $T_{\gamma(t)}M$ are the Fermi coordinates.

10.1.6. Jacobi fields. A Jacobi field is a vector field $Y(t)$ along a geodesic $\gamma(t)$ that satisfies the Jacobi equation

$$Y''(t) + R(X, Y)X = 0, \tag{10.1}$$

where $X(t) = \dot\gamma(t)$, $R(X, Y)$ is the curvature tensor associated with the vector fields X and Y, and $Y''(t)$ is the second covariant derivative of the vector field $Y(t)$. Using the Fermi coordinates $\{e_i(t)\}$, $i = 1, \dots, p$, one can rewrite equation (10.1) as a second-order linear differential equation along $\gamma(t)$,

$$\frac{d^2}{dt^2}Y(t) + K(t)Y(t) = 0, \tag{10.2}$$

where $Y(t) = (Y_1(t), \dots, Y_p(t))$ is a vector and $K(t) = (K_{ij}(t))$ is a matrix whose entry $K_{ij}(t)$ is the sectional curvature at $\gamma(t)$ in the direction of the plane given by vectors $e_i(t)$ and $e_j(t)$. It follows that every Jacobi field $Y(t)$ is uniquely determined by the values $Y(0)$ and $\frac{d}{dt}Y(0)$ and is well-defined for all t.

[2] More precisely, $R(v, w) = R(V, W)$ where V and W are any vector fields on M such that $V(x) = v$ and $W(x) = w$. One can show that the value $R(v, w)$ does not depend on the choice of the vector fields V and W.

The Jacobi fields arise in the formula for the second variation of the geodesic, which describes some infinitesimal properties of geodesics. More precisely, let $\gamma(t)$ be a geodesic and let $Z(s)$, $-\varepsilon \leq s \leq \varepsilon$, be a curve in TM such that $Z(0) = \dot{\gamma}(0)$. Consider the variation $r(t, s)$ of $\gamma(t)$ of the form

$$r(t, s) = \exp(tZ(s)), \quad t \geq 0, \ -\varepsilon \leq s \leq \varepsilon$$

(in other words, for a fixed $s \in [-\varepsilon, \varepsilon]$ the curve $r(t, s)$ is the geodesic through $\pi(Z(s))$ in the direction of $Z(s)$). One can show that the vector field

$$Y(t) = \frac{\partial}{\partial s} r(t, s)|_{s=0}$$

along the geodesic $\gamma(t)$ is the Jacobi field along $\gamma(t)$.

10.2. Definition and local properties of geodesic flows

The *geodesic flow* g_t is a flow on the tangent bundle TM that acts by the formula

$$g_t(v) = \dot{\gamma}_v(t).$$

See Figure 10.1. The geodesic flow generates a vector field V on TM given by

$$V(v) = \left.\frac{d(g_t(v))}{dt}\right|_{t=0}.$$

Since M is compact, the flow g_t is well-defined for all $t \in \mathbb{R}$ and is a smooth flow of class C^{r-1} where r is the class of smoothness of the Riemannian metric.

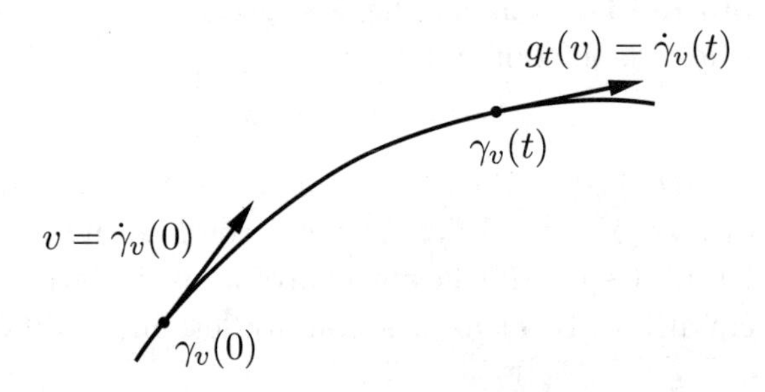

Figure 10.1. The geodesic flow

We present a different definition of the geodesic flow in terms of the special symplectic structure generated by the Riemannian metric. Consider the cotangent bundle T^*M and define the canonical 1-form ω on T^*M by the formula $\omega(x, q) = q(d\pi^*(x, q))$, where $x \in M$, $q \in T^*M$ is a 1-form and $\pi^*: T^*M \to M$ is the natural projection (i.e., $\pi^*(x, q) = x$; note that $d\pi^*(x, q) \in T_xM$). The canonical 2-form $\Omega_* = d\omega$ is nondegenerate and induces a symplectic structure on T^*M. The Riemannian metric allows one to identify the tangent and cotangent bundles via the map $\mathcal{L}\colon TM \to T^*M$ defined by

$$\mathcal{L}(x,v) = (x,q), \quad x \in M, \quad v \in T_xM,$$

where the 1-form q satisfies $q(w) = \langle v, w\rangle$ for any $w \in T_xM$. This identification produces the canonical 2-form Ω on TM by the formula

$$\Omega(Y,Z) = \Omega_*(\mathcal{L}(Y), \mathcal{L}(Z)), \quad Y, Z \in TM.$$

Consider the function $K\colon TM \to TM$ given by

$$K(x,v) = \frac{1}{2}\langle v, v\rangle = \frac{1}{2}\|v\|^2. \tag{10.3}$$

The vector field V, generating the geodesic flow g_t, is now defined as the vector field corresponding to the 1-form dK relative to the canonical 2-form Ω; that is, $\Omega(V,Z) = dK(Z)$ for any vector field $Z \in TM$.

Geodesic flows serve as mathematical models of classical mechanics, that is, as the phase flows of certain Hamiltonian systems. Here M is viewed as the configuration space and TM as the phase space of the mechanical system; points $x \in M$ are treated as "generalized" coordinates, vectors $v \in TM$ as "generalized" velocities, and 1-forms $q \in T^*M$ as impulses. According to Hamilton's principle, the trajectory of the mechanical system in the configuration space passing through the points x and y is an extremum of the energy functional $E = \frac{1}{2}\int_{t_0}^{t_1} \langle v(t), v(t)\rangle\, dt$.[3] One can show that these extrema are geodesics connecting x and y, that is, the Euler differential equation for the corresponding variational problem with fixed endpoints is the differential equation for the geodesics.

To describe the trajectories of the mechanical system in the phase space, we observe that the geodesic flow is a Hamiltonian flow with the Hamiltonian function $H(x,v) = K(x,v)$ given by (10.3).[4] Its vector field in local coordinates (x,v) is of the form $\left(\frac{\partial K}{\partial v}, -\frac{\partial K}{\partial x}\right)$.

Since the total energy of the system is a first integral, the hypersurfaces $K(x,v) = \text{const}$ (that is, $\|v\| = \text{const}$) are invariant under the geodesic flow; therefore, we always consider the geodesic flow as acting on the unit tangent bundle $SM = \{v \in TM : \|v\| = 1\}$.

---

[3] We assume that there is no potential energy.

[4] This is a Hamiltonian system of a special type. In general, a Hamiltonian system is defined on a symplectic manifold N (endowed with a symplectic 2-form Ω) by a Hamiltonian function $H\colon N \to \mathbb{R}$ so that the vector field of the Hamiltonian flow is a field corresponding to the 1-form dH relative to the 2-form Ω. In the local coordinate system (x,p) in which $\Omega = \sum dp_i \wedge dx^i$, this vector field has the form $(\partial H/\partial p, -\partial H/\partial x)$. For a conservative mechanical system with a configuration manifold M we have $N = T^*M$ and $H = K - U$ where K is the kinetic energy and U the potential energy of the system. By the Maupertuis–Lagrange–Jacobi principle, for a fixed value E of the Hamiltonian H the motion can be reduced to a geodesic flow on M by introducing a new Riemannian metric $\langle\cdot,\cdot\rangle'_x = (E - U(x))\langle\cdot,\cdot\rangle_x$ and making a certain time change. In our case everything simplifies since $U = 0$ and T^*M is naturally identified with TM.

The representation of a geodesic flow g_t as a Hamiltonian flow allows one to obtain an invariant measure μ for g_t; namely, by Liouville's theorem, this measure is given by $d\mu = d\sigma dm$ where $d\sigma$ is a surface element on the $(p-1)$-dimensional unit sphere and m the Riemannian volume on M.

In order to study the stability of trajectories of the geodesic flow, one should examine the system of variational equations (3.2). To this end, let us fix $v \in SM$ and $\xi \in T_v SM$ and consider the Jacobi equation (10.2) along the geodesic $\gamma_v(t)$. Let $Y_\xi(t)$ be the unique solution of this equation satisfying the initial conditions:

$$Y_\xi(0) = d_v\pi\xi, \quad \frac{d}{dt}Y_\xi(0) = K\xi. \tag{10.4}$$

One can show that the map $\xi \mapsto Y_\xi(t)$ is an isomorphism for which

$$Y_\xi(t) = d_{g_t v}\pi d_v g_t \xi, \quad \frac{d}{dt}Y_\xi(t) = K d_v g_t \xi. \tag{10.5}$$

This map establishes the identification between solutions of the system of variational equations (3.2) and solutions of the Jacobi equation (10.7).

10.3. Hyperbolic properties and Lyapunov exponents

From now on we assume that M is a compact surface endowed with a Riemannian metric of class C^3 and of *nonpositive curvature*, i.e., for any $x \in M$ the Riemannian curvature at x satisfies

$$K(x) \le 0. \tag{10.6}$$

Take $v \in TM$ and consider the geodesic $\gamma_v(t)$. The Jacobi equation along it is a scalar second-order differential equation of the form

$$\frac{d^2}{dt^2}Y(t) + K(t)Y(t) = 0, \tag{10.7}$$

where $K(t) = K(\gamma_v(t))$. Due to the requirement (10.6) and compactness of the manifold M, we have that $-k^2 \le K(t) \le 0$ for some $k > 0$ and all $t \in \mathbb{R}$. It follows that the boundary value problem for equation (10.7) has a unique solution, i.e., for any s_i and y_i, $i = 1, 2$, there exists a unique solution $Y(t)$ of (10.7) satisfying $Y(s_i) = y_i$ (see [**12**]).

Proposition 10.1. *Given $s \in \mathbb{R}$, let $Y_s(t)$ be the unique solution of equation (10.7) satisfying the boundary conditions: $Y_s(0) = 1$ and $Y_s(s) = 0$. Then there exists the limit*

$$\lim_{s\to\infty} \frac{d}{dt}Y_s(t)\Big|_{t=0} = Y^+. \tag{10.8}$$

Proof. Let $A(t)$ be the solution of equation (10.7) satisfying the initial conditions $A(0) = 0$ and $\frac{d}{dt}A(0) = 1$. For $t > 0$ consider the function

$$Z_s(t) = A(t) \int_t^s A^{-2}(u)\, du. \tag{10.9}$$

Exercise 10.2. Show that:

(1) $Z_s(t)$ is the solution of equation (10.7) satisfying the initial conditions $Z_s(s) = 0$ and $\frac{d}{dt}Z_s(s) = -A^{-1}(s)$;

(2) $Z_s(0) = 1$. Hint: Use the fact that the Wronskian of any two solutions of equation (10.7) is a constant and apply this to the Wronskian of the solutions $A(t)$ and $Z_s(t)$.

This implies that $Z_s(t) = Y_s(t)$. It is easy to verify that for any numbers $0 < q < s$ and all $t > 0$ we have

$$Y_s(t) - Y_q(t) = A(t) \int_q^s A^{-2}(u)\, du$$

and

$$\frac{d}{dt}Y_s(0) - \frac{d}{dt}Y_q(0) = \int_q^s A^{-2}(u)\, du.$$

Furthermore, one can show that the function $\int_q^s A^{-2}(u)\, du$ is monotonically increasing in t and hence, the limit

$$\lim_{s\to+\infty} \left(\frac{d}{dt}Y_s(0) - \frac{d}{dt}Y_q(0) \right)$$

exists. It follows that the limit $\lim_{s\to+\infty} \frac{d}{dt}Y_s(0)$ exists as well and the desired result follows. □

We define the *positive limit solution* $Y^+(t)$ of (10.7) as the solution that satisfies the initial conditions:

$$Y^+(0) = 1 \quad \text{and} \quad \frac{d}{dt}Y^+(t)\Big|_{t=0} = Y^+$$

(see (10.8)). Since solutions of equation (10.7) depend continuously on the initial conditions, by (10.9), we obtain that for every $t > 0$

$$Y^+(t) = \lim_{s\to+\infty} Y_s(t) = A(t) \int_t^\infty A^{-2}(u)\, du. \tag{10.10}$$

It follows that this solution is nondegenerate (i.e., $Y^+(t) \neq 0$ for every $t \in \mathbb{R}$).

Similarly, letting $s \to -\infty$, we can define the *negative limit solution* $Y^-(t)$ of equation (10.7).

For every $v \in SM$ set

$$\begin{aligned} E^+(v) &= \{\xi \in T_vSM : \langle \xi, V(v)\rangle = 0 \text{ and } Y_\xi(t) = Y^+(t)\|d_v\pi\xi\|\}, \\ E^-(v) &= \{\xi \in T_vSM : \langle \xi, V(v)\rangle = 0 \text{ and } Y_\xi(t) = Y^-(t)\|d_v\pi\xi\|\}, \end{aligned} \tag{10.11}$$

where V is the vector field generated by the geodesic flow and Y_ξ is the solution of equation (10.7) satisfying the initial conditions (10.4).

Using the relations (10.5) and (10.10) (and a similar relation for $Y^-(t)$), one can prove the following properties of the subspaces $E^-(v)$ and $E^+(v)$.

Proposition 10.3. *The following properties hold:*

(1) $E^-(v)$ *and* $E^+(v)$ *are one-dimensional linear subspaces of* T_vSM*;*

(2) $d_v\pi E^-(v) = d_v\pi E^+(v) = \{w \in T_{\pi v}M : w \text{ is orthogonal to } v\}$*;*

(3) *the subspaces* $E^-(v)$ *and* $E^+(v)$ *are invariant under the differential* d_vg_t*; i.e.,* $d_vg_tE^-(v) = E^-(g_t(v))$ *and* $d_vg_tE^+(v) = E^+(g_t(v))$*;*

(4) *if* $\tau\colon SM \to SM$ *is the involution defined by* $\tau v = -v$*, then*

$$E^+(-v) = d_v\tau E^-(v) \quad \text{and} \quad E^-(-v) = d_v\tau E^+(v);$$

(5) *if* $\xi \in E^+(v)$ *or* $\xi \in E^-(v)$*, then* $\|K\xi\| \le k\|d_v\pi\xi\|$*;*

(6) *if* $\xi \in E^+(v)$ *or* $\xi \in E^-(v)$*, then* $Y_\xi(t) \neq 0$ *for every* $t \in \mathbb{R}$*;*

(7) $\xi \in E^+(v)$ *(respectively,* $\xi \in E^-(v)$*) if and only if*

$$\langle \xi, V(v)\rangle = 0 \quad \text{and} \quad \|d_{g_t(v)}\pi d_vg_t\xi\| \le c$$

for every $t > 0$ *(respectively,* $t < 0$*) and some* $c > 0$*;*

(8) *if* $\xi \in E^+(v)$ *(respectively,* $\xi \in E^-(v)$*), then the function* $t \mapsto |Y_\xi(t)|$ *is nonincreasing (respectively, nondecreasing).*

In view of properties (5) and (7), we have $\xi \in E^+(v)$ (respectively, $\xi \in E^-(v)$) if and only if $\langle \xi, V(v)\rangle = 0$ and $\|d_vg_t\xi\| \le c$ for $t > 0$ (respectively, $t < 0$), for some constant $c > 0$.

The subspaces $E^-(v)$ and $E^+(v)$ are natural candidates for stable and unstable subspaces for the geodesic flows. However, in general, these subspaces may not span the whole second tangent space T_vSM; i.e., the intersection $E(v) = E^-(v) \cap E^+(v)$ may be a nontrivial subspace of T_vSM. If this is the case, then since the subspaces $E^-(v)$ and $E^+(v)$ are invariant under dg_t, the intersection $E(g_t(v)) = E^-(g_t(v)) \cap E^+(g_t(v))$ coincides with $dg_tE(v)$ and is a nontrivial subspace of $T_{g_t(v)}SM$. For every $\xi \in E(v)$ the vector field $Y_\xi(t)$ is parallel along the geodesic $\gamma_v(t)$.[5] Hence, for every $\xi \in E(v)$ the Lyapunov exponent $\chi(v,\xi) = 0$.

[5]Indeed, in view of statement (8) of Proposition 10.3, the function $|Y_\xi(t)|$ is both nonincreasing and nondecreasing.

Furthermore, one can show that if the subspaces $E^-(v)$ and $E^+(v)$ do span the space T_vSM (i.e., $T_vSM = E^-(v) \oplus E^+(v)$), then the geodesic flow is, indeed, Anosov (see [**32**]). This is the case when the curvature is strictly negative.[6] However, for manifolds of nonpositive curvature one can only expect that the geodesic flow is nonuniformly hyperbolic. To see this, consider the set

$$\Delta = \left\{ v \in SM : \limsup_{t\to\infty} \frac{1}{t} \int_0^t K(\gamma_v(s))ds < 0 \right\}. \tag{10.12}$$

It is easy to see that this set is measurable and invariant under the flow g_t. The following result from [**69**] (see also [**71**]) shows that the Lyapunov exponents are nonzero on the set Δ.

Theorem 10.4. *For every $v \in \Delta$ we have $\chi(v,\xi) < 0$ if $\xi \in E^+(v)$ and $\chi(v,\xi) > 0$ if $\xi \in E^-(v)$.*

Proof. Let $\psi\colon \mathbb{R}^+ \to \mathbb{R}$ be a continuous function. We need the following lemma.

Lemma 10.5. *Assume that $c = \sup_{t\geq 0} |\psi(t)| < \infty$. Then:*

(1) *if $\psi(t) \leq 0$ for all $t \geq 0$ and $\tilde{\psi} > 0$, then $\overline{\psi} < 0$;*
(2) *if $\psi(t) \geq 0$ for all $t \geq 0$ and $\tilde{\psi} > 0$, then $\underline{\psi} > 0$,*

where $\overline{\psi}$ and $\underline{\psi}$ are defined by (2.13), *and*

$$\tilde{\psi} = \liminf_{t\to\infty} \frac{1}{t} \int_0^t \psi(s)^2\, ds.$$

Proof of the lemma. Assume that $\psi(t) \leq 0$. Then $\overline{\psi} \leq 0$. On the other hand, if $c > 0$, then

$$-\frac{\overline{\psi}}{c} = \underline{\left|\frac{\psi}{c}\right|} \geq \widetilde{\left(\frac{\psi}{c}\right)} = \frac{\tilde{\psi}}{c^2} > 0.$$

This implies that $\overline{\psi} < 0$ and completes the proof of the first statement. The proof of the second statement is similar. □

We proceed with the proof of the theorem. Fix $v \in \Delta$, $\xi \in E^+(v)$, and consider the function $\varphi(t) = \frac{1}{2}|Y_\xi(t)|^2$. Using (10.7), we obtain

$$\frac{d^2}{dt^2}\varphi(t) = -K(t)\varphi(t) + \frac{d}{dt}|Y_\xi(t)|^2.$$

[6]Riemannian manifolds whose geodesic flows are Anosov are said to be of *Anosov type*; see [**32**]. They are closely related to manifolds of hyperbolic type that admit a metric of negative curvature. The latter have been a subject of intensive study in differential geometry: Morse [**64**] already understood that on surfaces of negative Euler characteristic (which admit a metric of constant negative curvature) the geodesic flow in any Riemannian metric inherits to some extent the properties of the geodesic flow in a metric of negative curvature.

It follows from Proposition 10.3 (see statements (6) and (8)) that $\varphi(t) \neq 0$ and $\frac{d}{dt}\varphi(t) \leq 0$ for all $t \geq 0$. Set

$$z(t) = (\varphi(t))^{-1} \frac{d}{dt}\varphi(t).$$

It is easy to check that the function $z(t)$ satisfies the Riccati equation

$$\frac{d}{dt}z(t) + z(t)^2 - (\varphi(t))^{-1}\frac{d}{dt}|Y_\xi(t)|^2 + K(t) = 0. \tag{10.13}$$

By Proposition 10.3,

$$\begin{aligned}\left|\frac{d}{dt}\varphi(t)\right| &= \left|\frac{1}{2}\frac{d}{dt}|Y_\xi(t)|^2\right| = |Y_\xi(t)| \cdot \left|\frac{d}{dt}|Y_\xi(t)|\right| \\ &= \|d_{g_t v}\pi d_v g_t \xi\| \cdot \|K d_v g_t \xi\| \leq a\|d_{g_t v}\pi d_v g_t \xi\|^2 = 2a\varphi(t).\end{aligned}$$

It follows that $\sup_{t\geq 0}|z(t)| \leq 2a$. Integrating the Riccati equation (10.13) on the interval $[0, t]$, we obtain that

$$z(t) - z(0) + \int_0^t z(s)^2\, ds = \int_0^t (\varphi(s))^{-1}\frac{d}{ds}|Y_\xi(s)|^2\, ds - \int_0^t K(s)\, ds.$$

It follows that for $v \in \Delta$ (see (10.12)) we have

$$\begin{aligned}\liminf_{t\to\infty} \frac{1}{t}\int_0^t z(s)^2\, ds \geq \liminf_{t\to\infty} \frac{1}{t}\int_0^t (\varphi(s))^{-1}\frac{d}{ds}|Y_\xi(s)|^2\, ds \\ - \limsup_{t\to\infty}\frac{1}{t}\int_0^t K(s)\, ds > 0.\end{aligned}$$

Therefore, in view of Lemma 10.5 we conclude that

$$\limsup_{t\to\infty}\frac{1}{t}\int_0^t z(s)\, ds < 0.$$

On the other hand, using Proposition 10.3, we find that

$$\begin{aligned}\chi(v, \xi) &= \limsup_{t\to\infty}\frac{1}{t}\log\|d_v g_t \xi\| = \limsup_{t\to\infty}\frac{1}{t}\log\|d_{g_t v}\pi d_v g_t \xi\| \\ &= \limsup_{t\to\infty}\frac{1}{t}\log|Y_\xi(t)| = \frac{1}{2}\limsup_{t\to\infty}\frac{1}{t}\int_0^t z(s)\, ds.\end{aligned}$$

This completes the proof of the first statement of the theorem. The second statement can be proved in a similar way. □

It follows from Theorem 10.4 that if the set Δ has positive Liouville measure, then the geodesic flow $g_t|\Delta$ is nonuniformly hyperbolic. It is, therefore, crucial to find conditions which guarantee that Δ has positive Liouville measure.

Theorem 10.6. *Let M be a smooth compact surface of nonpositive curvature $K(x)$ and genus greater than 1. Then $\mu(\Delta) > 0$.*

Proof. By the Gauss–Bonnet formula the Euler characteristic of M is

$$\frac{1}{2\pi}\int_M K(x)\,dm(x).$$

It follows from the condition of the theorem that

$$\int_M K(x)\,dm(x) < 0. \tag{10.14}$$

By the Birkhoff Ergodic Theorem, we obtain that for μ-almost every $v \in SM$ there exists the limit

$$\lim_{t\to\infty}\frac{1}{t}\int_0^t K(\pi(g_s v))\,ds = \Phi(v)$$

and that

$$\int_{SM}\Phi(v)\,d\mu(v) = \int_M K(x)\,dm(x).$$

The desired result follows from (10.14). □

10.4. Ergodic properties

As we saw in Section 10.3 the geodesic flow g_t on a compact surface M of nonpositive curvature and of genus greater than 1 is nonuniformly hyperbolic on the set Δ of positive Liouville measure which is defined by (10.12). Since the Liouville measure is invariant under the geodesic flow, the results of Section 9.1 apply and show that the ergodic components are of positive Liouville measure (see Theorem 9.1). In this section we show that, indeed, the geodesic flow on Δ is ergodic.

We wish to show that every ergodic component of $g_t|\Delta$ is open (mod 0). To achieve this, we construct one-dimensional foliations of SM, W^-, and W^+, such that $W^s(x) = W^-(x)$ and $W^u(x) = W^+(x)$ for almost every $x \in \Delta$ (W^- and W^+ are known as the stable and unstable horocycle foliations; see below). We then apply Theorem 9.18 to derive that the flow $g_t|\Delta$ is ergodic. In order to proceed in this direction, we need some more information on surfaces of nonpositive curvature (see [**33, 72**]).

We denote by H the universal Riemannian cover of M, i.e., a simply connected two-dimensional complete Riemannian manifold for which $M = H/\Gamma$ where Γ is a discrete subgroup of the group of isometries of H, isomorphic to $\pi_1(M)$. According to the Hadamard–Cartan theorem, any two points $x, y \in H$ are joined by a single geodesic which we denote by γ_{xy}. For any $x \in H$, the exponential map $\exp_x\colon \mathbb{R}^2 \to H$ is a diffeomorphism. Hence, the map

$$\varphi_x(y) = \exp_x\left(\frac{y}{1-\|y\|}\right) \tag{10.15}$$

is a homeomorphism of the open unit disk D onto H.

Two geodesics $\gamma_1(t)$ and $\gamma_2(t)$ in H are said to be *asymptotic* if

$$\sup_{t>0} \rho(\gamma_1(t), \gamma_2(t)) < \infty,$$

where ρ is the distance in H induced by the Riemannian metric. Given a point $x \in H$, there is a unique geodesic starting at x which is asymptotic to a given geodesic. The asymptoticity is an equivalence relation, and the equivalence class $\gamma(\infty)$ corresponding to a geodesic γ is called a *point at infinity.* The set of these classes is denoted by $H(\infty)$ and is called the *ideal boundary* of H. Using (10.15), one can extend the topology of the space H to $\overline{H} = H \cup H(\infty)$ so that $\overline{H}$ becomes a compact space.

The map φ_x can be extended to a homeomorphism (still denoted by φ_x) of the closed disk $\overline{D} = D \cup S^1$ onto $\overline{H}$ by the equality

$$\varphi_x(y) = \gamma_y(+\infty), \qquad y \in S^1.$$

In particular, φ_x maps S^1 homeomorphically onto $H(\infty)$.

For any two distinct points x and y on the ideal boundary there is a geodesic joining them. This geodesic is uniquely defined if the Riemannian metric is of strictly negative curvature (i.e., if inequality (10.6) is strict). Otherwise, there may exist a pair of distinct points $x, y \in H(\infty)$ which can be joined by more than one geodesic. More precisely, there exists a geodesically isometric embedding into H of an infinite strip of zero curvature which consists of geodesics joining x and y. Moreover, any two geodesics on the universal cover which are asymptotic both for $t > 0$ and for $t < 0$ (i.e., they join two distinct points on the ideal boundary) bound a flat strip. The latter means that there is a geodesically isometric embedding of a flat strip in $\mathbb{R}^2$ into the universal cover. This statement is know as the *flat strip theorem.*

The fundamental group $\pi_1(M)$ of the manifold M acts on the universal cover H by isometries. This action can be extended to the ideal boundary $H(\infty)$. Namely, if $p = \gamma_v(+\infty) \in H(\infty)$ and $\zeta \in \pi_1(M)$, then $\zeta(p)$ is the equivalence class of geodesics which are asymptotic to the geodesic $\zeta(\gamma_v(t))$.

We now describe the invariant foliations for the geodesic flow. We consider the distributions E^- and E^+ introduced in Section 10.2 (see (10.11)).

Proposition 10.7. *The distributions E^- and E^+ are integrable. Their integral manifolds form foliations of SM which are invariant under the flow g_t.*

Sketch of the proof. Consider the circle $S^1(\gamma_v(t), t)$ of radius t centered at $\gamma_v(t)$, for $t > 0$. The intersection

$$\Gamma(t) = D(x, R) \cap S^1(\gamma_v(t), t)$$

(here $D(x, R)$ is the disk centered at x of some radius R) is a smooth convex curve passing through x. One can show that the curvature of $\Gamma(t)$ is bounded uniformly over t. Moreover, any geodesic which is orthogonal to this curve passes through the center of the circle $\gamma_v(t)$. The family of curves $\Gamma(t)$ is monotone (by inclusion) and is compact. Therefore, the sequence of curves $\Gamma(t)$ converges as $t \to \infty$ to a convex smooth curve of bounded curvature. Moreover, any geodesic which is orthogonal to this curve is asymptotic to $\gamma_v(t)$. The framing of the limit curve is a local manifold at v which is a piece of the integral manifold for the distribution E^-. The integral manifolds corresponding to the distribution E^+ can be obtained in a similar fashion. □

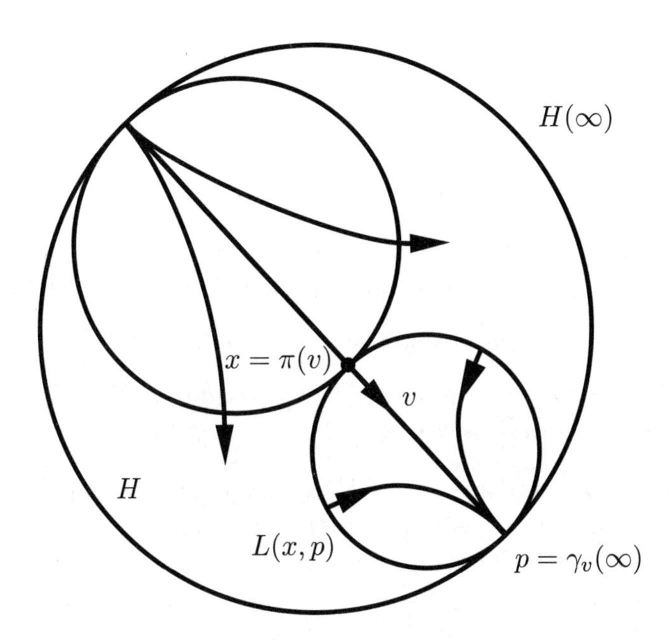

Figure 10.2. Horocycles

We denote by W^- and W^+ the foliations of SM corresponding to the invariant distributions E^- and E^+. These foliations can be lifted from SM to SH. We denote these lifts by $\tilde{W}^-$ and $\tilde{W}^+$, respectively.

Given $x \in H$ and $p \in H(\infty)$, set

$$L(x, p) = \pi(\tilde{W}^-(v))$$

where $x = \pi(v)$ and $p = \gamma_v(\infty)$. The set $L(x, p)$ is called the *horocycle* centered at p and passing through x. See Figure 10.2.

We summarize the properties of the foliations and horocycles in the following statement.

Proposition 10.8. *The following properties hold:*

(1) *for any $x \in H$ and $p \in H(\infty)$ there exists a unique horocycle $L(x,p)$ centered at p and passing through x; it is a limit in the C^1 topology of circles $S^1(\gamma(t),t)$ as $t \to +\infty$ where γ is the unique geodesic joining x and p;*

(2) *the leaf $W^-(v)$ is the framing of the horocycle $L(x,p)$ ($x = \pi(v)$ and $p = \gamma_v(+\infty)$) by orthonormal vectors which have the same direction as the vector v (i.e., they are "inside" the limit sphere). The leaf $W^+(v)$ is the framing of the horocycle $L(x,p)$ ($x = \pi(v)$ and $p = \gamma_v(-\infty) = \gamma_{-v}(+\infty)$) by orthonormal vectors which have the same direction as the vector v (i.e., they are "outside" the limit sphere);*

(3) *for every $\zeta \in \pi_1(M)$ we have*

$$\zeta(L(x,p)) = L(\zeta(x), \zeta(p)),$$

$$d_v\zeta\tilde{W}^-(v) = \tilde{W}^-(d_v\zeta v), \quad d_v\zeta\tilde{W}^+(v) = \tilde{W}^+(d_v\zeta v);$$

(4) *for every $v, w \in SH$, for which $\gamma_v(+\infty) = \gamma_w(+\infty) = p$, the geodesic $\gamma_w(t)$ intersects the horocycle $L(\pi(v),p)$ at some point.*

We now state a remarkable result by Eberlein [**31**].

Proposition 10.9. *The geodesic flow g_t on any surface M of nonpositive curvature and of genus greater that 1 is topologically transitive.*

Proof. We call two points $x, y \in H(\infty)$ *dual* if for any open sets U and V, containing x and y, respectively, there exists an element $\xi \in \pi_1(M)$ such that $\xi(\bar{H} \setminus U) \subset V$. Clearly, $\xi(\bar{H} \setminus U) \subset V$ if and only if $\xi^{-1}(\bar{H} \setminus V) \subset U$. One can show that on any surface M of nonpositive curvature and of genus greater that 1 any two points $x, y \in H(\infty)$ are dual (see [**31**]).

Lemma 10.10. *If the points $x, y \in H(\infty)$ are dual, then there exists a sequence $\xi_n \in \pi_1(M)$ such that $\xi_n^{-1}(p) \to x$ and $\xi_n(p) \to y$ as $n \to \infty$ for any $p \in H$.*

Proof of the lemma. Let $\{U_n\}$ and $\{V_n\}$ be local bases for the cone topology at x and y, respectively. For each $n > 0$ there exists $\xi_n \in \pi_1(M)$ such that $\xi_n(\bar{H} \setminus U_n) \subset V_n$ and $\xi_n^{-1}(\bar{H} \setminus V_n) \subset U_n$. For any $p \in H$ and any sufficiently large n we have $p \in (\bar{H} \setminus U_n) \cap (\bar{H} \setminus V_n)$ and hence, $\xi_n(p) \in V_n$ and $\xi_n^{-1}(p) \in U_n$. The lemma follows. □

Given $v, w \in SM$, let $\tilde{v}$ and $\tilde{w}$ be their lifts to SH. Denote by $\zeta : H \to M$ the covering map. Then $d\zeta$ maps SH onto SM.

Since the points $x = \gamma_{\tilde{v}}(+\infty)$ and $y = \gamma_{\tilde{w}}(-\infty)$ are dual, by the lemma, there exists a sequence $\xi_n \in \pi_1(M)$ such that $\xi_n^{-1}(p) \to x$ and $\xi_n(p) \to y$ as

$n \to \infty$ for any $p \in H$. Let $p = \pi\tilde{v}$ and $q = \pi\tilde{w}$ where $\pi\colon SM \to M$ is the projection, and set $t_n = d(\xi_n(p), q)$ and $\tilde{v}_n = \dot{\gamma}_{\xi_n(p),q}(0)$ (where $\gamma_{\xi_n(p),q}(t)$ is the geodesic connecting the points $\xi_n(p)$ and q). Then $g_{t_n}(\tilde{v}_n) = -\dot{\gamma}_{q,\xi_n(p)}(0)$. Since $\xi_n(p) \to y$, it follows that $g_{t_n}(\tilde{v}_n) \to \tilde{w}$. Observing that $\xi_n^{-1}(q) \to x$, we obtain that $d\xi_n^{-1}\tilde{v}_n = \dot{\gamma}_{p,\xi_n^{-1}(q)}(0) \to \tilde{v}$. It follows that

$$d\zeta\tilde{v}_n = d\zeta d\xi_n^{-1}\tilde{v}_n \to d\zeta\tilde{v} = v$$

and that

$$g_{t_n}(d\zeta\tilde{v}_n) = d\zeta g_{t_n}(\tilde{v}_n) \to d\zeta\tilde{w} = w.$$

This means that for any open neighborhoods U and V of v and w, respectively, the set $g_t(U)$ meets V for arbitrarily large positive values of t, implying topological transitivity of the geodesic flow g_t. □

We now state our main result.

Theorem 10.11. *Let M be a compact surface of nonpositive curvature and of genus greater than 1. Then the following properties hold:*

(1) *the set Δ defined by* (10.12) *has positive Liouville measure, is open* (mod 0), *and is everywhere dense;*

(2) *the geodesic flow $g_t|\Delta$ is ergodic.*

Proof. By Theorem 10.6, $\mu(\Delta) > 0$ where μ is the Liouville measure. We shall show that the set Δ is open (mod 0). Note that given $v \in \Delta$, there is a number t such that the curvature $K(x)$ of M at the point $x = \gamma_v(t)$ is strictly negative. Therefore, there is a disk $D(x,r)$ in M centered at x of radius r such that $K(y) < 0$ for every $y \in D(x,r)$. It follows that there is a neighborhood U of v in SM such that $\gamma_w(t) \in D(x,r)$ for every $w \in U$. We denote $\tilde{D}(x,r) = \{z \in SM : \pi(z) \in D(x,r)\}$. Note that the set $\tilde{D}(x,r)$ is open.

By Birkhoff's Ergodic Theorem, for almost every $w \in \tilde{D}(x,r)$ (with respect to the Liouville measure) the limit

$$\lim_{t\to\infty} \frac{\tilde{T}(w,t)}{t} \tag{10.16}$$

exists, where

$$\tilde{T}(w,t) = \int_0^t \chi_{\tilde{D}(x,r)}(\dot{\gamma}_w(\tau))\, d\tau$$

and $\chi_{\tilde{D}(x,r)}$ is the characteristic function of the set $\tilde{D}(x,r)$. In fact, using the ergodic decomposition of the measure μ, it is easy to show that the limit (10.16) is positive for almost every $w \in \tilde{D}(x,r)$. Since the set $\tilde{D}(x,r)$

is open, we obtain that the limit (10.16) is positive for almost every $w \in U$. This implies that the limit

$$\lim_{t \to \infty} \frac{T(w,t)}{t}$$

exists and is positive for almost every $w \in U$ where

$$T(w,t) = \int_0^t \chi_{D(x,r)}(\gamma_w(\tau))\, d\tau$$

and $\chi_{D(x,r)}$ is the characteristic function of the set $D(x,r)$. In view of (10.12), this implies that almost every $w \in U$ lies in Δ.

Note that since the geodesic flow is topologically transitive (see Proposition 10.9) and the set Δ is open (mod 0), this set is everywhere dense. This proves the first statement.

For μ-almost every $v \in \Delta$, consider one-dimensional local stable and unstable manifolds $V^s(v)$ and $V^u(v)$. We denote by $\tilde{V}^s(\tilde{v})$ and $\tilde{V}^u(\tilde{v})$ their lifts to SH (with $\tilde{v}$ being a lift of v). Given $\tilde{w} \in \tilde{V}^s(\tilde{v})$, we have

$$\rho(\pi(g_t(\tilde{v})), \pi(g_t(\tilde{w}))) \to 0 \quad \text{as } t \to \infty. \tag{10.17}$$

It follows that the geodesics $\gamma_{\tilde{v}}(t)$ and $\gamma_{\tilde{w}}(t)$ are asymptotic and hence, $\gamma_{\tilde{v}}(+\infty) = \gamma_{\tilde{w}}(+\infty)$. We wish to show that $\tilde{w} \in \tilde{W}^+(\tilde{v})$. Assuming the contrary, consider the horocycle $L(\pi(\tilde{v}), \gamma_{\tilde{v}}(+\infty))$. Let $\tilde{z}$ be the point of intersection of the geodesic $\gamma_{\tilde{w}}(t)$ and this horocycle (such a point exists by Proposition 10.8). We have

$$\rho(\pi(g_t(\tilde{v})), \pi(g_t(\tilde{w}))) \geq \rho(\pi(\tilde{w}), \tilde{z}) > 0,$$

which contradicts (10.17). It follows that $\tilde{V}^s(\tilde{v}) \subset \tilde{W}^-(\tilde{v})$ for every $v \in \Delta$ and every lift $\tilde{v}$ of v. Arguing similarly, one can show that $\tilde{V}^u(\tilde{v}) \subset \tilde{W}^-(\tilde{v})$ for every $v \in \Delta$ and every lift $\tilde{v}$ of v. By Proposition 10.9 and Theorem 9.18, we conclude that every ergodic component of $g_t|\Delta$ is open (mod 0). In view of Proposition 10.9 we conclude that $g_t|\Delta$ is ergodic. □

10.5. The entropy formula for geodesic flows

For geodesic flows on surfaces of negative curvature the entropy formula (see Section 9.3) can be transferred into a remarkable form that explicitly relates the metric entropy with the curvature of horocycles. More precisely, for $v \in SM$ consider the horocycle $L(\pi(v), \gamma_v(+\infty))$ through the point $x = \pi(v)$, which is a submanifold in H of class C^2 if the Riemannian metric is of class C^4.

Theorem 10.12. *Let g_t be the geodesic flow on a compact surface endowed with a C^4 Riemannian metric of nonpositive curvature. Then the metric* *entropy of g_1 with respect to the Liouville measure μ is*

$$h_\mu(g_1) = -\int_{SM} K(v)\, d\mu(v).$$

Proof. For $v \in SM$ and $\xi \in E^+(v)$, by statement (4) of Proposition 10.3, we have that for $t \geq 1$,

$$\|dg_t\xi\| \leq (1+a)\|d\pi \circ dg_t\xi\|.$$

This implies that for any LP-regular point $v \in SM$ and $\xi \in E^+(v)$,

$$\chi^-(v,\xi) = \lim_{t\to\infty} \frac{1}{t} \log \|dg_t\xi\| = \lim_{t\to\infty} \frac{1}{t} \log \|d\pi \circ dg_t\xi\|. \tag{10.18}$$

For any $s \neq 0$ let

$$\lambda_s(v) = \frac{\|d\pi \circ dg_s\xi\|}{\|d\pi\xi\|}$$

(note that $\lambda_s(v)$ does not depend on the choice of the vector ξ). By the entropy formula, Birkhoff's Ergodic Theorem, and (10.18), for every $s \neq 0$,

$$\begin{aligned} h_\mu(g_s) &= -\int_{SM} \chi^-(v,\xi)\, d\mu(v) \\ &= -\int_{SM} \lim_{n\to\infty} \frac{1}{sn} \log \|d\pi \circ dg_{sn}\xi\|\, d\mu(v) \\ &= -\int_{SM} \lim_{n\to\infty} \frac{1}{sn} \sum_{i=1}^{n-1} \log \lambda_s(g_{sn}(v))\, d\mu(v) \\ &= -\int_{SM} \log \lambda_s(v)\, d\mu(v). \end{aligned} \tag{10.19}$$

Since $h_\mu(g_s) = |s|h_\mu(g_1)$, we have by (10.19) that for $s > 0$,

$$\begin{aligned} h_\mu(g_1) &= \frac{1}{s} h_\mu(g_s) = \lim_{s\to 0} \frac{1}{s} h_\mu(g_s) \\ &= -\int_{SM} \lim_{s\to 0} \frac{1}{s} \log \lambda_s(v)\, d\mu(v). \end{aligned}$$

We shall compute the expression under the integral. For $v \in SM$, $\xi \in E^+(v)$, and $s > 0$ we have

$$\begin{aligned} |Y_\xi(s)|^2 &= |Y_\xi(0)|^2 + \int_0^s \frac{d}{du}|Y_\xi(u)|^2\, du \\ &= |Y_\xi(0)|^2 + 2\int_0^s |Y_\xi(u)| \cdot \frac{d}{du}|Y_\xi(u)|\, du. \end{aligned} \tag{10.20}$$

Let

$$a(s) = \frac{1}{s} \int_0^s |Y_\xi(u)| \cdot \frac{d}{du}|Y_\xi(u)|\, du.$$

For $s = 0$ we have

$$a(0) = 2|Y_\xi(s)| \cdot \frac{d}{ds}|Y_\xi(s)|\bigg|_{s=0} = 2K(v). \tag{10.21}$$

Using (10.20) and (10.21) and noting that $|Y_\xi(0)|^2 = 1$, we obtain that for $s \neq 0$,

$$\begin{aligned}\|d\pi \circ dg_s\xi\| &= \sqrt{\|d\pi \circ dg_s\xi\|^2} \\ &= \sqrt{|Y_\xi(s)|^2} = \sqrt{1 + sa(s)} = 1 + \frac{1}{2}sa(s) + O(s^2).\end{aligned}$$

It follows that

$$\lim_{s\to 0} \frac{1}{s} \log \lambda_s(v) = K(v)$$

and the desired result follows. □

Part 2

Selected Advanced Topics

Chapter 11

Cone Technics

As we have mentioned earlier in Chapter 4 one of the most effective ways to verify the conditions of nonuniform hyperbolicity for a diffeomorphism preserving a smooth measure ν is to show that the Lyapunov exponents are nonzero almost everywhere. While direct calculation of Lyapunov exponents is widely used in numerical studies of dynamical systems, the rigorous calculation may be difficult to carry out and the cone techniques come in handy in helping to verify that the exponents are nonzero though this does not allow us to actually compute them.

11.1. Introduction

The *cone* of size $\gamma > 0$ centered around $\mathbb{R}^{n-k}$ in the product space $\mathbb{R}^n = \mathbb{R}^k \times \mathbb{R}^{n-k}$ is defined by

$$C_\gamma = \big\{(v, w) \in \mathbb{R}^k \times \mathbb{R}^{n-k} : \|v\| < \gamma\|w\|\big\} \cup \{(0,0)\}.$$

Let $\mathcal{A}$ be a cocycle over an invertible measurable transformation $f\colon X \to X$ preserving a Borel probability measure ν in X and let $A\colon X \to GL(n, \mathbb{R})$ be its generator. The main idea underlying the cone techniques is the following. Let $Y \subset X$ be an f-invariant subset. Assume that there exist $\gamma > 0$ and $a > 1$ such that for every $x \in Y$:

(1) *invariance*: $A(x)C_\gamma \subset C_\gamma$;

(2) *expansion*: $\|A(x)v\| \geq a\|v\|$ for every $v \in C_\gamma$.

Exercise 11.1. Show that $\chi^+(x, v) > 0$ for every $x \in Y$ and $v \in C_\gamma \setminus \{0\}$.

This implies that the $n - k$ largest values of the Lyapunov exponent are positive.

We stress that positivity of Lyapunov exponents in the above exercise holds regardless of whether the set Y has positive ν-measure or not. A crucial observation made by Wojtkowski [**85**] is that in certain cases if condition (1) holds almost everywhere in X (or even on a set of positive measure), then this condition alone is sufficient to establish positivity of the Lyapunov exponents (almost everywhere or, respectively, on a set of positive measure) and no estimate on the growth of vectors inside the cone is necessary. We will present a result by Wojtkowski in this direction as well as describing another version of the cone techniques due to Burns and Katok [**46**].

We begin by stating a result of Wojtkowski establishing the positivity of the top Lyapunov exponent. Let $Q\colon \mathbb{R}^n \to \mathbb{R}$ be a nondegenerate quadratic form of type $(1, n-1)$. Without loss of generality we assume that

$$Q(v) = v_1^2 - v_2^2 - \cdots - v_n^2.$$

Consider the cone

$$C = \{v \in \mathbb{R}^n : Q(v) > 0\} \cup \{0\}$$

and the family of matrices

$$\mathcal{F} = \big\{A \in GL(n, \mathbb{R}) : |\det A| = 1,\ Q(Av) > 0 \text{ for } v \in \overline{C} \setminus \{0\}\big\}.$$

Clearly, the cone C is preserved by the action of the matrices in $\mathcal{F}$; i.e., $AC \subset C$ for every $A \in \mathcal{F}$.

Theorem 11.2 ([**85**]). *If* $\log^+\|A\| \in L^1(X, \nu)$ *and the cocycle only takes values in* $\mathcal{F}$*, then for* ν*-almost every* $x \in X$*,*

$$\limsup_{m\to\infty} \frac{1}{m} \log\|\mathcal{A}(x, m)\| > 0.$$

We outline the proof of this result. Consider the function $\rho\colon \mathcal{F} \to \mathbb{R}_0^+$ given by

$$\rho(A) = \inf_{v \in C\setminus\{0\}} \sqrt{\frac{Q(Av)}{Q(v)}}.$$

Exercise 11.3. Show that $\|A\| \geq \rho(A)$ and $\rho(AB) \geq \rho(A)\rho(B)$.

One can also show that the function $\rho(A)$ has the crucial property

$$\rho(A) > 1 \quad \text{for } A \in \mathcal{F}. \tag{11.1}$$

It follows from Exercise 11.3 that

$$\frac{1}{m} \log\|\mathcal{A}(x, m)\| \geq \frac{1}{m} \log \rho(\mathcal{A}(x, m)) \geq \frac{1}{m} \log \prod_{i=0}^{m-1} \rho(\mathcal{A}(f^i(x))).$$

On the other hand, by Birkhoff's Ergodic Theorem and (11.1), for ν-almost every $x \in X$ we have

$$\lim_{m\to\infty} \frac{1}{m} \log \prod_{i=0}^{m-1} \rho(A(f^i(x))) > 0.$$

This yields the desired result. It should be stressed that the above argument (in particular, the bound $\rho(A) > 1$ for $A \in \mathcal{F}$) cannot be used in the case of nondegenerate quadratic forms of type $(k, n-k)$ with $k > 1$ (as noted in [**85**]).

11.2. Lyapunov functions

We now describe an approach to establishing nonzero Lyapunov exponents which is based on Lyapunov functions. It was developed by Burns and Katok in [**46**].

Consider a measurable family of cones $C = \{C_x : x \in X\}$ in $\mathbb{R}^n$ and the complementary cones

$$\widehat{C}_x = \left(\mathbb{R}^n \setminus \overline{C}_x\right) \cup \{0\}.$$

The *rank* of a cone C_x is the maximal dimension of a linear subspace $L \subset \mathbb{R}^n$ contained in C_x. We denote it by $r(C_x)$ and we have that $r(C_x)+r(\widehat{C}_x) \le n$.

A pair of complementary cones C_x and $\widehat{C}_x$ is called *complete* if $r(C_x) + r(\widehat{C}_x) = n$. We say that the family of cones C is *complete* if the pair of complementary cones $(C_x, \widehat{C}_x)$ is complete for ν-almost every $x \in X$.

Let $\mathcal{A}$ be a measurable cocycle over a measurable transformation $f\colon X \to X$ preserving a finite measure ν. We say that the family of cones C is $\mathcal{A}$-*invariant* if for ν-almost every $x \in X$,

$$A(x)C_x \subset C_{f(x)} \quad \text{and} \quad A(f^{-1}(x))^{-1}\widehat{C}_x \subset \widehat{C}_{f^{-1}(x)}. \tag{11.2}$$

Let now $Q\colon \mathbb{R}^n \to \mathbb{R}$ be a continuous function that is homogeneous of degree one (i.e., $Q(av) = aQ(v)$ for every $v \in \mathbb{R}^n$) and takes both positive and negative values. The set

$$C^u(Q) := \{0\} \cup Q^{-1}(0, +\infty) \subset \mathbb{R}^n \tag{11.3}$$

is called the *positive cone of* Q, and the set

$$C^s(Q) := \{0\} \cup Q^{-1}(-\infty, 0) \subset \mathbb{R}^n \tag{11.4}$$

is called the *negative cone of* Q. The rank of $C^u(Q)$ (respectively, $C^s(Q)$) is called the *positive* (respectively, *negative*) *rank* of Q and is denoted by $r^u(Q)$ (respectively, $r^s(Q)$). Clearly, $r^u(Q) + r^s(Q) \le n$ and since Q takes both positive and negative values, we have $r^u(Q) \ge 1$ and $r^s(Q) \ge 1$. The function Q is said to be *complete* if the cones $C^u(Q)$ and $C^s(Q)$ form a complete pair of complementary cones, i.e., if

$$r^u(Q) + r^s(Q) = n.$$

A measurable function $Q\colon X \times \mathbb{R}^n \to \mathbb{R}$ is said to be a *Lyapunov function* for the cocycle $\mathcal{A}$ (with respect to ν) if there exist positive integers r^u and r^s such that for ν-almost every $x \in X$:

(1) the function $Q_x = Q(x, \cdot)$ is continuous, homogeneous of degree one, and takes both positive and negative values;

(2) Q_x is complete, $r^u(Q_x) = r^u$, and $r^s(Q_x) = r^s$;

(3) for every $x \in \mathbb{R}^n$ we have

$$Q_{f(x)}(A(x)v) \geq Q_x(v). \tag{11.5}$$

The numbers r^u and r^s are called, respectively, the *positive* and *negative ranks* of Q.

Exercise 11.4. Using (11.5), show that if Q is a Lyapunov function, then the two families of cones

$$C^u(Q_x) = \{v \in \mathbb{R}^n : Q_x(v) > 0\} \cup \{0\},$$
$$C^s(Q_x) = \{v \in \mathbb{R}^n : Q_x(v) < 0\} \cup \{0\}$$

are $\mathcal{A}$-invariant (see (11.2)).

A Lyapunov function is said to be *eventually strict* if for ν-almost every $x \in X$ there exists $m = m(x) \in \mathbb{N}$ depending measurably on x such that for every $v \in \mathbb{R}^n \setminus \{0\}$ we have

$$Q_{f^m(x)}(\mathcal{A}(x,m)v) > Q_x(v), \quad Q_{f^{-m}(x)}(\mathcal{A}(x,-m)v) < Q_x(v). \tag{11.6}$$

Theorem 11.5. *Assume that* $\log^+\|A\| \in L^1(X,\nu)$ *and that there exists an eventually strict Lyapunov function for the cocycle* $\mathcal{A}$*. Then for* ν*-almost every* $x \in X$ *the cocycle has* r^u *positive and* r^s *negative values of the Lyapunov exponent, counted with their multiplicities.*

Proof. Without loss of generality, we may assume that the measure ν is ergodic. Indeed, let Q be an eventually strict Lyapunov function and let $(\nu_\beta)_\beta$ be an ergodic decomposition of ν. Then $\log^+\|A\| \in L^1(X,\nu_\beta)$ for almost every β and Q is an eventually strict Lyapunov function for the cocycle $\mathcal{A}$ with respect to ν_β. Moreover, if the theorem holds for ergodic measures, then it also holds for arbitrary measures. From now on we assume that ν is ergodic.

For ν-almost every $x \in X$, we will construct subspaces $D^u_x, D^s_x \subset \mathbb{R}^n$, respectively, of dimensions r^u and r^s such that for every $m \in \mathbb{Z}$,

$$\mathcal{A}(x,m)D^u_x \subset C^u(Q_{f^m(x)}) \quad \text{and} \quad \mathcal{A}(x,m)D^s_x \subset C^s(Q_{f^m(x)})$$

and for every $v \in D^u_x \setminus \{0\}$ and $w \in D^s_x \setminus \{0\}$,

$$\lim_{m\to\infty} \frac{1}{m} \log\|\mathcal{A}(x,-m)v\| < 0 \quad \text{and} \quad \lim_{m\to\infty} \frac{1}{m} \log\|\mathcal{A}(x,m)w\| < 0.$$

This implies that $D^u_x = E^u(x)$ and $D^s_x = E^s(x)$ for ν-almost every $x \in X$. We shall establish the existence of the spaces D^s_x. The proof of the existence of the spaces D^u_x is entirely analogous.

Set $C_x = \overline{C^s(Q_x)}$ and for every $m \in \mathbb{N}$, define

$$C_{m,x} = \mathcal{A}(f^m(x),-m)C_{f^m(x)}.$$

By (11.5), we have $C_{1,x} \supset C_{2,x} \supset \cdots$. Each set $C_{m,x}$ contains a subspace of dimension r^s. By the compactness of the closed unit sphere, the intersection

$$C_{\infty,x} = \bigcap_{m=1}^{\infty} C_{m,x}$$

also contains a subspace D^s_x of dimension r^s. Fix $\varepsilon > 0$.

Exercise 11.6. Show that there exist a set $Y \subset X$ of measure $\nu(Y) > 1-\varepsilon$ and constants $c, d > 0$ such that for every $x \in Y$ and $v \in C_{\infty,x}$,

$$c\|v\| \le -Q_x(v) \le d\|v\|. \tag{11.7}$$

For each $m \in \mathbb{N}$ and $x \in X$, let

$$\kappa_m(x) = \sup\left\{ \frac{Q_{f^m(x)}(\mathcal{A}(x,v)x)}{Q_x(v)} : v \in C_{\infty,x} \setminus \{0\} \right\}.$$

By (11.5), we have $\kappa_m(x) \in (0,1]$. Moreover, by (11.6), for ν-almost every $x \in X$ there exists $m = m(x) \in \mathbb{N}$ such that $\kappa_m(x) < 1$. It follows from Luzin's theorem that there exist $N \in \mathbb{N}$ and a set $E \subset Y$ of measure $\nu(E) > 1-2\varepsilon$ such that $\kappa_N(x) < 1$ for every $x \in E$. Observe that for every $m, n \in \mathbb{N}$ and $x \in X$,

$$\kappa_{n+m}(x) \le \kappa_n(x)\kappa_m(f^n(x)).$$

Hence, $\kappa_n(x) < 1$ for every $n \ge N$ and $x \in E$. Consider the induced map $\bar{f}\colon E \to E$ on the set E, defined (mod 0) by $\bar{f}(x) = f^{\bar{n}(x)}(x)$, where

$$\bar{n}(x) = \min\{n \in \mathbb{N} : f^n(x) \in E\}$$

is the first return time. Consider also the induced cocycle $\bar{\mathcal{A}}$ whose generator is $\bar{A}(x) = \mathcal{A}(x, \bar{n}(x))$. For each $m \in \mathbb{N}$ and $x \in E$, let

$$\bar{\kappa}_m(x) = \sup\left\{ \frac{Q_{\bar{f}^m(x)}(\bar{\mathcal{A}}(x,m)v)}{Q_x(v)} : v \in C_{\infty,x} \setminus \{0\} \right\}.$$

We have $\bar{\kappa}_m(x) = \kappa_{\tau_m(x)}(x) \in (0,1]$, where

$$\tau_m(x) = \sum_{i=0}^{m-1} \bar{n}(\bar{f}^i(x)).$$

Exercise 11.7. Show that for every $m, n \in \mathbb{N}$ and $x \in E$ we have

$$\bar{\kappa}_{n+m}(x) \leq \bar{\kappa}_n(x)\bar{\kappa}_m(\bar{f}^n(x)). \tag{11.8}$$

Now let

$$K_m(x) = \log \bar{\kappa}_m(x).$$

By (11.8), the sequence of functions K_m is subadditive, that is,

$$K_{m+n}(x) \leq K_n(x) + K_m(\bar{f}^n(x)). \tag{11.9}$$

Since $K_m(x) \leq 0$ (and, hence, is bounded from above), it follows from Kingman's subadditive ergodic theorem [**53**] that for ν-almost every $x \in E$ the limit

$$F(x) := \lim_{m\to\infty} \frac{K_m(x)}{m} = \lim_{m\to\infty} \frac{1}{m} \log \bar{\kappa}_m(x)$$

exists and moreover, the function F is $\bar{f}$-invariant ν-almost everywhere in E. Since ν is ergodic, we have that

$$\begin{aligned} F(x) &= \lim_{m\to\infty} \frac{1}{m} \int_E K_m \, d\nu \\ &= \inf_{m\in\mathbb{N}} \frac{1}{m} \int_E K_m \, d\nu \\ &\leq \frac{1}{N} \int_E K_N \, d\nu < 0 \end{aligned} \tag{11.10}$$

for ν-almost every $x \in E$. Since $\tau_n(x) \geq n$, the last inequality follows from the fact that for every $x \in E$,

$$K_N(x) = \log \bar{\kappa}_N(x) = \log \kappa_{\tau_N(x)}(x) \leq \log \kappa_N(x) < 0.$$

On the other hand, it follows from (11.7) that for every $x \in E$, $m \in \mathbb{N}$, and $v \in C_{\infty,x}$,

$$\frac{c}{d}\|\bar{\mathcal{A}}(x,m)v\| \leq \bar{\kappa}_m(x)\|v\| \leq \frac{d}{c}\|\bar{\mathcal{A}}(x,m)v\|,$$

and hence,

$$\begin{aligned} \lim_{m\to\infty} \frac{1}{m} \log\|\mathcal{A}(x,m)v\| &= \lim_{m\to\infty} \frac{1}{\tau_m(x)} \log\|\bar{\mathcal{A}}(x,m)v\| \\ &\leq \lim_{m\to\infty} \frac{K_m(x)}{\tau_m(x)} = \frac{F(x)}{\tau(x)}, \end{aligned} \tag{11.11}$$

where $\tau(x) := \lim_{m\to\infty}(\tau_m(x)/m)$. By Kac's lemma and the ergodicity of the induced map $\bar{f}$ with respect to $\nu|E$, we have that for ν-almost every $x \in E$,

$$\tau(x) = \nu\left(\bigcup_{m\in\mathbb{N}} f^m(E)\right) > 0.$$

It follows from (11.10) and (11.11) that for ν-almost every $x \in E$,

$$\lim_{m\to\infty} \frac{1}{m} \log\|\mathcal{A}(x,m)v\| < 0.$$

Since $\nu(E) > 1 - 2\varepsilon$ and ε is arbitrary, the desired result follows. □

11.3. Cocycles with values in the symplectic group

Let $\mathcal{A}$ be a cocycle over an invertible measurable transformation $f\colon X \to X$ preserving a Borel probability measure ν in X and let $A\colon X \to GL(n,\mathbb{R})$ be its generator. For a homogeneous function Q of degree one and any $x \in X$ consider the two families of cones as in (11.3) and (11.4), that is,

$$C^u(Q_x) := \{0\} \cup Q_x^{-1}(0,+\infty) \subset \mathbb{R}^n$$

and

$$C^s(Q_x) := \{0\} \cup Q_x^{-1}(-\infty,0) \subset \mathbb{R}^n.$$

If Q_x is complete and

$$A(x)C^u(Q_x) \subset C^u(Q_{f(x)}), \quad A(f^{-1}(x))^{-1}C^s(Q_x) \subset C^s(Q_{f^{-1}(x)}) \tag{11.12}$$

for ν-almost every $x \in X$, then Q is a Lyapunov function and Theorem 11.5 applies. However, if condition (11.12) is satisfied only with respect to the family of cones $C^u(Q_x)$, then Q may not be a Lyapunov function. This situation occurs for some interesting classes of cocycles and families of cones. The most important case in applications involves cocycles with values in the symplectic group $Sp(2m,\mathbb{R})$ for $m \ge 1$ and the so-called symplectic cones that we define later.

Let C be an $\mathcal{A}$-invariant measurable family of cones. We say that:

(1) C is *strict* if for ν-almost every $x \in X$,

$$A(x)\overline{C_x} \subset C_{f(x)} \quad \text{and} \quad A(f^{-1}(x))^{-1}\overline{\widehat{C}_x} \subset \widehat{C}_{f^{-1}(x)};$$

(2) C is *eventually strict* if for ν-almost every $x \in X$ there exists $m = m(x) \in \mathbb{N}$ such that

$$\mathcal{A}(x,m)\overline{C_x} \subset C_{f^m(x)} \quad \text{and} \quad \mathcal{A}(x,-m)^{-1}\overline{\widehat{C}_x} \subset \widehat{C}_{f^{-m}(x)}. \tag{11.13}$$

One can show that for cocycles with values in the symplectic group $Sp(2m,\mathbb{R})$ the presence of just one eventually strict invariant family of cones guarantees existence of a Lyapunov function Q for the cocycle and hence, Theorem 11.5 applies.

We will only consider the simple case of $SL(2,\mathbb{R})$ cocycles and we refer the reader to [**9**] for the general case. We call a cone in $\mathbb{R}^n$ *connected* if its projection to the projective space $\mathbb{R}P^{n-1}$ is a connected set. A connected cone in $\mathbb{R}^2$ is simply the union of two opposite sectors bounded by two different straight lines intersecting at the origin plus the origin itself. By a linear coordinate change such a cone can always be reduced to the standard cone

$$S = \{(v, w) \in \mathbb{R}^2 : vw > 0\} \cup \{(0, 0)\}.$$

Theorem 11.8. *If a cocycle with values in $SL(2, \mathbb{R})$ has an eventually strictly invariant family of connected cones $C = \{C_x : x \in X\}$, then it has an eventually strict Lyapunov function Q such that for ν-almost every $x \in X$ the function Q_x has the form*

$$Q_x(v) = \operatorname{sgn} K_x(v, v) \cdot |K_x(v, v)|^{1/2}$$

for some quadratic form K_x of signature zero, and its zero set coincides with the boundary of the cone C_x.

Proof. First assume that $C_x = S$ for ν-almost every $x \in X$. Write

$$A(x) = \begin{pmatrix} a(x) & b(x) \\ c(x) & d(x) \end{pmatrix}.$$

By (11.2), the functions a, b, c, and d are nonnegative. Since $A(x) \in SL(2, \mathbb{R})$, we have $1 = a(x)d(x) - b(x)c(x)$. For each $(u, v) \in S$ put $K(u, v) = uv$. We obtain

$$\begin{aligned} K(A(x)(u, v)) &= a(x)c(x)u^2 + b(x)d(x)v^2 + (a(x)d(x) + b(x)c(x))uv \\ &\geq (a(x)d(x) + b(x)c(x))uv \\ &\geq (a(x)d(x) - b(x)c(x))uv = K(u, v). \end{aligned} \tag{11.14}$$

Arguing similarly, one can show that all entries of the matrix

$$\mathcal{A}(x, m) = \begin{pmatrix} a(x, m) & b(x, m) \\ c(x, m) & d(x, m) \end{pmatrix}$$

are nonnegative. Moreover, condition (11.13) implies that there exists $m = m(x) \geq 1$ such that $b(x, m) > 0$ and $c(x, m) > 0$. By (11.14), we conclude that $K(\mathcal{A}(x, m)(u, v)) > K(u, v)$.

In the case of an arbitrary family of connected cones, let us introduce a coordinate change $L\colon X \to SL(2, \mathbb{R})$ that takes the two lines bounding the cone C_x into the coordinate axes. Then $L(x)C_x = S$. For the cocycle $B\colon X \to SL(2, \mathbb{R})$ defined by $B(x) = L(f(x))A(x)L(x)^{-1}$, the constant family of cones S is eventually strictly invariant and hence by the previous argument, the function $Q_0(x, (u, v)) = \operatorname{sgn}(uv)|uv|^{1/2}$ is an eventually strict Lyapunov function. Hence, for the original cocycle A the function $Q(x, (u, v)) = Q_0(x, L(x)(u, v))$ has the same properties. □

Chapter 12

Partially Hyperbolic Diffeomorphisms with Nonzero Exponents

In this chapter we shall discuss a special class of dynamical systems which act as uniform contractions and/or uniform expansions in some directions in the tangent space while allowing nonuniform contractions and/or nonuniform expansions in some other directions, thus exhibiting hyperbolicity of a "mixed" type.

This form of hyperbolicity usually comes as a particular case of uniform partial hyperbolicity[1]. The latter is characterized by two transverse stable and unstable subspaces complemented by a central direction with weaker rates of contraction and expansion.

In general, some or all Lyapunov exponents in the central direction can be zero and the "mixed" hyperbolicity ensures that the Lyapunov exponents in the central direction are all nonzero so that the system is nonuniformly hyperbolic. In particular, a partially hyperbolic diffeomorphism with nonzero Lyapunov exponents in the central direction preserving a smooth measure can have at most countably many ergodic components of positive measure. Stronger results can be obtained if one assumes that the Lyapunov exponents in the central direction are all negative (or all positive).

[1]A more general setting deals with systems admitting a *dominated splitting* (see [**15**]) but this concept goes far beyond the scope of this book.

12.1. Partial hyperbolicity

In Section 4.4 we introduced nonuniformly partially hyperbolic diffeomorphisms. We now briefly discuss the stronger notion of uniform partial hyperbolicity and we refer the reader to the book [**73**] for a more detailed exposition of uniform partial hyperbolicity theory.

A diffeomorphism f is said to be *uniformly partially hyperbolic* if there exist (1) numbers λ, λ', μ, and μ' such that $0 < \lambda < 1 < \mu$ and $\lambda < \lambda' \leq \mu' < \mu$; (2) numbers $C > 0$ and $K > 0$; (3) subspaces $E^c(x)$, $E^s(x)$, and $E^u(x)$, for $x \in M$ such that:

(1) $dfE^\omega(x) = E^\omega(f(x))$ where $\omega = c$, s, or u and

$$T_xM = E^s(x) \oplus E^c(x) \oplus E^u(x);$$

(2) for $v \in E^s(x)$ and $n \geq 0$,

$$\|df^n v\| \leq C\lambda^n \|v\|;$$

(3) for $v \in E^u(x)$ and $n \leq 0$,

$$\|df^n v\| \leq C\mu^n \|v\|;$$

(4) for $v \in E^c(x)$ and $n \geq 0$,

$$C^{-1}(\lambda')^n \|v\| \leq \|df^n v\| \leq C(\mu')^n |v\|;$$

(5)

$$\angle(E^c(x), E^s(x)) \geq K, \quad \angle(E^c(x), E^u(x)) \geq K,$$
$$\angle(E^s(x), E^u(x)) \geq K.$$

In other words, f is uniformly partially hyperbolic if it is nonuniformly partially hyperbolic on M in the sense of Section 4.4 with the functions $\lambda(x)$, $\lambda'(x)$, $\mu(x)$, and $\mu'(x)$ constant, the function $C(x)$ bounded from above, and the function $K(x)$ bounded from below.

The subspaces $E^c(x)$, $E^s(x)$, and $E^u(x)$ are called, respectively, *central*, *stable*, and *unstable*.

Exercise 12.1. Show that the subspaces $E^c(x)$, $E^s(x)$, and $E^u(x)$ depend continuously on $x \in M$.

Using Theorem 4.13, one can strengthen this result and prove that the stable and unstable subspaces depend Hölder continuously on $x \in M$. Furthermore, one can extend the Stable Manifold Theorem 7.1 to partially hyperbolic diffeomorphisms and show that there are local stable $V^s(x)$ and unstable $V^u(x)$ manifolds through every point $x \in M$. Due to the uniform nature of hyperbolicity they depend continuously on $x \in M$ and their sizes are uniform in x. Now using Theorem 7.15, one can construct global stable $W^s(x)$ and unstable $W^u(x)$ manifolds through points in M that generate two f-invariant transverse stable and unstable foliations of M with smooth leaves—the integrable foliations for the stable and unstable distributions.

The Absolute Continuity Theorem 8.2 extends to partially hyperbolic diffeomorphisms ensuring that both stable and unstable foliations possess the absolute continuity property.

Unlike the stable and unstable distributions, the central distribution may not be integrable. Indeed, there is an example due to Smale of an Anosov diffeomorphism f of a three-dimensional manifold that is a factor of a nilpotent Lie group such that the tangent bundle admits an invariant splitting $TM = E_1 \oplus E_2 \oplus E_3$ where E_3 is the unstable distribution, $E_1 \oplus E_2$ is the stable distribution, and the "weakly stable" distribution E_2 is not integrable (see [**73**]). One can view f as a partially hyperbolic diffeomorphism for which E_2 is the nonintegrable central distribution (and E_1 is the stable distribution). This situation is robust: any diffeomorphism g that is sufficiently closed to f in the C^1 topology is a partially hyperbolic diffeomorphism whose central distribution is not integrable. The following result by Hirsch, Pugh, and Shub [**39**] describes the "opposite" situation.

Theorem 12.2. *Let f be a partially hyperbolic diffeomorphism whose central distribution is integrable to a smooth foliation. Then any diffeomorphism g that is sufficiently close to f in the C^1 topology is partially hyperbolic and its central distribution is integrable to a continuous foliation with smooth leaves.*[2]

Similarly to Anosov diffeomorphisms, partially hyperbolic diffeomorphisms form an open set in the space of C^1 diffeomorphisms of M.

In studying ergodic properties of uniformly partially hyperbolic diffeomorphisms an important role is played by the accessibility property that we now introduce.

We call two points $p, q \in M$ *accessible* if they can be connected by a path which consists of finitely many segments lying in *u*nstable and *s*table manifolds of some points in M; more precisely, if there are points $p = z_0, z_1, \dots, z_{\ell-1}, z_\ell = q$ such that $z_i \in W^u(z_{i-1})$ or $z_i \in W^s(z_{i-1})$ for $i = 1, \dots, \ell$. The collection of points $z_0, z_1, \dots, z_\ell$ is called an *us-path* connecting p and q and is denoted by $[p, q] = [z_0, z_1, \dots, z_\ell]$.

Exercise 12.3. Show that accessibility is an equivalence relation.

The diffeomorphism f is said to have the *accessibility property* if the partition into accessibility classes is trivial, that is, if any two points $p, q \in M$ are accessible. A weaker version is the *essential accessibility property*—the partition into accessibility classes is ergodic (i.e., a measurable union of

[2]In general, this foliation is not smooth.

equivalence classes must have zero or full measure). Finally, we call f *center bunched* if $\max\{\lambda, \mu\} < \lambda'/\mu'$. The following result provides sufficient conditions for ergodicity of partially hyperbolic diffeomorphisms with respect to smooth measures.

Theorem 12.4 (Burns–Wilkinson [**22**]). *Let f be a C^2 diffeomorphism of a compact smooth manifold preserving a smooth measure ν. Assume that f is uniformly partially hyperbolic, essentially accessible, and center bunched. Then f is ergodic.*

The assumption that f is center bunched is technical and can be dropped if $\dim E^c = 1$ (see [**22, 76**]).

We describe some examples of partially hyperbolic systems (see [**73**] for details and references).

1. Let M and N be compact smooth Riemannian manifolds and let $f\colon M \to M$ be a C^r Anosov diffeomorphism, $r \geq 1$. The diffeomorphism

$$F = f \times \mathrm{Id}_N\colon M \times N \to M \times N$$

is partially hyperbolic. Any sufficiently small perturbation $G \in \mathrm{Diff}^r(M \times N)$ of F is partially hyperbolic; the central distribution E^c_G is integrable and the corresponding central foliation W^c_G has compact smooth leaves, which are diffeomorphic to N. If f preserves a smooth measure ν on M, then the map F preserves the measure $\nu \times m$ where m is the Riemannian volume in N.

2. The time-t map of an Anosov flow on a compact smooth manifold M is a partially hyperbolic diffeomorphism with one-dimensional central direction. In particular, if this map is accessible, then, by Theorem 12.4, it is ergodic with respect to an invariant smooth measure. The time-t map of the geodesic flow on a negatively curved manifold is an example of such a map.

Another example is given by the special flow T^t over a C^r Anosov diffeomorphism $f\colon M \to M$ with a roof function $H(x)$ (see Section 1.2).

Theorem 12.5. *There exist $\varepsilon > 0$ and an open and dense set of C^r functions $\tilde{H}\colon M \to \mathbb{R}^+$ with $|\tilde{H}(x)| \leq 1$ such that the time-t map of the special flow T^t with the roof function*

$$H(x) = H_0 + \varepsilon\tilde{H}(x)$$

($H_0 > 0$ is a constant) is accessible (and, hence, is ergodic).

3. Let G be a compact Lie group and let $f\colon M \to M$ be a C^r Anosov diffeomorphism. Each C^r function $\varphi\colon M \to G$ defines a skew product transformation $f_\varphi\colon M \times G \to M \times G$, or *$G$-extension of f*, defined by the formula

$$F_\varphi(x, y) = (f(x), \varphi(x)y).$$

Left transformations are isometries of G in the bi-invariant metric, and therefore, F_φ is partially hyperbolic. If f preserves a smooth measure ν, then f_φ preserves the smooth measure $\nu \times \nu_G$ where ν_G is the (normalized) Haar measure on G.

Theorem 12.6 (Burns–Wilkinson [**21**]). *For every neighborhood $\mathcal{U} \subset C^r(M, G)$ of the function φ there exists a function $\psi \in \mathcal{U}$ such that the diffeomorphism F_ψ is accessible.*

12.2. Systems with negative central exponents

We describe an approach to the study of ergodic properties of uniformly partially hyperbolic systems preserving a smooth measure, which fits well with the spirit of this book: it takes into account the Lyapunov exponents along the central direction E^c. We consider the case when these central exponents are all negative on a subset of positive measure. The case when the central exponents are all positive on a subset of positive measure can be reduced to the previous one by switching to the inverse map.

Theorem 12.7 (Burns–Dolgopyat–Pesin [**20**]). *Let f be a C^2 diffeomorphism of a compact smooth Riemannian manifold preserving a smooth measure ν. Assume that f is uniformly partially hyperbolic, accessible, and has negative central exponents at every point x of a set A of positive measure. Then f is ergodic (indeed, has the Bernoulli property). In particular, A has full measure and hence, f has nonzero Lyapunov exponents almost everywhere.*

Proof. Since the Lyapunov exponents in the center direction are negative, the map $f|A$ has nonzero Lyapunov exponents. We denote by $V^{sc}(x)$ the local stable manifold tangent to $E^{sc}(x)$; by Theorem 7.1, $V^{sc}(x)$ is defined for almost every $x \in A$ and its size depends measurably on x. We also denote by $m^s(x)$, $m^u(x)$, $m^{sc}(x)$ the leaf volumes on $V^s(x)$, $V^u(x)$, and $V^{sc}(x)$, respectively.

Since the map $f|A$ has nonzero Lyapunov exponents, by Theorem 9.1, it has at most countably many ergodic components of positive measure. We shall show that every such component is open (mod 0) and hence, so is the set A.

Observe that for almost every $x \in A$, the manifold $V^{cs}(x)$ is transverse to the unstable foliation W^u. This and the uniform size of local unstable manifolds $V^u(z)$ ensure that the set

$$P(x) = \bigcup_{y \in V^{sc}(x)} V^u(y)$$

is an open neighborhood of x. Therefore, the set

$$Q(x) = \bigcup_{n\in\mathbb{Z}} f^n(P(x))$$

is open in M and by Lemma 9.3, f is ergodic on $Q(x)$.

We shall show that the map $f|A$ is topologically transitive and, hence, is ergodic. In fact, we prove the following stronger statement that will help us establish that $A = M \pmod 0$.

Lemma 12.8. *Almost every orbit of f is dense.*

Proof of the lemma. It suffices to show that if U is an open set, then the orbit of almost every point enters U. To this end, let us call a point *good* if it has a neighborhood in which the orbit of almost every point enters U. We wish to show that an arbitrary point p is good. Since f is accessible, there is an us-path $[z_0, \dots, z_k]$ with $z_0 \in U$ and $z_k = p$. We shall show by induction on j that each point z_j is good. This is obvious for $j = 0$.

Now suppose that z_j is good. Then z_j has a neighborhood N such that $\mathcal{O}(x) \cap U \neq \varnothing$ for almost every $x \in N$. Let S be the subset of N consisting of the points with this property that are also both forward and backward recurrent. It follows from Poincaré's Recurrence Theorem that S has full measure in N. If $x \in S$, any point $y \in W^s(x) \cup W^u(x)$ has the property that $\mathcal{O}(y) \cap U \neq \varnothing$. The absolute continuity of the foliations W^s and W^u ensures that the set

$$\bigcup_{x\in S} W^s(x) \cup W^u(x)$$

has full measure in the set

$$\bigcup_{x\in N} W^s(x) \cup W^u(x).$$

The latter is a neighborhood of z_{j+1}. Hence, z_{j+1} is good. □

We shall show that $A = M \pmod 0$. Otherwise the set $B = M \setminus A$ has positive ν-measure. By Lemma 12.8, we can choose a Lebesgue density point[3] $x \in B$ whose orbit is everywhere dense. Hence, there exists $n \in \mathbb{Z}$ such that $f^n(x) \in A$. We can choose $\varepsilon > 0$ so small that $B(f^n(x), \varepsilon) \subset A \pmod 0$ and then choose $\delta > 0$ such that $f^n(B(x, \delta)) \subset B(f^n(x), \varepsilon)$. Since B is invariant and $x \in B$ is a Lebesgue density point, the set $f^n(B(x, \delta))$ contains a subset in B of positive measure. Hence, B intersects A in a set of positive measure. This contradiction implies that $A = M \pmod 0$.

[3]Recall that a point $x \in B$ is a Lebesgue density point if $\lim_{r\to 0} \frac{\nu(B(x,r)\cap B)}{\nu(B(x,r))} = 1$.

To show that the map f has the Bernoulli property, observe that for every $n > 0$ the map f^n is accessible and has negative central exponents. Hence, it is ergodic and the desired result follows. □

12.3. Foliations that are not absolutely continuous

As we pointed out above, the stable and unstable foliations of a partially hyperbolic diffeomorphism possess the absolute continuity property. We also mentioned that the central distribution may not be integrable. Even if it is integrable, the central foliation may not be absolutely continuous. It is indeed the case when the Lyapunov exponents in the central direction are all negative or all positive.

To better illustrate this phenomenon, consider a linear hyperbolic automorphism A of the two-dimensional torus $\mathbb{T}^2$ and the map $F = A \times \mathrm{Id}$ of the three-dimensional torus $\mathbb{T}^3 = \mathbb{T}^2 \times S^1$. Any sufficiently small C^1 perturbation G of F is uniformly partially hyperbolic with one-dimensional central distribution. By Theorem 12.2, the latter is integrable and its integral curves form a continuous foliation W^c of M with smooth compact leaves that are diffeomorphic to the unit circle. One can show that if the perturbation G has nonzero Lyapunov exponent in the central direction, then the central foliation is not absolutely continuous: for almost every $x \in M$ the conditional measure (generated by the Riemannian volume) on the leaf $W^c(x)$ of the central foliation passing through x is atomic.

We describe a far more general version of this result which is due to Ruelle and Wilkinson [**79**] (see also [**9**]). Let (X, ν) be a probability space and let $f\colon X \to X$ be an invertible transformation, which preserves the measure ν and is ergodic with respect to ν. Also let M be a smooth compact Riemannian manifold and let $\varphi\colon X \to \mathrm{Diff}^{1+\alpha}(M)$ be a map. Consider the skew-product transformation $F\colon X \times M \to X \times M$ given by

$$F(x, y) = (f(x), \varphi_x(y))$$

and assume it is Borel measurable. We also assume that F possesses an invariant ergodic measure μ on $X \times M$ such that $\pi_*\mu = \nu$ where $\pi\colon X \times M \to X$ is the projection.

Fix $x \in X$. Letting $\varphi_x^{(0)}$ be the identity map, define the sequence of maps $\varphi_x^{(k)}$, $k \in \mathbb{Z}$, on M by the formula

$$\varphi_x^{(k+1)} = \varphi_{f^k(x)} \circ \varphi_x^{(k)}.$$

Since the tangent bundle to M is measurably trivial, the differential of the map φ along the M direction gives rise to a cocycle

$$\mathcal{A}\colon X \times M \times \mathbb{Z} \to GL(n, \mathbb{R}), \quad n = \dim M,$$

defined by $\mathcal{A}(x, y, k) = d_y\varphi_x^{(k)}$. If $\log^+ \|d\varphi\| \in L^1(X \times M, \mu)$, then the Multiplicative Ergodic Theorem 6.1 and the ergodicity of μ imply that the Lyapunov exponents of this cocycle are constant μ-almost everywhere. We write the distinct values in increasing order:

$$\chi_1 < \cdots < \chi_\ell, \quad 1 \le \ell \le n.$$

Theorem 12.9. *Assume that for some $\gamma > 0$ the function φ satisfies*

$$\log^+ \|d\varphi\|_\gamma \in L^1(X, \nu),$$

where $\|\cdot\|_\gamma$ is the γ-Hölder norm.[4] *Assume also that $\chi_\ell < 0$. Then there exists a set $S \subset X \times M$ of full measure and an integer $k \ge 1$ such that*

$$\operatorname{card}(S \cap (\{x\} \times M)) = k$$

for every $(x, y) \in S$.

Proof. We need the following adaptation of the construction of regular neighborhoods in Section 6.4.2 to the cocycle $\mathcal{A}$.

Lemma 12.10. *There exists a set $\Lambda_0 \subset X \times M$ of full measure such that for any $\varepsilon > 0$ the following statements hold:*

(1) *there exist a measurable function $r \colon \Lambda_0 \to (0, 1]$ and a collection of embeddings $\Psi_{x,y} \colon B(0, r(x, y)) \to M$ such that $\Psi_{x,y}(0) = y$ and*

$$\exp(-\varepsilon) < \frac{r(F(x, y))}{r(x, y)} < \exp(\varepsilon);$$

(2) *if*

$$\varphi_{(x,y)} = \Psi_{F(x,y)}^{-1} \circ \varphi_x \circ \Psi_{(x,y)} \colon B(0, r(x, y)) \to \mathbb{R}^n,$$

then

$$\exp(\chi_1 - \varepsilon) \le \|d_0\varphi_{(x,y)}^{-1}\|^{-1}, \|d_0\varphi_{(x,y)}\| \le \exp(\chi_\ell + \varepsilon);$$

(3) *the C^1 distance $d_{C^1}(\varphi_{(x,y)}, d_0\varphi_{(x,y)}) < \varepsilon$ in $B(0, r(x, y))$;*

(4) *there exist $K > 0$ and a measurable function $A \colon \Lambda_0 \to \mathbb{R}$ such that for any $z, w \in B(0, r(x, y))$,*

$$\begin{aligned} K^{-1} d(\Psi_{(x,y)}(z), \Psi_{(x,y)}(w)) &\le \|z - w\| \\ &\le A(x) d(\Psi_{(x,y)}(z), \Psi_{(x,y)}(w)) \end{aligned}$$

with

$$e^{-\varepsilon} < \frac{A(F(x, y))}{A(x, y)} < e^{\varepsilon}.$$

Exercise 12.11. Prove Lemma 12.10 applying the Reduction Theorem 6.10, which should be restated for the cocycle $\mathcal{A}$.

---

[4] The norm $\|d\varphi\|_\gamma$ is the smallest $C > 0$ for which $\sup_{x \in X} \sup_{y \in M} \|d_y\varphi_x(u) - d_y\varphi_x(v)\| \le C\|u - v\|^\gamma$ for any $u, v \in T_yM$.

Decomposing μ into a system of conditional measures

$$d\mu(x,y) = d\mu_x(y)d\nu(x)$$

and using the invariance of μ with respect to F, we obtain

$$(\varphi_x)_*\mu_x = \mu_{f(x)}$$

for ν-almost every $x \in X$.

Lemma 12.12. *There exist a set $\Lambda \subset \Lambda_0$ and real numbers $R > 0$, $C > 0$, and $0 < \alpha < 1$ such that*

(1) $\mu(\Lambda) > 1/2$ *and if $(x,y) \in \Lambda$, then $\mu_x(\Lambda_x) > 1/2$ where*

$$\Lambda_x = \{y \in M : (x,y) \in \Lambda\};$$

(2) *if $(x,y) \in \Lambda$ and $z \in M$ are such that the distance $d_M(y,z)$ between y and z in M is less than or equal to R, then for all $m \geq 0$,*

$$d_M(\varphi_x^{(m)}(y), \varphi_x^{(m)}(z)) \leq C\alpha^m d_M(y,z).$$

Exercise 12.13. Derive Lemma 12.12 from Lemma 12.10. Hint: Set $\alpha = e^{(\chi_\ell+\varepsilon)}$. Observe that $0 < \alpha < 1$ provided that $\varepsilon > 0$ is sufficiently small. Show that the set of points $(x,y) \in \Lambda_0$ for which statement (2) of the lemma holds for some $R > 0$ and $C > 0$ has positive measure in Λ_0.

To prove the theorem, it suffices to show that there is a set $B \subset X$ of positive ν-measure such that for any $x \in B$ the measure μ_x has an atom. To see this, for $x \in X$ let $d(x) = \sup_{y\in M} \mu_x(y)$. Clearly, the function $d(x)$ is measurable, f-invariant, and positive on B. Since f is ergodic, we have $d(x) = d > 0$ for almost every $x \in X$. Let

$$S = \{(x,y) \in X \times M : \mu_x(y) \geq d\}.$$

Note that S is F-invariant, has measure at least d, and, hence, has measure 1. The desired result follows.

To prove the above claim, let Λ be the set constructed in Lemma 12.12 and let $B = \pi(\Lambda)$. Clearly, $\nu(B) > 0$. We shall show that for any $x \in B$ the measure μ_x has an atom.

Let $\mathcal{U} = \{U_1, \ldots, U_N\}$ be a cover of M by N closed balls of radius $R/10$. For $x \in X$ set

$$m(x) = \inf \sum \operatorname{diam} U_j,$$

where the infimum is taken over all collections of closed balls $\{U_1, \ldots, U_k\}$ in M such that $k \leq N$ and $\mu_x(\bigcup_{j=1}^k U_j) \geq 1/2$. We also define the number $m = \operatorname{ess\,sup}_{x\in B} m(x)$. We will show that $m = 0$. Otherwise, there is a number $p > 0$ such that

$$C\Delta N\alpha^p < \frac{m}{2}, \tag{12.1}$$

where Δ is the diameter of M. For $x \in B$ let $U_1(x), \dots, U_{k(x)}(x)$ be those balls in the cover $\mathcal{U}$ that meet Λ_x. Since these balls cover Λ_x and $\mu_x(\Lambda_x) > 1/2$, we have that

$$\mu_x\left(\bigcup_{j=1}^{k(x)} U_j(x)\right) \geq \frac{1}{2}. \tag{12.2}$$

Taking into account that $(\varphi_x^{(i)})_*\mu_x = \mu_{f^i(x)}$, for all i we obtain

$$\mu_{f^i(x)}\left(\bigcup_{j=1}^{k(x)} \varphi_x^{(i)} U_j(x)\right) \geq \frac{1}{2}.$$

Since the balls $U_j(x)$ meet the set Λ_x and have diameter less than $R/10$, by Lemma 12.12 we obtain that

$$\operatorname{diam}(\varphi_x^{(i)} U_j(x)) \leq C\Delta\alpha^i. \tag{12.3}$$

Let $\tau(x)$ be the first return time of the point $x \in B$ to B under the map f^p and let $B_i = \{x \in B : \tau(x) = i\}$. We have that $B = \bigcup_{i=1}^{\infty} B_i$ and since f is invertible and its inverse f^{-1} preserves ν, we also have that

$$B' = \bigcup_{i=1}^{\infty} f^{pi}(B_i) = B \pmod 0.$$

If $z \in B'$, then $z = f^{pi}(x)$ for some $x \in B_i$ and some $i \geq 1$. It follows from the definition of $m(z)$ and inequalities (12.1), (12.2), and (12.3) that

$$m(z) \leq \sum_{j=1}^{k(x)} \operatorname{diam}(\varphi_x^{(pi)} U_j(x)) \leq Ck(x)\Delta\alpha^{pi} \leq CN\Delta\alpha^p \leq \frac{m}{2}.$$

This implies that

$$m = \operatorname{ess\,sup}_{x \in B} m(x) = \operatorname{ess\,sup}_{z \in B'} m(z) \leq \frac{m}{2},$$

contradicting the assumption that $m > 0$.

We have shown that $m = 0$ and hence, $m(x) = 0$ for ν-almost every $x \in B$. For such a point x there is a sequence of closed balls $U^1(x), U^2(x), \dots$ for which

$$\lim_{i \to \infty} \operatorname{diam} U^i(x) = 0$$

and $\mu_x(U^i(x)) \geq \frac{1}{2N}$ for all i. Take $z_i \in U^i(x)$. Any accumulation point of the sequence $\{z_i\}$ is an atom for μ_x. □

Corollary 12.14. *Let μ be an ergodic measure for a $C^{1+\alpha}$ diffeomorphism of a compact Riemannian manifold M. If μ has all of its exponents negative, then μ is concentrated on the orbit of a periodic sink.*

Proof. The result follows from Theorem 12.9 by taking $X = \{x\}$ and taking ν to be the point mass. □

Let f be a $C^{1+\alpha}$ diffeomorphism of a compact smooth Riemannian manifold M preserving a Borel probability measure μ. Assume that W is a foliation of M with smooth leaves, which is invariant under f. We say that f is *W-dissipative* if $L_W(x) \neq 0$ for μ-almost every $x \in M$, where $L_W(x)$ denotes the sum of the Lyapunov exponents of f at the point x along the subspace $T_xW(x)$.

For $x \in M$ we denote by $\mathrm{vol}(W(x))$ the leaf volume of the leaf $W(x)$ through x. We say that the foliation W has *finite volume leaves almost everywhere* if the set of those $x \in M$ for which $\mathrm{vol}(W(x)) < \infty$ has full volume. An example of a foliation whose leaves have finite volume almost everywhere is a foliation with smooth compact leaves. If W is such a foliation, then the function $M \ni x \mapsto \mathrm{vol}(W(x))$ is well-defined (finite) but may not be bounded (see an example in [**35**]).

Theorem 12.15 (Hirayama–Pesin [**38**]). *Let f be a $C^{1+\alpha}$ diffeomorphism of a compact smooth Riemannian manifold M preserving a smooth measure μ. Also let W be an f-invariant foliation of M with smooth leaves. Assume that W has finite volume leaves almost everywhere. If f is W-dissipative, then the foliation W is not absolutely continuous in the weak sense (see Chapter 8).*

Proof. Let $A^- \subset M$ be the set of points for which $L_W(x) < 0$ and let $A^+ \subset M$ be the set of points for which $L_W(x) > 0$. They both are f-invariant and either $m(A^-) > 0$ or $m(A^+) > 0$ or both (we use here the fact that the invariant measure μ is smooth and hence equivalent to volume). Without loss of generality we may assume that $m(A^+) > 0$. Then for a sufficiently small $\lambda > 0$, sufficiently large integer ℓ, and every small $\varepsilon > 0$ there exists a Borel set $A^+_{\lambda,\ell,\varepsilon} \subset A^+$ of positive μ-measure such that for every $x \in A^+_{\lambda,\ell,\varepsilon}$ and $n \geq 0$,

$$|\mathrm{Jac}(df^n|T_xW(x))| \geq \ell^{-1}e^{\lambda n}e^{-\varepsilon n}. \tag{12.4}$$

Given $V > 0$, consider the set

$$Y_V = \{y \in M : \mathrm{vol}(W(y)) \leq V\}.$$

Observe that the set $A^+_{\lambda,\ell,\varepsilon}$ has positive volume and let $x \in A^+_{\lambda,\ell,\varepsilon}$ be a Lebesgue density point of m. One can choose $V > 0$ such that the set

$$R = A^+_{\lambda,\ell,\varepsilon} \cap B(x,r) \cap Y_V$$

has positive volume.

Assume on the contrary that the foliation W is absolutely continuous in the weak sense. Then for almost every $y \in R$ the set $R_y = R \cap W(y)$ has positive volume in $W(y)$. Observe that $\mu(R) > 0$ (where μ is an invariant smooth measure for f). Therefore, the trajectory of almost every point $y \in R$ returns to R infinitely often. Let y be such a point and let $\{n_k\}$ be the sequence of successive returns to R. We also may assume that $m_{W(y)}(R_y) > 0$.

Since $f^{n_k}(y) \in R \subset Y_V$, we have that for every $k > 0$,

$$m_{W(f^{n_k}(y))}(f^{n_k}(R_y)) \le W(f^{n_k}(y)) \le V.$$

On the other hand, by (12.4),

$$\begin{aligned} \operatorname{vol}(W(f^{n_k}(y))) &\ge m_{W(f^{n_k}(y))}(f^{n_k}(R_y)) \\ &= \int_{R_y} \operatorname{Jac}(df^{n_k}|T_zW(z))\, dm(z) \\ &\ge \ell^{-1} e^{(\lambda-\varepsilon)n_k} m(R_y) > V \end{aligned}$$

if n_k is sufficiently large. This yields a contradiction and completes the proof of the theorem. □

Chapter 13

More Examples of Dynamical Systems with Nonzero Lyapunov Exponents

The goal of this chapter is to present some new examples of dynamical systems with nonzero Lyapunov exponents. The corresponding constructions use more sophisticated techniques and reveal some new and subtle properties of systems with nonzero exponents.

13.1. Hyperbolic diffeomorphisms with countably many ergodic components

Let f be a diffeomorphism with nonzero Lyapunov exponents with respect to a smooth invariant measure. According to Theorem 9.1, f can have countably many ergodic components. In Section 9.1 we presented an example of a diffeomorphism f that has finitely many ergodic components and here we will describe the example by Dolgopyat, Hu, and Pesin [**29**] of a diffeomorphism f that has countably many ergodic components, thus demonstrating that Theorem 9.1 cannot be improved. More precisely, the following statement holds.

Theorem 13.1. *There exists a C^∞ volume-preserving diffeomorphism f of the 3-torus $\mathbb{T}^3$ with nonzero Lyapunov exponents almost everywhere and with countably many ergodic components that are open* (mod 0).

Let $A: \mathbb{T}^2 \to \mathbb{T}^2$ be a linear hyperbolic automorphism with at least two fixed points p and p'. Consider the map $F = A \times \mathrm{Id}$ of the 3-torus $\mathbb{T}^3 = \mathbb{T}^2 \times S^1$. We will perturb F to obtain the desired diffeomorphism f.

Proposition 13.2. *Given $k \geq 2$ and $\delta > 0$, there exists a map g of the three-dimensional manifold $M = \mathbb{T}^2 \times I$ (where $I = [0,1]$) such that:*

(1) *g is a C^∞ volume-preserving diffeomorphism;*

(2) *$\|F - g\|_{C^k} \leq \delta$;*

(3) *$d^m g|\mathbb{T}^2 \times \{z\} = d^m F|\mathbb{T}^2 \times \{z\}$ for $z = 0, 1$ and all $0 \leq m < \infty$;*

(4) *g is ergodic and has nonzero Lyapunov exponents almost everywhere.*

We show how to complete the proof of the theorem by deducing it from the proposition. To this end, consider the countable collection of intervals $\{I_n\}_{n=1}^\infty$ in the circle S^1, given by

$$I_{2n} = [(n+2)^{-1}, (n+1)^{-1}], \quad I_{2n-1} = [1 - (n+1)^{-1}, 1 - (n+2)^{-1}].$$

Clearly, $\bigcup_{n=1}^\infty I_n = (0,1)$ and the interiors of I_n are pairwise disjoint. By the proposition, for each n one can construct a C^∞ volume-preserving ergodic diffeomorphism $f_n: \mathbb{T}^2 \times I \to \mathbb{T}^2 \times I$ satisfying:

(1) $\|F - f_n\|_{C^n} \leq e^{-n^2}$;

(2) $d^m f_n|\mathbb{T}^2 \times \{z\} = d^m F|\mathbb{T}^2 \times \{z\}$ for $z = 0, 1$ and all $0 \leq m < \infty$;

(3) f_n has nonzero Lyapunov exponents almost everywhere.

Let $L_n: I_n \to I$ be the affine onto map and let

$$\pi_n = (\mathrm{Id}, L_n): \mathbb{T}^2 \times I_n \to \mathbb{T}^2 \times I.$$

We define the map f by setting $f|\mathbb{T}^2 \times I_n = \pi_n^{-1} \circ f_n \circ \pi_n$ for all n and $f|\mathbb{T}^2 \times \{0\} = F|\mathbb{T}^2 \times \{0\}$. Note that for every $n > 0$ and $0 \leq m \leq n$ we have

$$\begin{aligned} \|d^m F|\mathbb{T}^2 \times I_n - \pi_n^{-1} \circ d^m f_n \circ \pi_n\|_{C^n} &\leq \|\pi_n^{-1} \circ (d^m F - d^m f_n) \circ \pi_n\|_{C^n} \\ &\leq e^{-n^2} \cdot (n+1)^n \to 0 \end{aligned}$$

when $n \to \infty$. It follows that f is a C^∞ diffeomorphism of M. It is easy to see that f is volume-preserving, has nonzero Lyapunov exponents almost everywhere, and that $f|\mathbb{T}^2 \times I_n$ is ergodic.

Proof of the proposition. Note that the map F is uniformly partially hyperbolic with one-dimensional center E_F^c, stable E_F^s and unstable E_F^u distributions. We obtain the desired result by arranging two C^∞ volume-preserving perturbations of F that ensure the essential accessibility property and positivity of the central Lyapunov exponent:

(1) The first perturbation is the time-t map (for a sufficiently small t) of a flow h_t generated by a vector field X, which vanishes outside a small neighborhood of a fixed point; moreover the map dh_t preserves the E_F^{uc}-planes.

(2) The second perturbation acts as small rotations in the E_F^{uc}-planes in a small neighborhood of another fixed point.

Let $\eta > 1$ and η^{-1} be the eigenvalues of A. Choose a small number $\varepsilon_0 > 0$ such that $d(p, p') \geq 3\varepsilon_0$ and consider the local stable and unstable one-dimensional manifolds $V^s(p)$, $V^u(p)$, $V^s(p')$, and $V^u(p')$ of "size" ε_0.

Let us choose the smallest positive number n_1 such that the intersection $A^{-n_1}(V^s(p')) \cap V^u(p) \cap B(p, \varepsilon_0)$ consists of a single point, which we denote by q_1. Similarly, let n_2 be the smallest positive number such that the intersection $A^{n_2}(V^u(p')) \cap V^s(p) \cap B(p, \varepsilon_0)$ consists of a single point, which we denote by q_2. See Figure 13.1.

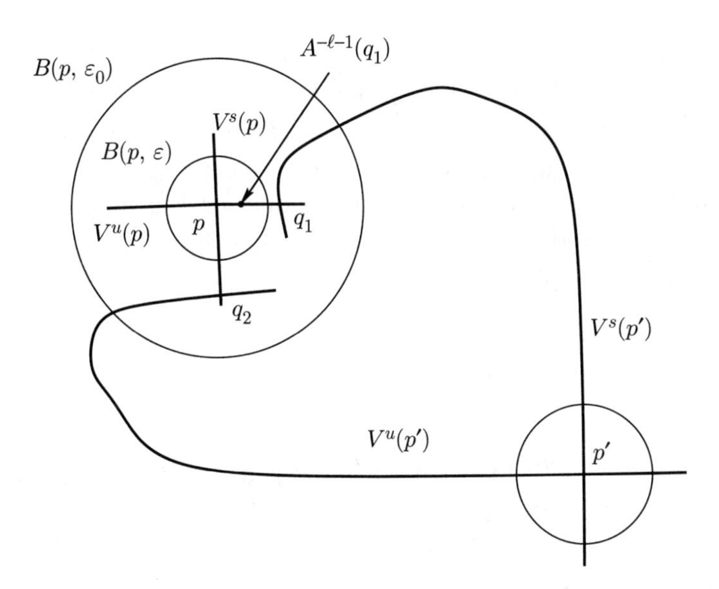

Figure 13.1

Take a sufficiently small $\varepsilon \in (0, \varepsilon_0)$ satisfying

$$\varepsilon \leq \frac{1}{2} \min\{d(p, q_1), d(p, q_2)\}.$$

There exists $\ell \geq 2$ such that (see Figure 13.1)

$$A^{-\ell}(q_1) \notin B(p, \varepsilon), \quad A^{-\ell-1}(q_1) \in B(p, \varepsilon). \tag{13.1}$$

We can choose ε so small that

$$B(p,\varepsilon)\cap(A^{-n_1}(V^s(p'))\cup A^{n_2}(V^u(p')))=\varnothing.$$

We can further reduce ε if necessary so that for some $q\in\mathbb{T}^2$, some number $N>0$, which will be determined later, and for every $i=1,\dots,N$,

$$A^i(B(q,\varepsilon))\cap B(q,\varepsilon)=\varnothing,\quad A^i(B(q,\varepsilon))\cap B(p,\varepsilon)=\varnothing.$$

Note that $\varepsilon=\varepsilon(N)$. Finally, we choose $\varepsilon'\in(0,\varepsilon)$ such that $A^{-\ell-1}(q_1)\in B(p,\varepsilon')$.

We define two disjoint open domains

$$\Omega_1=B(p,\varepsilon_0)\times I\quad\text{and}\quad\Omega_2=B^{uc}(\bar{q},\varepsilon_0)\times B^s(\bar{q},\varepsilon_0),\tag{13.2}$$

where $\bar{q}=(q,1/2)$ and the sets $B^{uc}(\bar{q},\varepsilon_0)\subset V^u(q)\times I$ and $B^s(\bar{q},\varepsilon_0)\subset V^s(q)$ are balls of radius ε_0.

The desired map g is obtained as the result of two perturbations H_1 and H_2 of F so that $g=H_1\circ F\circ H_2$, where each H_i is a C^∞ volume-preserving diffeomorphism of M which coincides with F outside of Ω_i, $i=1,2$. We construct the perturbation H_1 as a time-t map (for sufficiently small t) of a divergence-free vector field in Ω_1 and we construct the perturbation H_2 by applying a small rotation in the E_F^{uc}-plane at every point in Ω_2.

In order to construct the perturbation H_1, consider the coordinate system in Ω_1 with the origin at $(p,0)\in M$ and the x-, y-, z-axes, respectively, to be unstable, stable, and central directions for the map F. If a point $w=(x,y,z)\in\Omega_1$ and $F(w)\in\Omega_1$, then $F(w)=(\eta x,\eta^{-1}y,z)$.

We choose a C^∞ function $\xi\colon I\to\mathbb{R}^+$ satisfying (see Figure 13.2):

(1) $\xi(z)>0$ on $(0,1)$;

(2) $d^i\xi(0)=d^i\xi(1)=0$ for $i=0,1,\dots,k$;

(3) $\|\xi\|_{C^k}\le\delta$

and then two other C^∞ functions $\varphi=\varphi(x)$ and $\psi=\psi(y)$ on the interval $(-\varepsilon_0,\varepsilon_0)$ such that (see Figures 13.3 and 13.4):

(4) $\varphi(x)=\varphi_0$ for $x\in(-\varepsilon',\varepsilon')$ and $\psi(y)=\psi_0$ for $y\in(-\varepsilon',\varepsilon')$, where φ_0 and ψ_0 are positive constants;

(5) $\varphi(x)=0$ whenever $|x|\ge\varepsilon$; $\psi(y)\ge0$ for any y and $\psi(y)=0$ whenever $|y|\ge\varepsilon$;

(6) $\|\varphi\|_{C^k}\le\delta$, $\|\psi\|_{C^k}\le\delta$;

(7) $\int_0^{\pm\varepsilon}\varphi(s)\,ds=0$.

We now define a vector field X in Ω_1 by the formula

$$X(x,y,z)=\left(-\psi(y)\xi'(z)\int_0^x\varphi(s)\,ds,0,\psi(y)\xi(z)\varphi(x)\right).\tag{13.3}$$

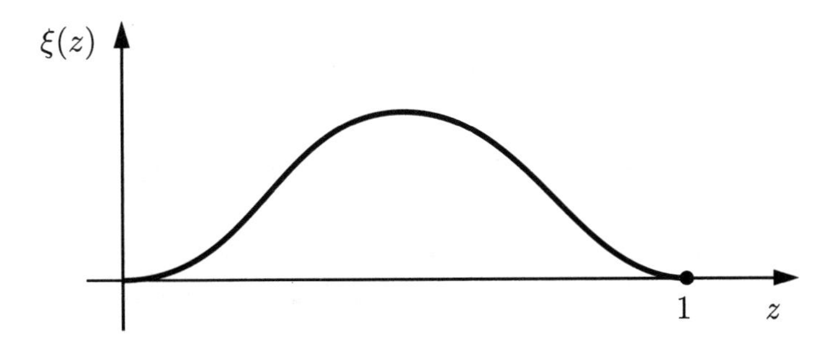

Figure 13.2. The function ξ

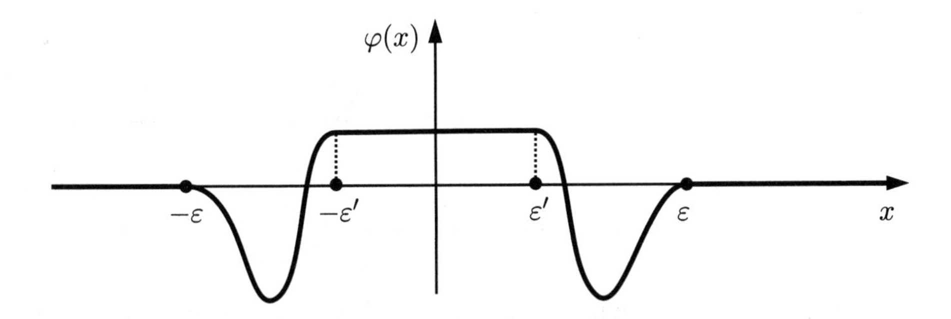

Figure 13.3. The function φ

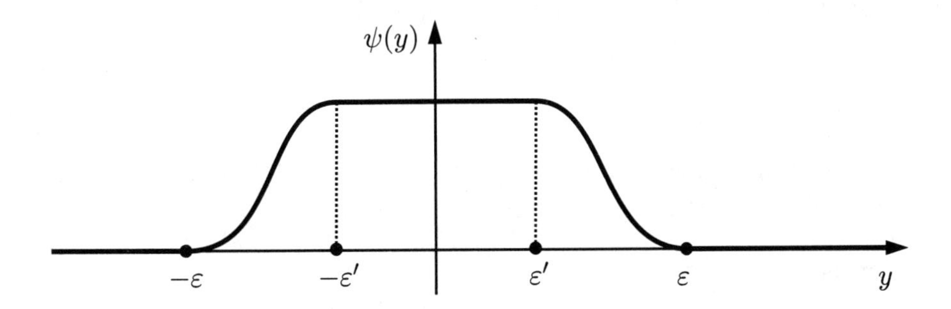

Figure 13.4. The function ψ

Exercise 13.3. Show that X is a divergence-free vector field supported on $(-\varepsilon, \varepsilon) \times (-\varepsilon, \varepsilon) \times I$.

We define a diffeomorphism h_t in Ω_1 to be the time-t map of the flow generated by X and we set $h_t = \mathrm{Id}$ on the complement of Ω_1. Fixing some sufficiently small $t > 0$, consider the map $H_1 = h_t$.

Exercise 13.4. Show that H_1 is a C^∞ volume-preserving diffeomorphism of M that preserves the y coordinate, that is, $H_1(\mathbb{R}\times\{y\}\times\mathbb{R}) \subset \mathbb{R}\times\{y\}\times\mathbb{R}$ for every $y \in \mathbb{R}$.

In order to construct the perturbation H_2, consider the coordinate system in Ω_2 with the origin at $(q, 1/2)$ and the x-, y-, z-axes, respectively, to be unstable, stable, and central directions. We then switch to the cylindrical coordinate system (r, θ, y), where $x = r\cos\theta$, $y = y$, and $z = r\sin\theta$.

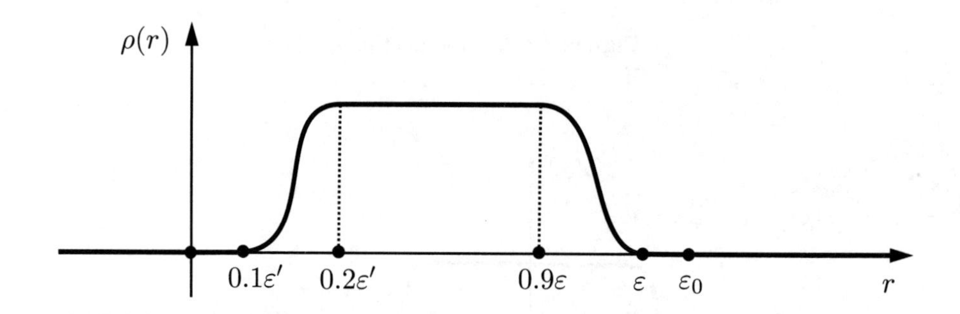

Figure 13.5. The function ρ

We choose a C^∞ function $\rho\colon (-\varepsilon_0, \varepsilon_0) \to \mathbb{R}^+$ satisfying (see Figure 13.5):

(8) $\rho(r) > 0$ for $0.2\varepsilon' \le r \le 0.9\varepsilon$ and $\rho(r) = 0$ for $r \le 0.1\varepsilon'$ and $r \ge \varepsilon$;

(9) $\|\rho\|_{C^k} \le \delta$.

For a sufficiently small $\tau > 0$, define a map $\tilde{h}_\tau$ in Ω_2 by

$$\tilde{h}_\tau(r, \theta, y) = (r, \theta + \tau\psi(y)\rho(r), y) \tag{13.4}$$

and set $\tilde{h}_\tau = \mathrm{Id}$ on $M \setminus \Omega_2$.

Exercise 13.5. Show that the map $H_2 = \tilde{h}_\tau$ is a C^∞ volume-preserving diffeomorphism of M that preserves the y coordinate.

Let us set

$$g = g_{t\tau} = h_t \circ F \circ \tilde{h}_\tau.$$

Exercise 13.6. Show that for all sufficiently small $t > 0$ and $\tau > 0$, the map $g_{t\tau}$ is a C^∞ diffeomorphism of M close to F in the C^1 topology and, hence, is partially hyperbolic. Show also that it is center bunched and preserves volume in M.

By Lemma 13.7 below, the map g has the essential accessibility property and hence, by Theorem 12.4, it is ergodic. It remains to explain that g has nonzero Lyapunov exponents almost everywhere.

Denote by $E^s_{t\tau}(w)$, $E^u_{t\tau}(w)$, and $E^c_{t\tau}(w)$ the stable, unstable, and central subspaces at a point $w \in M$ for the map $g_{t\tau}$. Set $\kappa_{t\tau}(w) = dg_{t\tau}|E^u_{t\tau}(w)$, $w \in M$. By Lemma 13.9 below, for all sufficiently small $\tau > 0$,

$$\int_M \log \kappa_{0\tau}(w)\, dw < \log \eta.$$

The subspace $E^u_{t\tau}(w)$ depends continuously on t and τ (for a fixed w) and hence, so does $\kappa_{t\tau}$. It follows that there are $t_0 > 0$ and $\tau_0 > 0$ such that for all $0 \le t \le t_0$ and $0 \le \tau \le \tau_0$ we have

$$\int_M \log \kappa_{t\tau}(w)\, dw < \log \eta.$$

Denote by $\chi^s_{t\tau}(w)$, $\chi^u_{t\tau}(w)$, and $\chi^c_{t\tau}(w)$ the Lyapunov exponents of $g_{t\tau}$ at the point $w \in M$ in the stable, unstable, and central directions, respectively (since these directions are one-dimensional the Lyapunov exponents do not depend on the vector). By ergodicity of $g_{t\tau}$ and Birkhoff's Ergodic Theorem, we have that for almost every $w \in M$,

$$\chi^u_{t\tau}(w) = \lim_{n\to\infty} \frac{1}{n} \log \prod_{i=0}^{n-1} \kappa_{t\tau}(g^i_{t\tau}(w)) = \int_M \log \kappa_{t\tau}(w)\, dw < \log \eta.$$

Since $E^s_{t\tau}(w) = E^s_{00}(w) = E^s_F(w)$ for every t and τ, we conclude that $\chi^s_{t\tau}(w) = -\log \eta$ for almost every $w \in M$. Since $g_{t\tau}$ is volume-preserving,

$$\chi^s_{t\tau}(w) + \chi^u_{t\tau}(w) + \chi^c_{t\tau}(w) = 0$$

for almost every $w \in M$. It follows that $\chi^c_{t\tau}(w) > 0$ for almost every $w \in M$ and hence, $g_{t\tau}$ has nonzero Lyapunov exponents almost everywhere. This completes the proof of the proposition. □

We now state and prove the two technical lemmas that we referred to in the proof of the theorem.

Lemma 13.7. *For every sufficiently small $t > 0$ and $\tau > 0$ the map $g_{t\tau}$ has the essential accessibility property.*

Proof of the lemma. Set $I_p = \{p\} \times (0, 1)$ (recall that p is one of the two fixed points of A; see Figure 13.1).

Exercise 13.8. Show that given a point $x \in M$, which does not lie on the boundary of M, there is a point $x' \in I_p$ such that x and x' are accessible.

It follows that in order to prove the lemma it suffices to show that any two points in I_p are accessible. Indeed, given two points $x, y \in M$, which do not lie on the boundary of M, we can find points $x', y' \in I_p$ such that the points x and x' are accessible and so are the points y and y'. If the points x', y' are accessible, the desired result follows, since accessibility is a transitive relation.

Given a point $w \in M$, denote by $\mathcal{A}(w)$ the accessibility class of w (i.e., the set of points $q \in M$ such that w and q are accessible). We use the coordinate system (x, y, z) in Ω_1 described above. Since the map h_t preserves the center leaf I_p, we have that for $z \in (0, 1)$,

$$h_t(0,0,z) = (h_t^{(1)}(0,0,z), h_t^{(2)}(0,0,z), h_t^{(3)}(0,0,z)) = (0,0,h_t^{(3)}(0,0,z)).$$

We claim that the accessibility property for points in I_p will follow if we show that for every $z \in (0, 1)$,

$$n\mathcal{A}(p,z) \supset \{(p,a) : a \in [(h_t^{-\ell})^{(3)}(p,z), z]\}, \tag{13.5}$$

where ℓ is chosen by (13.1). Indeed, since accessibility is a transitive relation and since $h_t^{-n}(p, z) \to (p, 0)$ for any $z \in (0, 1)$, condition (13.5) implies that $\mathcal{A}(p, z) \supset \{(p, a) : a \in (0, z]\}$ for all $z \in (0, 1)$ and the claim follows. We now proceed with the proof of (13.5).

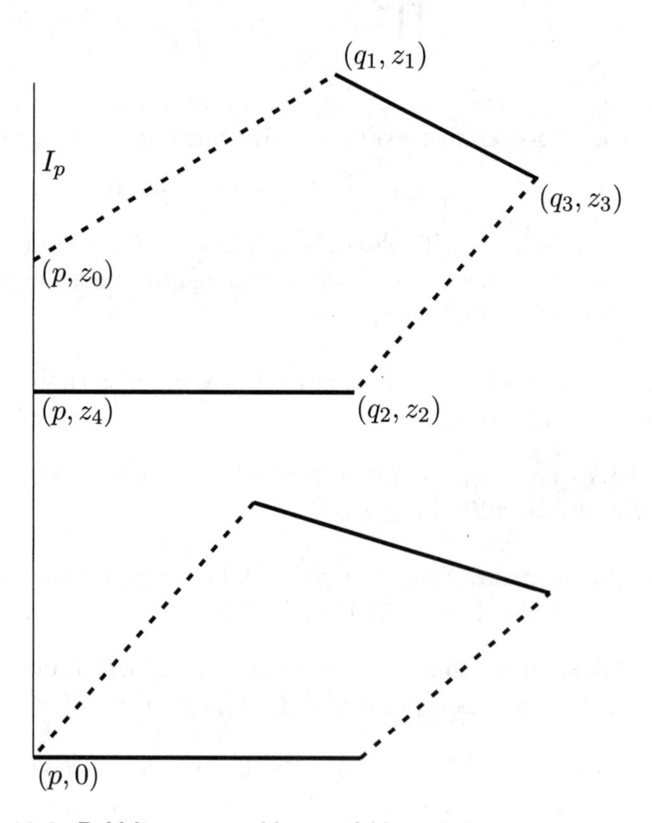

Figure 13.6. Bold lines are stable manifolds and dotted lines are unstable manifolds.

Denote by $V^s_{t\tau}(p)$, $V^u_{t\tau}(p)$ the local stable and unstable manifolds at p for the map $g_{t\tau}$. Let $q_1 \in V^u_{t\tau}(p)$ and $q_2 \in V^s_{t\tau}(p)$ be the two points constructed above (see Figure 13.1). The intersection $V^s_{t\tau}(q_1) \cap V^u_{t\tau}(q_2)$ is nonempty and consists of a single point q_3. We will show that for any $z_0 \in (0,1)$, there exist $z_i \in (0,1)$, $i = 1,2,3,4$, such that

$$(q_1, z_1) \in V^u_{t\tau}((p, z_0)), \quad (q_3, z_3) \in V^s_{t\tau}((q_1, z_1)),$$
$$(q_2, z_2) \in V^u_{t\tau}((q_3, z_3)), \quad (p, z_4) \in V^s_{t\tau}((q_2, z_2))$$

and

$$z_4 \le (h_t^{-\ell})^{(3)}(p, z_0) \tag{13.6}$$

(see Figure 13.6). This will imply that $(p, z_4) \in \mathcal{A}(p, z_0)$. By continuity, we conclude that

$$\{(p, a) : a \in [z_4, z_0]\} \subset \mathcal{A}(p, z_0)$$

and (13.5) will follow.

Since $g_{t\tau}$ preserves the (x,z)-plane, we have $V^{uc}_{t\tau}((p, z_0)) = V^{uc}_F((p, z_0))$. Hence, there is a unique $z_1 \in (0,1)$ such that $(q_1, z_1) \in V^u_{t\tau}((p, z_0))$. Notice that for $n \le \ell$,

$$g^{-n}_{t\tau}(p, z_0) = (p, h_t^{-n}((p, z_0))), \quad g^{-n}_{t\tau}(q_1, z_1) = (A^{-n} q_1, z_1).$$

This is true because the points $A^{-n} q_1$, $n = 0, 1, \ldots, \ell$, lie outside the ε-neighborhood of I_p, where the perturbation map $h_t = \mathrm{Id}$. Similarly, since the points $A^{-n} q_1$, $n > \ell$, lie inside the ε'-neighborhood of I_p and the third component of h_t depends only on the z-coordinate, we have

$$g^{-n}_{t\tau}(q_1, z_1) = (A^{-n} q_1, h_t^{-n+\ell} z_1).$$

Since

$$d(g^{-n}_{t\tau}((p, z_0)), g^{-n}_{t\tau}((q_1, z_1))) \to 0$$

as $n \to \infty$, we have

$$d(h_t^{-n}((p, z_0)), h_t^{-n+\ell}((p, z_1))) \to 0$$

as $n \to \infty$. It follows that $z_1 = (h_t^{-\ell})^{(3)}((p, z_0))$.

Since $h_t = \mathrm{Id}$ outside Ω_1, the sets $A^{-n_1} V^s_{t\tau}(p')$ and $A^{n_2} V^u_{t\tau}(p')$ are pieces of horizontal lines and hence, $z_2 = z_3 = z_1$. Since the third component of h_t does not decrease when a point moves from (q_2, z_2) to (p, z_4) along $V^s_{t\tau}(p)$, we conclude that $z_4 \le z_3 = z_1 = (h_t^{-\ell})^3(p, z_0)$ and thus (13.6) holds, completing the proof of the lemma. □

Lemma 13.9. *For any sufficiently small $\tau > 0$,*

$$\int_M \log \kappa_{0\tau}(w) dw < \log \eta.$$

Proof of the lemma. For any $w \in M$, we introduce the coordinate system in T_wM associated with the splitting $T_wM = E_F^u(w) \oplus E_F^s(w) \oplus E_F^c(w)$. Given $\tau > 0$ and $w \in M$, there exists a unique number $\alpha_\tau(w)$ such that the vector $v_\tau(w) = (1, 0, \alpha_\tau(w))^\perp$ lies in $E_{0\tau}^u(w)$ (where $\perp$ denotes the transpose). Since the map $\tilde{h}_\tau$ preserves the y-coordinate, by the definition of the function $\alpha_\tau(w)$, one can write the vector $dg_{0\tau}(w)v_\tau(w)$ in the form

$$dg_{0\tau}(w)v_\tau(w) = (\bar{\kappa}_\tau(w), 0, \bar{\kappa}_\tau(w)\alpha_\tau(g_{t0}(w)))^\perp \tag{13.7}$$

for some $\bar{\kappa}_\tau(w) > 1$. Taking into account that the expansion rate of $dg_{0\tau}(w)$ along its unstable direction is $\kappa_{0\tau}(w)$, we obtain that

$$\kappa_{0\tau}(w) = \bar{\kappa}_\tau(w)\frac{\sqrt{1 + \alpha_\tau(g_{0\tau}(w))^2}}{\sqrt{1 + \alpha_\tau(w)^2}}.$$

Since $E_{0\tau}^u(w)$ is close to $E_{00}^u(w)$, the function $\alpha_\tau(w)$ is uniformly bounded. Using the fact that the map $g_{0\tau}$ preserves volume, we find that

$$L_\tau = \int_M \log \kappa_{0\tau}(w)\, dw = \int_M \log \bar{\kappa}_\tau(w)\, dw. \tag{13.8}$$

Consider the map $\tilde{h}_\tau$. Since it preserves the y-coordinate, using (13.4), we can write that

$$\tilde{h}_\tau(x, y, z) = (r\cos\sigma, y, r\sin\sigma),$$

where $\sigma = \sigma(\tau, r, \theta, y) = \theta + \tau\psi(y)\rho(r)$. Therefore, the differential

$$d\tilde{h}_\tau \colon E_F^u(w) \oplus E_F^c(w) \to E_F^u(g_{0\tau}(w)) \oplus E_F^c(g_{0\tau}(w))$$

can be written in the matrix form

$$\begin{aligned} d\tilde{h}_\tau(w) &= \begin{pmatrix} A(\tau, w) & B(\tau, w) \\ C(\tau, w) & D(\tau, w) \end{pmatrix} \\ &= \begin{pmatrix} r_x\cos\sigma - r\sigma_x\sin\sigma & r_y\cos\sigma - r\sigma_y\sin\sigma \\ r_x\sin\sigma + r\sigma_x\cos\sigma & r_y\sin\sigma + r\sigma_y\cos\sigma \end{pmatrix}, \end{aligned}$$

where

$$r_x = \frac{\partial r}{\partial x} = \frac{x}{r} = \cos\theta, \quad r_z = \frac{\partial r}{\partial z} = \frac{y}{r} = \sin\theta,$$

$$\sigma_x = \frac{\partial \sigma}{\partial x} = \frac{-z}{r^2} + \frac{z}{r}\tau\tilde{\rho}_r(y, r) = \frac{\sin\theta}{r} + \tau\tilde{\rho}_r(y, r)\cos\theta,$$

$$\sigma_z = \frac{\partial \sigma}{\partial z} = \frac{x}{r^2} + \frac{x}{r}\tau\tilde{\rho}_r(y, r) = \frac{\cos\theta}{r} + \tau\tilde{\rho}_r(y, r)\sin\theta,$$

and $\tilde{\rho}(y, r) = \psi(y)\rho(r)$.

Exercise 13.10. Show that

$$\begin{aligned}
A &= A(\tau, w) = 1 - \tau r \tilde{\rho}_r \sin\theta\cos\theta - \frac{\tau^2\tilde{\rho}^2}{2} - \tau^2 r \tilde{\rho}\tilde{\rho}_r \cos^2\theta + O(\tau^3),\\
B &= B(\tau, w) = -\tau\tilde{\rho} - \tau r \tilde{\rho}_r \sin^2\theta - \tau^2 r \tilde{\rho}\tilde{\rho}_r \sin\theta\cos\theta + O(\tau^3),\\
C &= C(\tau, w) = \tau\tilde{\rho} + \tau r \tilde{\rho}_r \cos^2\theta - \tau^2 r \tilde{\rho}\tilde{\rho}_r \sin\theta\cos\theta + O(\tau^3),\\
D &= D(\tau, w) = 1 + \tau r \tilde{\rho}_r \sin\theta\cos\theta - \frac{\tau^2\tilde{\rho}^2}{2} - \tau^2 r \tilde{\rho}\tilde{\rho}_r \sin^2\theta + O(\tau^3).
\end{aligned}$$

We now obtain the formula for L_τ:

$$L_\tau = \log\eta - \int_M \log(D(\tau, w) - \eta B(\tau, w)\alpha_\tau(g_{0\tau}(w)))dw. \tag{13.9}$$

Indeed, since $g_{0\tau} = h_0 \circ F \circ \tilde{h}_\tau = F \circ \tilde{h}_\tau$, we have that

$$\mathcal{D}_\tau(w) = dg_{0\tau}(w)|E^u_{0\tau}(w) \oplus E^c_{0\tau}(w) = \begin{pmatrix} \eta A(\tau, w) & \eta B(\tau, w) \\ C(\tau, w) & D(\tau, w) \end{pmatrix}.$$

By (13.7),

$$\begin{aligned}
\mathcal{D}_\tau(w)\begin{pmatrix} 1 \\ \alpha_\tau(w) \end{pmatrix} &= \begin{pmatrix} \eta A(\tau, w) + \eta B(\tau, w)\alpha_\tau(w) \\ C(\tau, w) + D(\tau, w)\alpha_\tau(w) \end{pmatrix}\\
&= \begin{pmatrix} \kappa_\tau(w) \\ \kappa_\tau(w)\alpha_\tau(g_{0\tau}(w)) \end{pmatrix}.
\end{aligned} \tag{13.10}$$

Since $\tilde{h}_\tau$ is volume-preserving, $AD - BC = 1$ and therefore,

$$A + B\alpha = \frac{1}{D} + \frac{B}{D}(C + D\alpha).$$

Comparing the components in (13.10), we obtain

$$\begin{aligned}
\kappa_\tau(w) &= \eta(A(\tau, w) + B(\tau, w)\alpha_\tau(w))\\
&= \eta\left(\frac{1}{D(\tau, w)} + \frac{B(\tau, w)}{D(\tau, w)}(C(\tau, w) + D(\tau, w)\alpha_\tau(w))\right)\\
&= \eta\left(\frac{1}{D(\tau, w)} + \frac{B(\tau, w)}{D(\tau, w)}(\kappa_\tau(w)\alpha_\tau(g_{0\tau}(w)))\right).
\end{aligned}$$

Solving for $\kappa_\tau(w)$, we get

$$\kappa_\tau(w) = \frac{\eta}{D(\tau, w) - \eta B(\tau, w)\alpha_\tau(g_{0\tau}(w))}.$$

The desired relation (13.9) follows from (13.8).

One can deduce from (13.9) by a straightforward calculation that

$$\left.\frac{dL_\tau}{d\tau}\right|_{\tau=0} = 0 \quad \text{and} \quad \frac{d^2L_\tau}{d\tau^2}|_{\tau=0} > 0 \tag{13.11}$$

(we stress that the argument is not quite trivial and we refer the reader to [**9**] for details). To conclude the proof of the lemma, it remains to notice that by (13.11), $L_\tau < \log \eta$ for all sufficiently small τ. $\square$

13.2. The Shub–Wilkinson map

We construct an example of a uniformly partially hyperbolic diffeomorphism with positive central exponents. Let A be a linear hyperbolic automorphism of the two-dimensional torus $\mathbb{T}^2$. Consider the map $F = A \times \mathrm{Id}$ of the three-dimensional torus $\mathbb{T}^3 = \mathbb{T}^2 \times S^1$ where Id is the identity map. This map preserves volume and is uniformly partially hyperbolic. It has zero Lyapunov exponent in the central direction. Any sufficiently small C^1 perturbation G of F is also uniformly partially hyperbolic with one-dimensional central direction. The following result due to Shub and Wilkinson (see [**81**] and also [**28**]) shows that this perturbation can be arranged in such a way to ensure positive Lyapunov exponents in the central direction.

Theorem 13.11. *For any $k \geq 2$ and $\delta > 0$, there exists a volume-preserving C^∞ perturbation G of F such that G is δ-close to F in the C^k topology and has positive central exponents almost everywhere.*

Proof. We follow the approach described in the previous section. Without loss of generality we may assume that the linear hyperbolic automorphism A has at least two fixed points that we denote by p and p'. For a sufficiently small ε_0 consider the two disjoint open domains Ω_1 and Ω_2 given by (13.2). Define a diffeomorphism h_t in Ω_1 to be the time-t map of the flow generated by the vector field X given by (13.3) and set $h_t = \mathrm{Id}$ outside Ω_1. Now define a diffeomorphism $\tilde{h}_\tau$ in Ω_2 by (13.4) and set $\tilde{h}_\tau = \mathrm{Id}$ outside Ω_2. Finally, set $G = h_t \circ F \circ \tilde{h}_\tau$. Repeating the arguments in the proof of Proposition 13.2, one can show that for sufficiently small $t > 0$ and $\tau > 0$ the map G is a C^∞ volume-preserving diffeomorphism, which has the essential accessibility property and positive central exponents almost everywhere. The desired result follows from Theorem 12.4. $\square$

One can show that any sufficiently small C^∞ volume-preserving perturbation of G in the C^1 topology has the essential accessibility property and positive central exponents almost everywhere. Hence, it is ergodic and has the Bernoulli property. In particular, we obtain an open set in the C^1 topology in $\mathbb{T}^3$ of volume-preserving non-Anosov diffeomorphisms with nonzero Lyapunov exponents.

Chapter 14

Anosov Rigidity

In every example of volume-preserving nonuniformly hyperbolic diffeomorphisms that we have constructed in the book, the Lyapunov exponents were nonzero on a set of full volume but *not* everywhere. On the other hand, the Lyapunov exponents of an Anosov diffeomorphism are nonzero at *every* point on the manifold. This observation leads to a natural problem of whether *nonuniform hyperbolicity everywhere on a compact manifold implies uniform hyperbolicity*—a phenomenon that we call *Anosov rigidity*. We describe two versions of the Anosov rigidity phenomenon requiring two quite different approaches. The first one deals with the situation when a diffeomorphism f is nonuniformly hyperbolic on a *compact* and *invariant* subset K, in which case the problem is to show that $f|K$ is uniformly hyperbolic. The second one requires the weaker hypothesis that Lyapunov exponents are nonzero on a set of total measure, i.e., off a set that is a null set with respect to every f-invariant Borel probability measure.

14.1. The Anosov rigidity phenomenon. I

We describe the approach to the Anosov rigidity phenomenon developed in [**36**]. It can be expressed in the following two statements. Let f be a C^1 diffeomorphism of a compact smooth Riemannian manifold M and let K be a compact invariant subset.

Theorem 14.1. *Assume that there is a continuous invariant cone family $C(x) \subset T_xM$ on K such that:*

(1) *it is invariant; i.e., $dfC(x) \subset C(f(x))$;*

(2) *for all* $x \in K$,

$$\varphi(x) = \liminf_{n\to\infty} \frac{1}{n} \min_{v\in C(x), \|v\|=1} \log \|d_x f^n(v)\| > 0. \tag{14.1}$$

Then there exist $c > 0$ *and* $\chi > 0$ *such that for every* $x \in K$, $v \in C(x)$, $\|v\| = 1$, *and* $n \in \mathbb{N}$,

$$\|d_x f^n(v)\| \geq c e^{\chi n}.$$

Furthermore, for every $x \in K$ *there is a subspace* $E(x) \subset T_x M$ *such that*

$$df(E(x)) = E(f(x)) \quad \text{and} \quad \|d_x f^n(v)\| \geq c e^{\chi n}$$

for every $v \in E_x$, $\|v\| = 1$, *and* $n \in \mathbb{N}$.

Theorem 14.2. *Assume that there are two continuous transverse cone families* $C(x)$ *and* $D(x)$ *on* K *such that:*

(1) *they are invariant; i.e.,*

$$dfC(x) \subset C(f(x)), \quad df^{-1}D(x) \subset D(f^{-1}(x));$$

(2) *for all* $x \in K$,

$$\liminf_{n\to\infty} \frac{1}{n} \min_{v\in C(x), \|v\|=1} \log \|d_x f^n(v)\| > 0,$$

$$\liminf_{n\to-\infty} \frac{1}{|n|} \min_{v\in D(x), \|v\|=1} \log \|d_x f^n(v)\| > 0.$$

Then K *is a uniformly hyperbolic set for* f. *In particular, if* $K = M$, *then* f *is an Anosov diffeomorphism.*

Exercise 14.3. Deduce Theorem 14.2 from Theorem 14.1.

Theorem 14.2 was first proved by Mañé in [**61**]. Its continuous-time version is due to Sacker and Sell [**80**]. We present a proof of Theorem 14.1 following closely the approach developed by Hasselblatt, Pesin, and Schmeling in [**36**]. It exploits some ideas from descriptive set theory.

14.1.1. Transfinite hierarchy of set filtrations. We present a set-theoretic construction providing a detailed study of representations of a compact space as a nested union of compact subsets. The main idea can be highlighted by examining the well-known proof that a positive continuous function φ on a compact space has a positive minimum: the open cover by sets $\varphi^{-1}(\frac{1}{n}, \infty)$ has a finite subcover. Attempting to extend this proof to Baire functions, one might try to cover the space with the interiors of the sets $\varphi^{-1}[\frac{1}{n}, \infty)$. If the compact set A that remains after deleting all these interiors is nonempty, then one repeats the entire process on the set A with respect to its subspace topology. We develop a transfinite process of this sort for the function $\varphi(x)$ given by (14.1). We split our construction into three steps.

14.1.1.1. *Set filtrations.* Let (X, d) be a compact separable metric space. A *set filtration* of X is a collection of compact subsets $X_n \subset X$ such that:

(1) they exhaust X; i.e., $\bigcup_{n\in\mathbb{N}} X_n = X$;
(2) they are nested; i.e., $X_n \subseteq X_{n+1}$ for $n \in \mathbb{N}$;
(3) if $X_{n+1} \neq X$, then $X_n \subsetneqq X_{n+1}$.

We say that X is *uniform* with respect to this filtration if $X = X_n$ for some $n \in \mathbb{N}$.

Lemma 14.4. *We have that* $X = \mathrm{Cl} \bigcup_{n\in\mathbb{N}} \operatorname{int} X_n$, *where* Cl *denotes closure of the corresponding set.*[1]

Proof. We need to show that given a closed ball B centered at some point x and of some radius r, the intersection $B \cap \bigcup_{n\in\mathbb{N}} \operatorname{int} X_n$ is nonempty. To this end, note that

$$B = B \cap X = B \cap \bigcup_{n\in\mathbb{N}} X_n = \bigcup_{n\in\mathbb{N}} B \cap X_n$$

is a complete metric space and, hence, not a countable union of sets of first category. Thus, there exists $N \in \mathbb{N}$ such that $X_N \cap B$ is of second category and, hence, not nowhere dense. This means that

$$\varnothing \neq \operatorname{int}_B(\mathrm{Cl}\, X_N) = \operatorname{int}_B X_N \subset B \cap \operatorname{int} X_N,$$

where int_B denotes the interior in the subspace topology of B. This implies the desired result. □

The set $\Gamma = X \setminus \bigcup_{n\in\mathbb{N}} \operatorname{int} X_n$ is clearly compact.

Lemma 14.5. *We have that*

$$\Gamma = \{x \in X : \textit{there exists } x_n \to x \textit{ such that } x_n \notin X_n\}.$$

Proof. For $x \in \Gamma$ there exists a sequence $y_n \to x$ such that $y_n \notin \operatorname{int} X_n$. By the definition of interior, one can find $x_n \notin X_n$ such that $d(x_n, y_n) < 1/n$. Thus,

$$\Gamma \subset \{x \in X : \text{there exists } x_n \to x \text{ such that } x_n \notin X_n\}.$$

The reverse inclusion is clear because $x_n \notin X_n$ implies that $x_n \notin \operatorname{int} X_n$. □

[1] We use different notation for closure for reasons that will be clear later (see Section 14.1.1.2).

14.1.1.2. *The hierarchy.* In view of Lemma 14.4 we wish to exhaust the set X with the interiors of sets X_n from the filtration. This leaves uncovered the compact set Γ, and we now describe how to continue this process recursively in a transfinite way.

Set $X_n^{(0)} = X_n$, $F^{(0)} = X$, and $\Gamma^{(0)} = \Gamma$. Given an ordinal β such that we already have sets $\Gamma^{(\alpha)}$ for all $\alpha < \beta$, we inductively define

$$F^{(\beta)} = \bigcap_{\alpha<\beta} \Gamma^{(\alpha)}, \quad X_n^{(\beta)} = F^{(\beta)} \cap X_n,$$

and

$$\Gamma^{(\beta)} = \mathrm{Cl}_{F^{(\beta)}} \left(\bigcup_{n\in\mathbb{N}} \mathrm{int}_{F^{(\beta)}} X_n^{(\beta)} \right) \setminus \bigcup_{n\in\mathbb{N}} \mathrm{int}_{F^{(\beta)}} X_n^{(\beta)} \subset F^{(\beta)},$$

where $\mathrm{Cl}_{F^{(\beta)}}$ denotes the closure in the subspace topology of $F^{(\beta)}$. Our next statement implies that taking the ambient closure gives the same set.

Lemma 14.6. *The sets $\Gamma^{(\beta)}$, $F^{(\beta)}$, and $X_n^{(\beta)}$ are compact.*

Proof. For $\beta = 0$ this is the compactness of Γ, X, and X_n. We proceed by induction assuming that $\Gamma^{(\alpha)}$ is compact for all $\alpha < \beta$. Then $F^{(\beta)}$ is compact because it is defined as an intersection of compact sets. Since X_n is compact, this implies compactness of $X_n^{(\beta)}$. Finally, $\Gamma^{(\beta)}$ is a closed subset of $F^{(\beta)}$, hence also compact. □

Lemma 14.7. *We have that:*

(1) *the sets $F^{(\beta)}$, $X_n^{(\beta)}$, and $\Gamma^{(\beta)}$ are nested; i.e., for $\alpha < \beta$,*

$$F^{(\beta)} \subseteq F^{(\alpha)}, \quad X_n^{(\beta)} \subseteq X_n^{(\alpha)}, \quad \Gamma^{(\beta)} \subseteq \Gamma^{(\alpha)};$$

(2) *the sets $X_n^{(\beta)}$ form a filtration of the set $F^{(\beta)}$; i.e.,*

$$\bigcup_{n\in\mathbb{N}} X_n^{(\beta)} = F^{(\beta)}, \quad X_n^{(\beta)} \subseteq X_{n+1}^{(\beta)};$$

(3) $F^{(\beta)} = \mathrm{Cl} \bigcup_{n\in\mathbb{N}} \mathrm{int}_{F^{(\beta)}} X_n^{(\beta)}$ *and hence,*

$$\Gamma^{(\beta)} = F^{(\beta)} \setminus \bigcup_{n\in\mathbb{N}} \mathrm{int}_{F^{(\beta)}} X_n^{(\beta)};$$

(4) *if $\alpha < \beta$ and $F^{(\alpha)} \neq \varnothing$, then $F^{(\beta)} \subsetneq F^{(\alpha)}$; i.e., the transfinite induction for sets $F^{(\beta)}$ can stabilize only at $\varnothing$;*

(5) $F^{(\alpha+1)} = \Gamma^{(\alpha)}$ *and hence,* $X_n^{(\alpha+1)} = \Gamma^{(\alpha)} \cap X_n$.

Proof. (1) For the sets $F^{(\beta)}$ and $X_n^{(\beta)}$ the statement follows from the definitions and for the sets $\Gamma^{(\beta)}$ it follows from

$$\Gamma^{(\beta)} \subseteq F^{(\beta)} = \bigcap_{\tau<\beta} \Gamma^{(\tau)} \subseteq \Gamma^{(\alpha)}.$$

(2) We have that

$$F^{(\beta)} = F^{(\beta)} \cap X = F^{(\beta)} \cap \bigcup_{n\in\mathbb{N}} X_n = \bigcup_{n\in\mathbb{N}} F^{(\beta)} \cap X_n = \bigcup_{n\in\mathbb{N}} X_n^{(\beta)}$$

and

$$X_n^{(\beta)} = F^{(\beta)} \cap X_n \subset F^{(\beta)} \cap X_{n+1} = X_{n+1}^{(\beta)}.$$

(3) The result follows from (2) by applying Lemmas 14.6 and 14.4 to the set $F^{(\beta)} = \bigcup_{n\in\mathbb{N}} X_n^{(\beta)}$.

(4) The set $F^{(\alpha)} = \bigcup_{n\in\mathbb{N}} X_n^{(\alpha)}$ is nonempty, compact, and, hence, complete. Therefore, there exists $n_0 \in \mathbb{N}$ such that $X_{n_0}^{(\alpha)}$ is of second category in the induced topology of $F^{(\alpha)}$. Then $\operatorname{int} X_{n_0}^{(\alpha)} \neq \varnothing$ because $X_{n_0}^{(\alpha)}$ is compact and nonempty. It follows that

$$\begin{aligned} F^{(\beta)} &= \bigcap_{\gamma<\beta} \Gamma^{(\gamma)} \subset \bigcap_{\alpha\leq\gamma<\beta} \Gamma^{(\gamma)} \subset \Gamma^{(\alpha)} \\ &= F^{(\alpha)} \setminus \bigcup_{n\in\mathbb{N}} \operatorname{int} X_n^{(\alpha)} \subset F^{(\alpha)} \setminus \operatorname{int} X_{n_0}^{(\alpha)} \subsetneqq F^{(\alpha)}. \end{aligned}$$

(5) By statement (1) the sets $\Gamma^{(\tau)}$ are nested and hence,

$$F^{(\alpha+1)} = \bigcap_{\tau<\alpha+1} \Gamma^{(\tau)} = \bigcap_{\tau\leq\alpha} \Gamma^{(\tau)} = \Gamma^{(\alpha)}.$$

This completes the proof of the lemma. □

14.1.1.3. *Termination of the process.* We now describe what happens in the end of the induction process. The following statement reflects the tacit assumption that $F^{(0)} = X \neq \varnothing$.

Lemma 14.8. *There is a countable ordinal ξ such that $F^{(\xi)} \neq \varnothing = F^{(\xi+1)}$.*

Proof. Since the space X is second countable, it has a countable base $\mathcal{U}$. If $F^{(\alpha)} \neq \varnothing$, then $F^{(\alpha)} \setminus F^{(\alpha+1)} \neq \varnothing$ by Lemma 14.7(4). Since this is open in the subspace topology of $F^{(\alpha)}$, there exists $\mathcal{O}_\alpha \in \mathcal{U}$ such that $\mathcal{O}_\alpha \cap F^{(\alpha)} \neq \varnothing$ and $\mathcal{O}_\alpha \cap F^{(\alpha+1)} = \varnothing$. Such sets $\mathcal{O}_\alpha$ are pairwise distinct, so there are only countably many α for which $F^{(\alpha)} \neq \varnothing$. Thus, $F^{(\alpha_0)} = \varnothing$ for a countable ordinal.

The set $\{\alpha < \omega_1 \colon F^{(\alpha)} = \varnothing\}$, where ω_1 is the first uncountable ordinal, contains α_0 and, hence, is a nonempty subset of the well-ordered set of countable ordinals. Therefore, it contains a minimal element η.

If η is a limit ordinal, i.e., is not of the form $\xi+1$ for any ordinal ξ, then there is an increasing sequence $(\alpha_n)_{n\in\mathbb{N}}$ of ordinals such that for all $\tau<\eta$ there exists $n\in\mathbb{N}$ for which $\tau<\alpha_n<\eta$. Hence,

$$F^{(\eta)}=\bigcap_{\tau<\eta}\Gamma^{(\tau)}=\bigcap_{\tau<\eta}F^{(\tau+1)}=\bigcap_{\tau<\eta}F^{(\tau)}=\bigcap_{n\in\mathbb{N}}F^{(\alpha_n)}\neq\varnothing,$$

since $\varnothing\neq F^{(\alpha_{n+1})}\subset F^{(\alpha_n)}$. This is a contradiction, so we can write $\eta=\xi+1$. □

Lemma 14.9. *If ξ is as in Lemma* 14.8, *i.e., $F^{(\xi)}\neq\varnothing=F^{(\xi+1)}$, then the following statements hold:*

(1) *$F^{(\xi)}\subset X_R$ for some $R\in\mathbb{N}$; in particular, if $\xi=0$, then X is uniform with respect to the filtration $(X_n)_{n\in\mathbb{N}}$;*

(2) *if $\tau<\xi$, then $\bigcup_{n=1}^{\infty}\operatorname{int}X_n^{(\tau)}\subsetneqq F^{(\tau)}$.*

(3) *if $\xi>0$, then for every $\varepsilon>0$ there exist $\tau<\xi$ and $N\in\mathbb{N}$ such that*

$$F^{(\tau)}\setminus U_\varepsilon(F^{(\xi)})\subset\bigcup_{n=1}^{N}\operatorname{int}X_n^{(\tau)}\subset X_N^{(\tau)}\subset X_N.$$

Proof. (1) Statements (3)–(5) of Lemma 14.7 yield

$$\varnothing=F^{(\xi+1)}=\Gamma^{(\xi)}=F^{(\xi)}\setminus\bigcup_{n\in\mathbb{N}}\operatorname{int}_{F^{(\xi)}}X_n^{(\xi)}.$$

Hence, $F^{(\xi)}\subset\bigcup_{n\in\mathbb{N}}\operatorname{int}_{F^{(\xi)}}X_n^{(\xi)}$. The desired result follows since this open cover has a finite subcover.

(2) If $F^{(\tau)}=\bigcup_{n=1}^{\infty}\operatorname{int}_{F^{(\tau)}}X_n^{(\tau)}$, then by statements (3), (4), and (5) of Lemma 14.7, we have $\varnothing=\Gamma^{(\tau)}=F^{(\tau+1)}$ and $\tau\geq\xi$.

(3) We have that

$$\bigcap_{\alpha<\xi}\Gamma^{(\alpha)}=F^{(\xi)}\neq\varnothing.$$

The sets $\Gamma^{(\alpha)}$ are nested and compact and hence, $\inf_{\alpha<\xi}d_H(F^{(\xi)},\Gamma^{(\alpha)})=0$, where d_H is the Hausdorff distance. That means that there exists $\tau<\xi$ such that $\Gamma^{(\alpha)}\subseteq U_\varepsilon(F^{(\xi)})$ whenever $\tau\leq\alpha<\xi$; in particular, for $\alpha=\tau$ we have

$$F^{(\tau)}\setminus\bigcup_{n\in\mathbb{N}}\operatorname{int}X_n^{(\tau)}=\Gamma^{(\tau)}\subseteq U_\varepsilon(F^{(\xi)}),$$

and hence

$$F^{(\tau)}\setminus U_\varepsilon(F^{(\xi)})\subseteq\bigcup_{n\in\mathbb{N}}\operatorname{int}X_n^{(\tau)}.$$

This is an open cover of a compact set. Hence, the claim follows. □

14.1.2. Proof of Theorem 14.1. Set

$$\varphi_n = \frac{1}{n} \min_{v \in C(x), \|v\|=1} \log \|d_x f^n v\|$$

and consider the filtration of X by the sets

$$X_n = \left\{ x \in X \colon \varphi_k(x) \geq \frac{1}{n} \text{ for } k \geq n \right\} \subset \left\{ x \in X \colon \varphi(x) \geq \frac{1}{n} \right\}.$$

The main idea of the proof is the following. Assume that we can find two compact sets $K_1 \subset K_2 \subset X$ such that the set K_1 is uniform with respect to the filtration X_n (i.e., $K_1 \subset X_N$ for some $N > 0$). Assume also that $K_2 \setminus O$ is known to be uniform whenever the set O is open and $K_1 \subset O$. If there is a uniform neighborhood U of K_1, then we conclude that $K_2 \subset U \cup K_2 \setminus U$ is uniform as well.

To implement this idea, we first show that $K_1 = F^{(\xi)}$, which is uniform by statement (1) of Lemma 14.9, has a uniform ε-neighborhood U_ε (see Lemma 14.10). This is the main step in the proof. Now we observe that if $\xi > 0$ in Lemma 14.8, then we can take $\tau < \xi$ as in statement (3) of Lemma 14.9 and conclude from the above that $K_2 = F^{(\tau)}$ is uniform. Since this implies that $F^{(\tau+1)} = \varnothing$, we conclude that $\tau + 1 > \xi$ after all, a contradiction. Consequently, $\xi = 0$, and $X = F^{(0)}$ is uniform by statement (1) of Lemma 14.9, as claimed.

Lemma 14.10. *There exist $C > 0$, $\varepsilon > 0$, and $\lambda > 1$ such that if $f^n(x) \in U_\varepsilon(F^{(\xi)})$ whenever $0 \leq n \leq K$ for some $K \in \mathbb{N}$, then*

$$\min_{v \in C(x), \|v\|=1} \|d_x f^n v\| \geq C\lambda^n \quad \textit{whenever} \quad 0 \leq n \leq K.$$

Proof. By statement (1) of Lemma 14.9, there exists $R \in \mathbb{N}$ (which depends on ξ) such that $F^{(\xi)} \subset X_R$. Thus for all $n \geq R$ and $y \in F^{(\xi)}$ we have

$$\frac{1}{R} \leq \varphi_n(y) = \frac{1}{n} \min_{v \in C_y^1, \|v\|=1} \log \|d_x f^n v\|;$$

hence,

$$\min_{v \in C(y), \|v\|=1} \|d_x f^n v\| \geq e^{n/R}.$$

Now take $L \in \mathbb{N}$ so large that if $y \in F^{(\xi)}$, then

$$\min_{v \in C(y), \|v\|=1} \|d_x f^L v\| \geq 3 \max_{x \in X} \max_{v \in C(x), \|v\|=1} \|v\| = 3$$

and write $g = d_x f^L$. Note that L depends only on R and hence only on ξ. If $v \in C(y)$ and $\|v\| = 1$, then

$$\begin{aligned} \|d_x g^n v\| = \frac{\left\| g\left(g^{n-1}v / \|g^{n-1}v\|\right) \right\|}{\left\| \left(g^{n-1}v / \|g^{n-1}v\|\right) \right\|} \|g^{n-1}v\| \geq 3\|g^{n-1}v\| \geq \cdots \\ \geq 3^{n-1}\|gv\| \geq 3^n. \end{aligned}$$

Thus, for $n \in \mathbb{N}$ and $y \in F^{(\xi)}$ we have

$$\min_{v\in C(y),\|v\|=1} \|g^n(v)\| \geq 3^n.$$

If $K \leq L$, then the conclusion of Lemma 14.10 is obtained by taking

$$C \leq \min_{1\leq n\leq L} \min_{v\in C(x),\|v\|=1} \|d_x f^n v\| \lambda^{-n},$$

where $\lambda > 1$ can be chosen arbitrarily (its particular choice will be determined below). For $K > L$ we continue as follows. For any $x \in U_\varepsilon(F^{(\xi)})$ we can choose $y \in F^{(\xi)}$ such that $d(x, y) < \varepsilon$. Then

$$\min_{v\in C(x),\|v\|=1} \|gv\| = \min_{v\in C(y),\|v\|=1} \|gv\| \frac{\min_{v\in C(x),\|v\|=1} \|gv\|}{\min_{v\in C(y),\|v\|=1} \|gv\|}.$$

By continuity of the cone family $C(x)$, one can choose ε so small that the last fraction is bounded from below by $2/3$.[2] This yields

$$\min_{v\in C(x),\|v\|=1} \|gv\| \geq \frac{2}{3} \min_{w\in C(y),\|w\|=1} \|gw\| \geq 2. \tag{14.2}$$

Thus, for any $n \in \mathbb{N}$ such that $nL \leq K$, by (14.2) we find that

$$\begin{aligned}\min_{v\in C(x),\|v\|=1} \|g^n v\| &= \min_{v\in C(x),\|v\|=1} a\|g(g^{n-1}v)\| \\ &\geq 2 \min_{v\in C(x),\|v\|=1} \|g^{n-1}v\| \geq \cdots \\ &\geq 2^{n-1} \min_{v\in C(x),\|v\|=1} \|gv\| \geq 2^n.\end{aligned}$$

Writing $n = kL + r$, we find that for every $v \in C_x^1$,

$$\begin{aligned}\|d_x f^n v\| &= \|d_x f^{kL+r} v\| \\ &= \frac{\|d_x f^n v\|}{\|d_x f^{n-1} v\|} \cdots \frac{\|d_x f^{n-r+1} v\|}{\|d_x f^{n-r} v\|} \|g^k v\| \\ &\geq C' 2^k = C'(2^{k/n})^n \geq C\lambda^n\end{aligned}$$

for suitable $\lambda > 1$ and $C > 0$. □

We now conclude the proof of Theorem 14.1. Recall that we chose ξ as in Lemma 14.8, which determines R via statement (1) of Lemma 14.9, and these parameters in turn determine the choice of ε in Lemma 14.10.

Suppose that $\xi > 0$ and choose $\tau < \xi$ and N as in statement (3) of Lemma 14.9.

[2] Thus, ε depends on L and R and hence ultimately only on ξ. Note also that this is the only place where the continuity of the cone family $C(x)$ is used.

Consider any $x \in F^{(\tau)}$. If there is a $k_0 \in \mathbb{N}_0$ such that $f^k(x) \in U_\varepsilon(F^{(\xi)})$ for $k < k_0$ and $f^{k_0}(x) \notin U_\varepsilon(F^{(\xi)})$, then $f^{k_0}(x) \in X_N^{(\tau)}$. Thus for any $v \in C(x)$, $\|v\| = 1$, and all $n \in \mathbb{N}$ we have

$$\begin{aligned}\|d_x f^n v\| &= \|d_x f^{\max(0,n-k_0)/N} d_x f^{\min(n,k_0)} v\| \\ &\geq e^{\max(0,n-k_0)/N} \|d_x f^{\min(n,k_0)} v\| \\ &\geq C\lambda^{\min(n,k_0)} e^{\max(0,n-k_0)/N} \geq C\gamma^n,\end{aligned}$$

where $\gamma = \min(\lambda, e^{1/N}) > 1$. Note that the same estimate holds if $f^n(x) \in U_\varepsilon(F^{(\xi)})$ for all $n \in \mathbb{N}$, so it holds for all $x \in F^{(\tau)}$.

It is easy to check that this implies that $F^{(\tau)} \subset X_{2\max\{1,-\log C\}/\log\gamma}$. By statement (3) of Lemma 14.7, we conclude that $F^{(\tau+1)} = \varnothing$, and hence, $\tau \geq \xi$, which is contrary to our choice of τ.

14.2. The Anosov rigidity phenomenon. II

We describe another approach to the Anosov rigidity phenomenon due to Cao [24] (see also [1, 25]).

Let f be a C^1 diffeomorphism of a compact smooth Riemannian manifold M and let $K \subset M$ be a compact invariant subset. We say that a subset $A \subset K$ has *total measure* if $\mu(A) = 1$ for any invariant Borel probability measure μ.[3]

Theorem 14.11. *Assume that there is a continuous invariant distribution $E(x) \subset T_xM$ on K such that the relation*

$$\psi(x) = \limsup_{n\to\infty} \frac{1}{n} \log \|d_x f^{-n}|E(x)\| < 0$$

holds for every x in a set A of total measure. Then there exist $c > 0$ and $\chi > 0$ such that for every $x \in K$ and $n \in \mathbb{N}$,

$$\|d_x f^{-n}|E(x)\| \leq ce^{-\chi n}.$$

Proof. We denote by $\mathcal{K}(f)$ the space of invariant measures on K, endowed with the weak* topology.

Lemma 14.12. *Let φ be a continuous function on K and let $\lambda \in \mathbb{R}$. If $\int_K \varphi\, d\mu < \lambda$ for every $\mu \in \mathcal{K}(f)$, then:*

(1) *for every $x \in K$, there exists a number $n(x) > 0$ such that*

$$\frac{1}{n(x)} \sum_{i=0}^{n(x)-1} \varphi(f^i(x)) < \lambda;$$

[3]Of course, the set A itself has total measure but we are interested in the smallest such set.

(2) *there exists $N > 0$ such that for all $n \geq N$, we have*

$$\frac{1}{n}\sum_{i=0}^{n-1} \varphi(f^i(x)) < \lambda.$$

Proof of the lemma. Assuming the contrary, we obtain that for some $x \in K$ and every $n > 0$,

$$\frac{1}{n}\sum_{i=0}^{n-1} \varphi(f^i(x)) \geq \lambda.$$

Consider the sequence of probability measures

$$\mu_n = \frac{1}{n}\sum_{i=0}^{n-1} \delta(f^i(x)), \quad n \geq 1,$$

where $\delta(f^i(x))$ is the Dirac measure at $f^i(x)$. The set $\mathcal{K}(f)$ is compact. Let μ be an accumulation measure and μ_{n_k} a subsequence, which converges to μ. By the Bogolubov–Krilov theorem, the measure μ is invariant. Since the function φ is continuous, we obtain that

$$\int_K \varphi \, d\mu = \lim_{k\to\infty} \frac{1}{n_k}\sum_{i=0}^{n_k-1} \varphi(f^i(x)) \geq \lambda.$$

This contradiction proves the first statement. To show the second statement, observe that for every $x \in K$ there are numbers $n(x) > 0$ and $0 < a(x) < \lambda$ such that

$$\frac{1}{n(x)}\sum_{i=0}^{n(x)-1} \varphi(f^i(x)) < a(x).$$

Thus, by the continuity of φ, for each $x \in K$ there is a neighborhood V_x of x such that for every $y \in V_x$,

$$\frac{1}{n(x)}\sum_{i=0}^{n(x)-1} \varphi(f^i(y)) < a(x).$$

Since K is compact, there is a finite cover of K by open neighborhoods $V_{x_1}, \ldots, V_{x_p}$. Set

$$\bar{N} = \max\{n(x_1), \ldots, n(x_p)\}, \quad a = \max\{a(x_1), \ldots, a(x_p)\}$$

and observe that $a < \lambda$. For $x \in K$ let

$$N_0(x) = 0, \quad N_1(x) = \min\{n(x_k)\colon x \in V_{x_i}, i = 1, \ldots, p\}$$

and define the sequence of functions $N_k(x)$ by

$$N_{k+1}(x) = N_k(x) + N_1(f^{N_k(x)}(x)).$$

For every $x \in K$ and $n > 0$ there exists $k > 0$ such that $N_k \le n \le N_{k+1}$. Letting $\alpha = \max_{x\in K} \|\varphi(x)\|$, we find that

$$\sum_{i=0}^{n-1} \varphi(f^i(x)) \le aN_k + \alpha\bar{N} \le an + (|a| + \alpha)\bar{N}.$$

Taking $N = \frac{2(|a|+\alpha)\bar{N}}{\lambda - a}$, we obtain the desired estimate. □

For $n > 0$ consider the function on K

$$\varphi_n(x) = \log \|df^{-n}(x)|E(x)\|.$$

Since f is a C^1 diffeomorphism, the function $\varphi_n(x)$ is continuous.

Exercise 14.13. Show that the sequence of functions φ_n is subadditive, i.e., $\varphi_{m+n}(x) \le \varphi_n(x) + \varphi_m(f^n(x))$ (see (11.9)).

By Kingman's subadditive ergodic theorem [**53**], the limit

$$\lim_{n\to\infty} \frac{\varphi_n(x)}{n} = \bar{\varphi}(x)$$

exists for almost every x and every measure $\mu \in \mathcal{K}(f)$ and the function $\bar{\varphi}$ is invariant and integrable. By the dominated convergence theorem, for every $\mu \in \mathcal{K}(f)$,

$$\lim_{n\to\infty} \int_K \frac{\varphi_n}{n}\, d\mu = \int_K \bar{\varphi}\, d\mu. \tag{14.3}$$

The assumption that the relation (14.1) holds on a set of total measure implies that $\bar{\varphi} < 0$ for almost every x and any $\mu \in \mathcal{K}(f)$. It follows that $\int_K \bar{\varphi}\, d\mu < 0$ for every invariant measure μ.

We need the following statement.

Lemma 14.14. *There exist $L > 0$ and $\lambda < 0$ such that for any $\mu \in \mathcal{K}(f)$,*

$$\frac{1}{L} \int_K \varphi_L\, d\mu < \lambda. \tag{14.4}$$

Proof of the lemma. Fix $\mu \in \mathcal{K}(f)$. Since $\int_K \bar{\varphi}\, d\mu$, by (14.3), there exists $n_\mu > 0$ such that for all $n \ge n_k$,

$$\int_K \frac{\varphi_n}{n}\, d\mu < \frac{1}{2} \int_K \bar{\varphi}\, d\mu.$$

Since $\varphi_{n_\mu}(x)/n_\mu$ is a continuous function on K, there exists an open neighborhood $O(\mu) \in \mathcal{K}(f)$ of μ such that for all $\mu' \in O(\mu)$,

$$\frac{1}{n_\mu} \int_K \varphi_{n_\mu}\, d\mu' < \frac{1}{4} \int_K \bar{\varphi}\, d\mu.$$

The collection of sets $O(\mu)$ forms an open cover of $\mathcal{K}(f)$ and since it is compact, there is a finite subcover $O(\mu_1), \dots, O(\mu_\ell)$. Setting $n_j = n_{\mu_j}$ and

$\lambda = \max\{\frac{1}{4}\int_K \bar{\varphi}\, d\mu_j\}$, we find that $\lambda < 0$ and for any $\mu \in \mathcal{K}(f)$ there is a number i such that $\mu \in O(\mu_i)$ and

$$\frac{1}{n_i}\int_K \varphi_{n_i}\, d\mu_i < \lambda.$$

Using the subadditivity of the sequence of functions φ_n repeatedly, we find that for any $k > 0$ and $x \in K$,

$$\varphi_{kn_j}(x) \le \sum_{i=0}^{k-1} \varphi_{n_j}(f^{in_j}(x)).$$

Since the measure μ is invariant, we have that

$$\begin{aligned}\frac{1}{kn_j}\int_K \varphi_{kn_j}\, d\mu &\le \frac{1}{k}\sum_{i=0}^{k-1}\frac{1}{n_j}\int_K \varphi_{n_j}(f^{in_j}(x))\, d\mu \\ &= \frac{1}{k}\sum_{i=0}^{k-1}\frac{1}{n_j}\int_K \varphi_{n_j}(x)\, d\mu < \lambda.\end{aligned}$$

Therefore, setting $L = n_1 \cdots n_\ell$, we obtain the desired inequality (14.4). $\square$

We proceed with the proof of the theorem. Since φ_L is a continuous function on K, by statement (2) of Lemma 14.12, there exists $\bar{N} > 0$ such that for all $n \ge \bar{N}$ and $x \in K$,

$$\frac{1}{n}\sum_{i=0}^{n-1}\frac{\varphi_L(f^i(x))}{L} < \lambda.$$

Using subadditivity of the sequence of functions $\varphi_n(x)$, we have that for any $x \in K$ and $k > 0$,

$$\varphi_{kL}(x) \le \sum_{i=0}^{k-1} \varphi_L(f^{iL}(x)).$$

It follows that for any $0 \le j < L$,

$$\varphi_{kL}(x) \le \varphi_j(x) + \sum_{i=0}^{k-2} \varphi_L(f^{Li+j}(x)) + \varphi_{L-j}(f^{L(k-1)+j}(x)).$$

Summing over $j = 0, \dots, L-1$ and dividing by L, we obtain that

$$\varphi_{kL}(x) \le \frac{1}{L}\sum_{j=0}^{L-1}\sum_{i=0}^{k-2} \varphi_L(f^{Li+j}(x)) + \frac{1}{L}\sum_{j=0}^{L-1}[\varphi_j(x) + \varphi_{L-j}(f^{L(k-1)+j}(x))]. \tag{14.5}$$

Setting $C_1 = \max_{1\le i\le L}\max_{x\in K}\varphi_i(x)$, we obtain from (14.5) that

$$\varphi_{kL}(x) \le \sum_{j=0}^{(k-1)L-1}\frac{\varphi_L(f^j(x))}{L} + 2C_1$$

and hence, using Lemma 14.14, we find that for all k for which $L(k-1) \geq \bar{N}$,

$$\varphi_{kL}(x) \leq L(k-1)\lambda + 2C_1. \tag{14.6}$$

Fix $n \geq \bar{N} + 2L$ and write it in the form $n = kL + j$ where $0 \leq j < L$. We have that $(k-1)L = n - L + j > \bar{N}$. Again using subadditivity of the sequence of functions $\varphi_n(x)$, we have that for $x \in M$,

$$\varphi_n(x) \leq \varphi_{kL}(x) + \varphi_j(f^{kL}(x)).$$

Therefore, by (14.6), we have that

$$\varphi_n(x) \leq L(k-1)\lambda + 3C_1.$$

Since $L(k-1) < n$, we conclude that

$$\frac{1}{n}\varphi_n(x) \leq \lambda + \frac{3}{n}C_1.$$

Hence, setting $P = \max\{\bar{N} + 2L, \frac{6C_1}{-\lambda}\}$, we obtain that $\varphi_n(x) \leq \frac{\lambda}{2}$ for every $x \in K$ and every $n \geq P$. This implies that for all $x \in K$ and all $n > 0$

$$\|df_x^{-n}|E(x)\| \leq ce^{-\chi n},$$

where

$$\chi = -\frac{\lambda}{2} > 0 \quad \text{and} \quad c = \max_{1 \leq i \leq P-1}\{\|df_x^{-i}|E(x)\|, 1\} > 0.$$

The desired result follows. □

As an immediate corollary of Theorem 14.11 one obtains a different version of the Anosov rigidity phenomenon.

Theorem 14.15. *Let $K \subset M$ be a compact f-invariant subset admitting two continuous distributions $E_1(x)$ and $E_2(x)$ on K such that $T_xM = E_1(x) \oplus E_2(x)$ and the relations*

$$\liminf_{n\to\infty} \frac{1}{n} \log \|d_xf^n|E_1(x)\| > 0, \quad \limsup_{n\to\infty} \frac{1}{n} \log \|d_xf^{-n}|E_2(x)\| < 0$$

hold for every x in a set A of total measure. Then K is a uniformly hyperbolic set for f. In particular, if $K = M$, then f is an Anosov diffeomorphism.

Theorem 14.11 is a slight modification of the result in [**24**]. One can construct an example that illustrates that the continuity requirement for the distribution $E(x)$ is essential for this theorem to hold true (see [**26**]).

Chapter 15

C^1 Pathological Behavior: Pugh's Example

We illustrate that the requirement in the Stable Manifold Theorem 7.1 that the diffeomorphism f is of class $C^{1+\alpha}$ for some $\alpha > 0$ is crucial. Namely, we outline a construction due to Pugh [**74**] of a nonuniformly hyperbolic C^1 diffeomorphism (which is not $C^{1+\alpha}$ for any $\alpha > 0$) of a four-dimensional manifold with the following property: there exists no manifold tangent to $E^s(x)$ for which (7.1) holds on some open neighborhood of x (see [**9**]).

Consider the sphere S^2 and denote by S^2_0 its equator and by S^2_- and S^2_+ the southern and northern hemispheres, respectively. Let $\rho_-\colon \mathbb{R}^2 \to S^2_-$ and $\rho_+\colon \mathbb{R}^2 \to S^2_+$ be the central projections with the south pole at the origin.[1] Clearly, $\rho_+ = i \circ \rho_-$ where i is the antipodal map of S^2. Any map $f\colon \mathbb{R}^2 \to \mathbb{R}^2$ gives rise to two maps $f_\pm\colon S^2_\pm \to S^2_\pm$ such that the diagrams

$$\begin{array}{ccc} \mathbb{R}^2 & \xrightarrow{\ f\ } & \mathbb{R}^2 \\ {\scriptstyle\rho_\pm}\downarrow & & \downarrow{\scriptstyle\rho_\pm} \\ S^2_\pm & \xrightarrow{\ f_\pm\ } & S^2_\pm \end{array}$$

are commutative, i.e., $f_\pm(x) = (\rho_\pm \circ f \circ \rho_\pm^{-1})(x)$. Thus we obtain a map $\rho_- f \cup \rho_+ f\colon S^2 \setminus S^2_0 \to S^2 \setminus S^2_0$ defined as $(\rho_- f \cup \rho_+ f)(x) = f_\pm(x)$ for $x \in S^2_\pm$.

[1] The central projection associates to every point A on the plane (that is tangent to the sphere at the south pole) the point B on the southern (respectively, northern) hemisphere that is the point of intersection of the sphere with the line passing through the center of the sphere and the point A.

We wish to choose a map f in such a way that it can be extended to a map $\rho_\sharp f$ which is well-defined on the whole sphere S^2.

To this end, we need the following statement. Its proof while somewhat straightforward is technically involved and is omitted. We refer the reader to the proof of Lemma 4 in [**74**] for a complete argument.

Proposition 15.1. *Let $A = \left(\begin{smallmatrix} a & c \\ 0 & b \end{smallmatrix}\right)$ with $ab \neq 0$, and let $h\colon \mathbb{R} \to \mathbb{R}$ be a C^1 function with compact support. Then the map*

$$f(x, y) = (ax + cy + h(y), by)$$

lifts to a unique continuous map $\rho_\sharp f$ of S^2, which agrees with $(\rho_- f \cup \rho_+ f)(x)$ on $S^2 \setminus S_0^2$. Moreover, $\rho_\sharp f$ is a C^1 diffeomorphism whose values and derivatives at the equator S_0^2 are the same as those of $\rho_\sharp A$.

We proceed with the construction of Pugh's example. Let $g\colon (0, \infty) \to (0, \infty)$ be a smooth function such that

$$g(u) = \frac{u}{\log(1/u)} \quad \text{for } 0 < u < 1/e,$$

$g'(u) > 1$ for $u \geq 1/e$, and $g'(u) = c > 1$ is a constant for $u \geq 1$. We extend g to $\mathbb{R}$ by setting

$$g_o(u) = \begin{cases} g(u), & u > 0, \\ 0, & u = 0, \\ -g(u), & u < 0. \end{cases} \tag{15.1}$$

We have that $g_o'(u) = c > 1$ provided $|u| \geq 1$. Choose constants a and b such that $0 < a < ab < 1 < b$ and consider the maps

$$f_\pm(x, y) = (ax \pm g_o(y), by).$$

Since the function $h(y) = g_o(y) - c(y)$ has compact support, the maps $f_\pm$ satisfy the hypotheses of Proposition 15.1 and, hence, can be lifted to S^2 as $\rho_\sharp f_\pm$.

We divide S^2 into two hemispheres $H_\pm$ along the x-axis longitude L_x; that is, $H_\pm$ is the hemisphere containing the quarter sphere

$$\rho_-\{(x, y) \in \mathbb{R}^2 : \pm y > 0\}.$$

Define a map $F_S\colon S^2 \to S^2$ by

$$F_S = \begin{cases} \rho_\sharp f_+(z), & z \in H_+, \\ \rho_\sharp f_-(z), & z \in H_-. \end{cases} \tag{15.2}$$

Clearly, F_S is a C^1 diffeomorphism.

Let $h\colon M \to M$ be a diffeomorphism of a compact surface M having a hyperbolic invariant set Λ on which h is topologically conjugate to the full shift on two symbols and let $T_\Lambda M = E^s \oplus E^u$ be the invariant splitting into one-dimensional stable and unstable subspaces. Then there are numbers $0 < \lambda < 1 < \mu$ such that for every $x \in \Lambda$,

$$\|d_x hv\| \leq \lambda \quad \text{for every } v \in E^s(x), \|v\| = 1,$$

$$\|d_x hv\| \geq \mu \quad \text{for every } v \in E^u(x), \|v\| = 1.$$

One can choose the map h such that the numbers λ and μ satisfy

$$\lambda < \min_{u \in TS^2, \|u\|=1} \|dF_S(u)\|, \quad \mu > \max_{u \in TS^2, \|u\|=1} \|dF_S(u)\|. \tag{15.3}$$

Let Λ_i, $i = 0, 1$, be the ith "cylinder", i.e., the compact set of points in Λ that corresponds to sequences of symbols with i in the initial position. Choose a smooth bump function $\mu\colon M \to [0, \pi/2]$ such that $\Lambda_0 = \mu^{-1}(0) \cap \Lambda$, $\Lambda_1 = \mu^{-1}(\pi/2) \cap \Lambda$, and $\mu^{-1}(\{0, \pi/2\})$ is a neighborhood of Λ. Now let R_θ be the rotation of S^2 by angle θ that fixes the poles.

We are now ready to define the desired map F. It acts on the four-dimensional manifold $M \times S^2$ by the formula

$$F(w, z) = \big(h(w), (R_{\mu(w)} \circ F_S \circ R_{-\mu(w)})(z)\big),$$

where R_θ is the rotation of S^2 by angle θ that fixes the poles. It is easy to see that F is a C^1 diffeomorphism and it leaves invariant the foliation $\mathcal{F}$ by 2-spheres $\{w\} \times S^2$, $w \in M$. Furthermore, by (15.3), F is normally hyperbolic to $\mathcal{F}$. This means that for every $w \in M$, $z \in S^2$ we have the splitting

$$T_{(w,z)}(M \times S^2) = E^s(w, z) \oplus T_{(w,z)}\mathcal{F} \oplus E^u(w, z)$$

into stable $E^s(w, z)$, unstable $E^u(w, z)$, and central $T_{(w,z)}\mathcal{F}$ subspaces which are invariant under dF. Moreover, dF contracts vectors in $E^s(w, z)$ and expands vectors in $E^u(w, z)$ with uniform rates, which are strictly bigger than the rates of contraction and expansion along $T_{(w,z)}\mathcal{F}$. The subspaces $E^s(w, z)$ and $E^u(w, z)$ are locally integrable to local stable $V^s(w, z)$ and unstable $V^u(w, z)$ manifolds.

Proposition 15.2. *There is a point $P \in M \times S^2$, which is LP-regular with nonzero Lyapunov exponents for the diffeomorphism F and for which the stable set*

$$V^{sc}(P) = \left\{x \in M \times S^2 : \lim_{n\to\infty} \frac{1}{n} \log d(F^n(x), F^n(P)) < 0\right\}$$

is not an injectively immersed submanifold tangent to $E^s(P) \oplus T_P\mathcal{F}$.

Proof. We consider another extension of the function g to $\mathbb{R}$,

$$\tilde{g}(u) = \begin{cases} g(u), & u > 0, \\ 0, & u = 0, \\ g(u), & u < 0. \end{cases}$$

Then $\tilde{g}$ is of class C^1 on $\mathbb{R}$ and of class C^∞ on $\mathbb{R}\setminus\{0\}$. Consider the C^1 maps f_S and f_T in $\mathbb{R}^2$ defined by

$$f_S(x,y)=(ax+g(y),by) \quad \text{and} \quad f_T(x,y)=(bx,g(x)+ay).$$

One can easily verify that f_S and f_T are C^1 diffeomorphisms of $\mathbb{R}^2$ onto itself. Moreover, F_S lifts f_S to S^2 (but not as $\rho_- f_S\cup\rho_+ f_S$; in fact the canonical lift $\rho_\sharp f_S$ fails to be C^1 at the equator). This follows from the fact that $f_\pm(x,y)=f_S(x,y)$ for $\pm y\ge 0$ (see (15.1)) since F_S is constructed from lifts $\rho_\sharp f_\pm$, respectively, of $f_\pm$ (see (15.2)).

Set $L_k=k(k+1)/2$ and for each $n\in\mathbb{N}$,

$$f_n=\begin{cases} f_S, & n\in\mathcal{S},\\ f_T, & n\in\mathcal{T},\end{cases} \tag{15.4}$$

where

$$\mathcal{S}=\bigl\{n\in\mathbb{N}: L_{k-1}<n\le L_k \text{ for some odd } k\bigr\},$$
$$\mathcal{T}=\bigl\{n\in\mathbb{N}: L_{k-1}<n\le L_k \text{ for some even } k\bigr\}.$$

Notice that the sets $\mathcal{S}$ and $\mathcal{T}$ partition the set $\mathbb{N}$ of positive integers.

Since $h|\Lambda$ is topologically conjugate to the full 2-shift, there exists $p\in\Lambda$ such that $h^n(p)\in\Lambda_0$ whenever $f_n=f_S$ and $h^n(p)\in\Lambda_1$ whenever $f_n=f_T$. Let $P=(p,z_0)$ where z_0 is the south pole of S^2. Since z_0 is fixed under both F_S and R_θ, the F-orbit of P is $\{(h^n(p),z_0)\}$. We have that

$$d_{(h^n(p),z_0)}F=\begin{pmatrix} d_{h^n(p)}h & 0\\ 0 & d_0 f_n\end{pmatrix}. \tag{15.5}$$

Indeed, since μ is constant near Λ, F_S is a lift of f_S, and $f_T=R_{\pi/2}\circ f_S\circ R_{-\pi/2}$, one can show that the diagram

$$\begin{array}{ccc} \{h^n(p)\}\times S^2 & \xrightarrow{F} & \{h^n(p)\}\times S^2\\ \downarrow & & \downarrow\\ S^2 & \xrightarrow{F_n} & S^2 \end{array} \tag{15.6}$$

is commutative, where $F_n=F_S$ when $f_n=f_S$ and $F_n=F_T$ when $f_n=f_T$. It follows from (15.5) that P has one positive Lyapunov exponent corresponding to $dh|E^u$ and three negative Lyapunov exponents: one corresponds to $dh|E^s$ and the other two, being $\frac{1}{2}\log(ab)$, correspond to the products of the matrices $d_0 f_n$. Let $E^-(P)$ denote the space of vectors with negative Lyapunov exponents. Taking into account that $dF^n|E^s_P$ is diagonal preserving the decomposition $E^s\oplus(x\text{-axis})\oplus(y\text{-axis})$, we conclude that the F-orbit of P is LP-regular.

Since F is normally hyperbolic to $\mathcal{F}$, a point $(w, z) \in M \times S^2$ satisfies

$$\lim_{n\to\infty} \frac{1}{n} \log d(F^n(w, z), F^n(P)) < 0$$

if and only if (w, z) lies in the strongly stable manifold of some point $(p, z') \in V^{sc}(P) \cap (\{p\} \times S^2)$. This means that

$$V^{sc}(P) = V^s(V^{sc}(P) \cap (\{p\} \times S^2)).$$

By (15.6), the set $V^{sc}(P) \cap (\{p\} \times S^2)$ coincides with

$$\left\{z \in S^2 : \lim_{n\to\infty} \frac{1}{n} \log \|(f_n \circ \cdots \circ f_0)(z)\| < 0\right\}.$$

To complete the proof, we need the following statement. Its proof exploits the choice of maps f_n (see (15.4)) and is technically involved; we refer the reader to the proof of Theorem 1 in [**74**].

Lemma 15.3. *If $z = (x, y) \in \mathbb{R}^2$ is such that $x > 0$ and $y > 0$, then $\|f^n(z)\| \to \infty$ as $n \to \infty$.*

It follows from the lemma that $V^{sc}(P)$ is contained in $V^s(\{p\} \times S^2)$ but does not include any neighborhood of the point P, since it misses the entire first quadrant. Hence it cannot be an immersed manifold tangent to $E^-(P)$. The desired result follows. □

Despite Pugh's construction, there are some affirmative results due to Barreira and Valls [**10**] on constructing stable manifolds for diffeomorphisms with nonzero Lyapunov exponents of class C^1.

Bibliography

[1] J. Alves, V. Araújo, and B. Saussol, *On the uniform hyperbolicity of some nonuniformly hyperbolic systems*, Proc. Amer. Math. Soc. **131** (2003), 1303–1309.

[2] D. Anosov, *Tangential fields of transversal foliations in Y-systems*, Math. Notes **2** (1967), 818–823.

[3] D. Anosov, *Geodesic flows on closed Riemann manifolds with negative curvature*, Proc. Steklov Inst. Math. **90** (1969), 1–235.

[4] D. Anosov and Ya. Sinai, *Certain smooth ergodic systems*, Russian Math. Surveys **22** (1967), 103–167.

[5] L. Arnold, *Random Dynamical Systems*, Monographs in Mathematics, Springer, 1998.

[6] A. Ávila and R. Krikorian, *Reducibility or nonuniform hyperbolicity for quasiperiodic Schrödinger cocycles*, Ann. of Math. (2) **164** (2006), 911–940.

[7] L. Barreira and Ya. Pesin, *Lyapunov Exponents and Smooth Ergodic Theory*, University Lecture Series 23, American Mathematical Society, 2002.

[8] L. Barreira and Ya. Pesin, *Smooth ergodic theory and nonuniformly hyperbolic dynamics*, with appendix by O. Sarig, in Handbook of Dynamical Systems 1B, edited by B. Hasselblatt and A. Katok, Elsevier, 2006, pp. 57–263.

[9] L. Barreira and Ya. Pesin, *Nonuniform Hyperbolicity: Dynamics of Systems with Nonzero Lyapunov Exponents*, Encyclopedia of Mathematics and its Applications 115, Cambridge University Press, 2007.

[10] L. Barreira and C. Valls, *Smoothness of invariant manifolds for nonautonomous equations*, Comm. Math. Phys. **259** (2005), 639–677.

[11] L. Barreira and C. Valls, *Stability of Nonautonomous Differential Equations*, Lect. Notes in Math. 1926, Springer, 2008.

[12] G. Birkhoff and G.-C. Rota, *Ordinary Differential Equations*, John Wiley & Sons, Inc., 1989.

[13] J. Bochi, *Genericity of zero Lyapunov exponents*, Ergodic Theory Dynam. Systems **22** (2002), 1667–1696.

[14] J. Bochi and M. Viana, *The Lyapunov exponents of generic volume-preserving and symplectic maps*, Ann. of Math. (2) **161** (2005), 1423–1485.

[15] C. Bonatti, L. Díaz, and M. Viana, *Dynamics beyond uniform hyperbolicity. A global geometric and probabilistic perspective*, Encyclopaedia of Mathematical Sciences 102, Mathematical Physics III, Springer, 2005.

[16] J. Bourgain, *Positivity and continuity of the Lyapounov exponent for shifts on* $\mathbb{T}^d$ *with arbitrary frequency vector and real analytic potential*, J. Anal. Math. **96** (2005), 313–355.

[17] J. Bourgain and S. Jitomirskaya, *Continuity of the Lyapunov exponent for quasi-periodic operators with analytic potential*, J. Stat. Phys. **108** (2002), 1203–1218.

[18] M. Brin, *Hölder continuity of invariant distributions*, in Smooth Ergodic Theory and its Applications, edited by A. Katok, R. de la Llave, Ya. Pesin, and H. Weiss, Proc. Symp. Pure Math. 69, Amer. Math. Soc., 2001, pp. 99–101.

[19] M. Brin and G. Stuck, *Introduction to Dynamical Systems*, Cambridge University Press, 2002.

[20] K. Burns, D. Dolgopyat, and Ya. Pesin, *Partial hyperbolicity, Lyapunov exponents and stable ergodicity*, J. Statist. Phys. **108** (2002), 927–942.

[21] K. Burns and A. Wilkinson, *Stable ergodicity of skew products*, Ann. Sci. École Norm. Sup. (4) **32** (1999), 859–889.

[22] K. Burns and A. Wilkinson, *On the ergodicity of partially hyperbolic systems*, Annals of Math. (2) **171** (2010), 451–489.

[23] D. Bylov, R. Vinograd, D. Grobman, and V. Nemyckii, *Theory of Lyapunov exponents and its application to problems of stability*, Izdat. "Nauka", Moscow, 1966, in Russian.

[24] Y. Cao, *Non-zero Lyapunov exponents and uniform hyperbolicity*, Nonlinearity **16** (2003), 1473–1479.

[25] Y. Cao, S. Luzzatto, and I. Rios, *A minimum principle for Lyapunov exponents and a higher-dimensional version of a theorem of Mañé*, Qual. Theory Dyn. Syst. **5** (2004), 261–273.

[26] Y. Cao, S. Luzzatto, and I. Rios, *Some non-hyperbolic systems with strictly non-zero Lyapunov exponents for all invariant measures: horseshoes with internal tangencies*, Discrete Contin. Dyn. Syst. **15** (2006), 61–71.

[27] I. Cornfeld, S. Fomin, and Ya. Sinai, *Ergodic Theory*, Springer, 1982.

[28] D. Dolgopyat, *On dynamics of mostly contracting diffeomorphisms*, Comm. Math. Phys. **213** (2000), 181–201.

[29] D. Dolgopyat, H. Hu, and Ya. Pesin, *An example of a smooth hyperbolic measure with countably many ergodic components*, in Smooth Ergodic Theory and its Applications, edited by A. Katok, R. de la Llave, Ya. Pesin, and H. Weiss, Proc. Symp. Pure Math. 69, Amer. Math. Soc., 2001, pp. 102–115.

[30] D. Dolgopyat and Ya. Pesin, *Every compact manifold carries a completely hyperbolic diffeomorphism*, Ergodic Theory Dynam. Systems **22** (2002), 409–435.

[31] P. Eberlein, *Geodesic flows on negatively curved manifolds I*, Ann. of Math. (2) **95** (1972), 492–510.

[32] P. Eberlein, *When is a geodesic flow of Anosov type? I*, J. Differential Geom. **8** (1973), 437–463; *II*, J. Differential Geom. **8** (1973), 565–577.

[33] P. Eberlein, *Geodesic flows in manifolds of nonpositive curvature*, in Smooth Ergodic Theory and its Applications, edited by A. Katok, R. de la Llave, Ya. Pesin, and H. Weiss, Proc. Symp. Pure Math. 69, Amer. Math. Soc., 2001, pp. 525–571.

[34] L. Eliasson, *Reducibility and point spectrum for linear quasi-periodic skew-products*, Proceedings of the International Congress of Mathematicians (Berlin, 1998), Doc. Math. **Extra Vol. II** (1998), 779–787.

[35] D. Epstein, *Foliations with all leaves compact*, Ann. Inst. Fourier (Grenoble) **26** (1976), 265–282.

[36] B. Hasselblatt, Ya. Pesin, and J. Schmeling, *Pointwise hyperbolicity implies uniform hyperbolicity*, Discrete Contin. Dyn. Syst., to appear.

[37] G. Hedlund, *The dynamics of geodesic flows*, Bull. Amer. Math. Soc. **45** (1939), 241–260.

[38] M. Hirayama and Ya. Pesin, *Non-absolutely continuous foliations*, Israel J. Math. **160** (2007), 173–187.

[39] M. Hirsch, C. Pugh, and M. Shub, *Invariant Manifolds*, Lect. Notes. in Math. 583, Springer, 1977.

[40] E. Hopf, *Statistik der geodätischen Linien in Mannigfaltigkeiten negativer Krümmung*, Ber. Verh. Sächs. Akad. Wiss. Leipzig **91** (1939), 261–304.

[41] H. Hu, Ya. Pesin, and A. Talitskaya, *Every compact manifold carries a hyperbolic Bernoulli flow*, in Modern Dynamical Systems and Applications, Cambridge University Press, 2004, pp. 347–358.

[42] M. Jakobson and G. Świątek, *One-dimensional maps*, in Handbook of Dynamical Systems 1A, edited by B. Hasselblatt and A. Katok, Elsevier, 2002, pp. 599–664.

[43] R. Johnson, *The recurrent Hill's equation*, J. Differential Equations **46** (1982), 165–193.

[44] R. Johnson and G. Sell, *Smoothness of spectral subbundles and reducibility of quasiperiodic linear differential systems*, J. Differential Equations **41** (1981), 262–288.

[45] A. Katok, *Bernoulli diffeomorphism on surfaces*, Ann. of Math. (2) **110** (1979), 529–547.

[46] A. Katok and K. Burns, *Infinitesimal Lyapunov functions, invariant cone families and stochastic properties of smooth dynamical systems*, Ergodic Theory Dynam. Systems **14** (1994), 757–785.

[47] A. Katok and B. Hasselblatt, *Introduction to the Modern Theory of Dynamical Systems*, Encyclopedia of Mathematics and its Applications 54, Cambridge University Press, 1995.

[48] A. Katok and L. Mendoza, *Dynamical systems with nonuniformly hyperbolic behavior*, in *Introduction to the modern theory of dynamical systems* by A. Katok and B. Hasselblatt, Cambridge University Press, 1995.

[49] A. Katok and V. Nitica, *Rigidity in Higher Rank Abelian Group Actions*, Cambridge Tracts in Mathematics 185, Cambridge University Press, 2011.

[50] A. Katok and J.-M. Strelcyn, with the collaboration of F. Ledrappier and F. Przytycki, *Invariant Manifolds, Entropy and Billiards; Smooth Maps with Singularities*, Lect. Notes. in Math. 1222, Springer, 1986.

[51] Yu. Kifer, *Ergodic Theory of Random Transformations*, Progress in Probability and Statistics 10, Birkhäuser, 1986.

[52] Yu. Kifer and P.-D. Liu, *Random dynamics*, in Handbook of Dynamical Systems 1B, edited by B. Hasselblatt and A. Katok, Elsevier, 2006, pp. 379–499.

[53] J. Kingman, *Subadditive processes*, in École d'Été de Probabilités de Saint-Flour V–1975, Lect. Notes. in Math. 539, Springer, 1976, pp. 167–223.

[54] G. Knieper, *Hyperbolic dynamics and Riemannian geometry*, in Handbook of Dynamical Systems 1A, edited by B. Hasselblatt and A. Katok, Elsevier, 2002, pp. 453–545.

[55] R. Krikorian, *Réductibilité des Systèmes Produits-Croisés à Valeurs dans des Groupes Compacts*, Astérisque 259, 1999.

[56] P.-D. Liu and M. Qian, *Smooth Ergodic Theory of Random Dynamical Systems*, Lect. Notes in Math. 1606, Springer, 1995.

[57] S. Luzzatto, *Stochastic-like behaviour in nonuniformly expanding maps*, in Handbook of Dynamical Systems 1B, edited by B. Hasselblatt and A. Katok, Elsevier, 2006, pp. 265–326.

[58] S. Luzzatto and M. Viana, *Parameter exclusions in Hénon-like systems*, Russian Math. Surveys **58** (2003), 1053–1092.

[59] A. Lyapunov, *The General Problem of the Stability of Motion*, Taylor & Francis, 1992.

[60] I. Malkin, *A theorem on stability via the first approximation*, Dokladi, Akademii Nauk USSR **76** (1951), 783–784.

[61] R. Mañé, *Quasi-Anosov diffeomorphisms and hyperbolic manifolds*, Trans. Amer. Math. Soc. **229** (1977), 351–370.

[62] R. Mañé, *A proof of Pesin's formula*, Ergodic Theory Dynam. Systems **1** (1981), 95–102; errata in **3** (1983), 159–160.

[63] J. Milnor, *Fubini foiled: Katok's paradoxical example in measure theory*, Math. Intelligencer **19** (1997), 30–32.

[64] M. Morse, *Instability and transitivity*, J. Math. Pures Appl. **40** (1935), 49–71.

[65] V. Oseledets, *A multiplicative ergodic theorem. Liapunov characteristic numbers for dynamical systems*, Trans. Moscow Math. Soc. **19** (1968), 197–221.

[66] O. Perron, *Die Ordnungszahlen linearer Differentialgleichungssyteme*, Math. Z. **31** (1930), 748–766.

[67] Ya. Pesin, *An example of a nonergodic flow with nonzero characteristic exponents*, Func. Anal. and its Appl. **8** (1974), 263–264.

[68] Ya. Pesin, *Families of invariant manifolds corresponding to nonzero characteristic exponents*, Math. USSR-Izv. **40** (1976), 1261–1305.

[69] Ya. Pesin, *Characteristic Ljapunov exponents, and smooth ergodic theory*, Russian Math. Surveys **32** (1977), 55–114.

[70] Ya. Pesin, *A description of the π-partition of a diffeomorphism with an invariant measure*, Math. Notes **22** (1977), 506–515.

[71] Ya. Pesin, *Geodesic flows on closed Riemannian manifolds without focal points*, Math. USSR-Izv. **11** (1977), 1195–1228.

[72] Ya. Pesin, *Geodesic flows with hyperbolic behaviour of the trajectories and objects connected with them*, Russian Math. Surveys **36** (1981), 1–59.

[73] Ya. Pesin, *Lectures on Partial Hyperbolicity and Stable Ergodicity*, Zürich Lectures in Advanced Mathematics, European Mathematical Society, 2004.

[74] C. Pugh, *The $C^{1+\alpha}$ hypothesis in Pesin theory*, Inst. Hautes Études Sci. Publ. Math. **59** (1984), 143–161.

[75] C. Pugh and M. Shub, *Ergodic attractors*, Trans. Amer. Math. Soc. **312** (1989), 1–54.

[76] F. Rodriguez Hertz, M. A. Rodriguez Hertz, and R. Ures, *Accessibility and stable ergodicity for partially hyperbolic diffeomorphisms with 1D-center bundle*, Invent. Math. **172** (2008), 353–381.

[77] D. Ruelle, *An inequality for the entropy of differentiable maps*, Bol. Soc. Brasil. Mat. **9** (1978), 83–87.

[78] D. Ruelle, *Analyticity properties of the characteristic exponents of random matrix products*, Adv. Math. **32** (1979), 68–80.

[79] D. Ruelle and A. Wilkinson, *Absolutely singular dynamical foliations*, Comm. Math. Phys. **219** (2001), 481–487.

[80] R. Sacker and G. Sell, *Existence of dichotomies and invariant splittings for linear differential systems. I*, J. Differential Equations **15** (1974), 429–458.

[81] M. Shub and A. Wilkinson, *Pathological foliations and removable zero exponents*, Invent. Math. **139** (2000), 495–508.

[82] Ya. Sinai, *Dynamical systems with countably-multiple Lebesgue spectrum II*, Amer. Math. Soc. Trans. (2) **68** (1966), 34–88.

[83] S. Smale, *Differentiable dynamical systems*, Bull. Amer. Math. Soc. **73** (1967), 747–817.

[84] A. Tahzibi, *C^1-generic Pesin's entropy formula*, C. R. Math. Acad. Sci. Paris **335** (2002), 1057–1062.

[85] M. Wojtkowski, *Invariant families of cones and Lyapunov exponents*, Ergodic Theory Dynam. Systems **5** (1985), 145–161.

Index

图字：01-2023-0921号

光滑遍历理论导论

Guanghua Bianli Lilun Daolun

图书在版编目 (CIP) 数据

光滑遍历理论导论 = Introduction to Smooth Ergodic Theory : 英文 / (葡) 路易斯 ·巴雷拉 (Luis Barreira) , (美) 亚科夫 ·B ·佩辛 (Yakov Pesin) 著. -- 影印本. —北京 : 高等教育出版社 , 2024.2
ISBN 978-7-04-061199-1
Ⅰ. ①光… Ⅱ. ①路… ②亚… Ⅲ. ①遍历性理论—英文 Ⅳ. ①O177.99
中国国家版本馆 CIP 数据核字 (2023) 第 179608 号

策划编辑 吴晓丽　　责任编辑 吴晓丽
封面设计 张申申　　责任印制 高 峰

出版发行 高等教育出版社
社址 北京市西城区德外大街4号
邮政编码 100120
购书热线 010-58581118
咨询电话 400-810-0598
网址 http://www.hep.edu.cn
http://www.hep.com.cn
网上订购 http://www.hepmall.com.cn
http://www.hepmall.com
http://www.hepmall.cn
印刷 固安县铭成印刷有限公司

开本 787mm × 1092mm 1/16
印张 18.75
字数 480千字
版次 2024年2月第1版
印次 2024年2月第1次印刷
定价 135.00元

[物 料 号 61199-00]

美国数学会经典影印系列

1	**Lars V. Ahlfors**, Lectures on Quasiconformal Mappings, Second Edition	9787040470109
2	**Dmitri Burago, Yuri Burago, Sergei Ivanov**, A Course in Metric Geometry	9787040469080
3	**Tobias Holck Colding, William P. Minicozzi II**, A Course in Minimal Surfaces	9787040469110
4	**Javier Duoandikoetxea**, Fourier Analysis	9787040469011
5	**John P. D'Angelo**, An Introduction to Complex Analysis and Geometry	9787040469981
6	**Y. Eliashberg, N. Mishachev**, Introduction to the *h*-Principle	9787040469028
7	**Lawrence C. Evans**, Partial Differential Equations, Second Edition	9787040469356
8	**Robert E. Greene, Steven G. Krantz**, Function Theory of One Complex Variable, Third Edition	9787040469073
9	**Thomas A. Ivey, J. M. Landsberg**, Cartan for Beginners: Differential Geometry via Moving Frames and Exterior Differential Systems	9787040469172
10	**Jens Carsten Jantzen**, Representations of Algebraic Groups, Second Edition	9787040470086
11	**A. A. Kirillov**, Lectures on the Orbit Method	9787040469103
12	**Jean-Marie De Koninck, Armel Mercier**, 1001 Problems in Classical Number Theory	9787040469998
13	**Peter D. Lax, Lawrence Zalcman**, Complex Proofs of Real Theorems	9787040470000
14	**David A. Levin, Yuval Peres, Elizabeth L. Wilmer**, Markov Chains and Mixing Times	9787040469943
15	**Dusa McDuff, Dietmar Salamon**, *J*-holomorphic Curves and Symplectic Topology	9787040469936
16	**John von Neumann**, Invariant Measures	9787040469974
17	**R. Clark Robinson**, An Introduction to Dynamical Systems: Continuous and Discrete, Second Edition	9787040470093
18	**Terence Tao**, An Epsilon of Room, I: Real Analysis: pages from year three of a mathematical blog	9787040469004
19	**Terence Tao**, An Epsilon of Room, II: pages from year three of a mathematical blog	9787040468991
20	**Terence Tao**, An Introduction to Measure Theory	9787040469059
21	**Terence Tao**, Higher Order Fourier Analysis	9787040469097
22	**Terence Tao**, Poincaré's Legacies, Part I: pages from year two of a mathematical blog	9787040469950
23	**Terence Tao**, Poincaré's Legacies, Part II: pages from year two of a mathematical blog	9787040469967
24	**Cédric Villani**, Topics in Optimal Transportation	9787040469219
25	**R. J. Williams**, Introduction to the Mathematics of Finance	9787040469127
26	**T. Y. Lam**, Introduction to Quadratic Forms over Fields	9787040469196

27	**Jens Carsten Jantzen**, Lectures on Quantum Groups	9 787040 469141
28	**Henryk Iwaniec**, Topics in Classical Automorphic Forms	9 787040 469134
29	**Sigurdur Helgason**, Differential Geometry, Lie Groups, and Symmetric Spaces	9 787040 469165
30	**John B.Conway**, A Course in Operator Theory	9 787040 469158
31	**James E. Humphreys**, Representations of Semisimple Lie Algebras in the BGG Category O	9 787040 468984
32	**Nathanial P. Brown, Narutaka Ozawa**, C^*-Algebras and Finite-Dimensional Approximations	9 787040 469325
33	**Hiraku Nakajima**, Lectures on Hilbert Schemes of Points on Surfaces	9 787040 501216
34	**S. P. Novikov, I. A. Taimanov, Translated by Dmitry Chibisov**, Modern Geometric Structures and Fields	9 787040 469189
35	**Luis Caffarelli, Sandro Salsa**, A Geometric Approach to Free Boundary Problems	9 787040 469202
36	**Paul H. Rabinowitz**, Minimax Methods in Critical Point Theory with Applications to Differential Equations	9 787040 502299
37	**Fan R. K. Chung**, Spectral Graph Theory	9 787040 502305
38	**Susan Montgomery**, Hopf Algebras and Their Actions on Rings	9 787040 502312
39	**C. T. C. Wall, Edited by A. A. Ranicki**, Surgery on Compact Manifolds, Second Edition	9 787040 502329
40	**Frank Sottile**, Real Solutions to Equations from Geometry	9 787040 501513
41	**Bernd Sturmfels**, Gröbner Bases and Convex Polytopes	9 787040 503081
42	**Terence Tao**, Nonlinear Dispersive Equations: Local and Global Analysis	9 787040 503050
43	**David A. Cox, John B. Little, Henry K. Schenck**, Toric Varieties	9 787040 503098
44	**Luca Capogna, Carlos E. Kenig, Loredana Lanzani**, Harmonic Measure: Geometric and Analytic Points of View	9 787040 503074
45	**Luis A. Caffarelli, Xavier Cabré**, Fully Nonlinear Elliptic Equations	9 787040 503067
46	**Teresa Crespo, Zbigniew Hajto,** Algebraic Groups and Differential Galois Theory	9 787040 510133
47	**Barbara Fantechi, Lothar Göttsche, Luc Illusie, Steven L. Kleiman, Nitin Nitsure, Angelo Vistoli,** Fundamental Algebraic Geometry: Grothendieck's FGA Explained	9 787040 510126
48	**Shinichi Mochizuki,** Foundations of p-adic Teichmüller Theory	9 787040 510089
49	**Manfred Leopold Einsiedler, David Alexandre Ellwood, Alex Eskin, Dmitry Kleinbock, Elon Lindenstrauss, Gregory Margulis, Stefano Marmi, Jean-Christophe Yoccoz,** Homogeneous Flows, Moduli Spaces and Arithmetic	9 787040 510096
50	**David A. Ellwood, Emma Previato,** Grassmannians, Moduli Spaces and Vector Bundles	9 787040 510393
51	**Jeffery McNeal, Mircea Mustaţă,** Analytic and Algebraic Geometry: Common Problems, Different Methods	9 787040 510553
52	**V. Kumar Murty,** Algebraic Curves and Cryptography	9 787040 510386
53	**James Arthur, James W. Cogdell, Steve Gelbart, David Goldberg, Dinakar Ramakrishnan, Jiu-Kang Yu,** On Certain L-Functions	9 787040 510409

54	**Rick Miranda,** Algebraic Curves and Riemann Surfaces	9 787040 469066
55	**Hershel M. Farkas, Irwin Kra,** Theta Constants, Riemann Surfaces and the Modular Group	9 787040 469042
56	**Fritz John,** Nonlinear Wave Equations, Formation of Singularities	9 787040 517262
57	**Henryk Iwaniec, Emmanuel Kowalski,** Analytic Number Theory	9 787040 517231
58	**Jan Malý, William P. Ziemer,** Fine Regularity of Solutions of Elliptic Partial Differential Equations	9 787040 517248
59	**Jin Hong, Seok-Jin Kang,** Introduction to Quantum Groups and Crystal Bases	9 787040 517255
60	**V. I. Arnold,** Topological Invariants of Plane Curves and Caustics	9 787040 517057
61	**Dusa McDuff, Dietmar Salamon,** *J*-Holomorphic Curves and Quantum Cohomology	9 787040 517040
62	**James Eells, Luc Lemaire,** Selected Topics in Harmonic Maps	9 787040 517279
63	**Yuri I. Manin,** Frobenius Manifolds, Quantum Cohomology, and Moduli Spaces	9 787040 517026
64	**Bernd Sturmfels,** Solving Systems of Polynomial Equations	9 787040 517033
65	**Liviu I. Nicolaescu,** Notes on Seiberg-Witten Theory	9 787040 469035
66	**I. G. Macdonald,** Symmetric Functions and Orthogonal Polynomials	9 787040 535938
67	**Craig Huneke,** Tight Closure and Its Applications	9 787040 535945
68	**Victor Guillemin, Viktor Ginzburg, Yael Karshon,** Moment Maps, Cobordisms, and Hamiltonian Group Actions	9 787040 535952
69	**Gerald B. Folland,** Fourier Analysis and Its Applications	9 787040 535969
70	**Luis Barreira, Yakov B. Pesin,** Lyapunov Exponents and Smooth Ergodic Theory	9 787040 534962
71	**Gilles Pisier,** Factorization of Linear Operators and Geometry of Banach Spaces	9 787040 535037
72	**Richard Elman, Nikita Karpenko, Alexander Merkurjev,** The Algebraic and Geometric Theory of Quadratic Forms	9 787040 534955
73	**William Fulton,** Introduction to Intersection Theory in Algebraic Geometry	9 787040 534870
74	**S. K. Lando,** Lectures on Generating Functions	9 787040 535006
75	**Frank D. Grosshans, Gian-Carlo Rota, Joel A. Stein,** Invariant Theory and Superalgebras	9 787040 534979
76	**Lawrence C. Evans,** Weak Convergence Methods for Nonlinear Partial Differential Equations	9 787040 534993
77	**Harish-Chandra,** Admissible Invariant Distributions on Reductive *p*-adic Groups	9 787040 535020
78	**Marc Levine,** Mixed Motives	9 787040 535013
79	**Pavel I. Etingof, Igor B. Frenkel, Alexander A. Kirillov, Jr.,** Lectures on Representation Theory and Knizhnik-Zamolodchikov Equations	9 787040 534986
80	**Nicholas M. Katz, Peter Sarnak,** Random Matrices, Frobenius Eigenvalues, and Monodromy	9 787040 534948
81	**Kenneth R. Davidson,** C*-Algebras by Example	9 787040 534924
82	**Max-Albert Knus, Alexander Merkurjev, Markus Rost, Jean-Pierre Tignol,** The Book of Involutions	9 787040 534931

83 **Judy L. Walker,** Codes and Curves — 9787040534917

84 **J. Amorós, M. Burger, K. Corlette, D. Kotschick, D. Toledo,** Fundamental Groups of Compact Kähler Manifolds — 9787040536331

85 **V. I. Arnold,** Experimental Mathematics — 9787040556261

86 **Leslie Jane Federer Vaaler, James W. Daniel,** Mathematical Interest Theory, Second Edition — 9787040556278

87 **Paul R. Halmos,** Linear Algebra Problem Book — 9787040556285

88 **Luis Barreira, Claudia Valls,** Ordinary Differential Equations: Qualitative Theory — 9787040556315

89 **Terence Tao,** Topics in Random Matrix Theory — 9787040556476

90 **Gerald Teschl,** Ordinary Differential Equations and Dynamical Systems — 9787040556483

91 **Lawrence C. Evans,** An Introduction to Stochastic Differential Equations — 9787040556490

92 **Gerald Teschl,** Mathematical Methods in Quantum Mechanics: With Applications to Schrödinger Operators, Second Edition — 9787040556506

93 **András Vasy,** Partial Differential Equations: An Accessible Route through Theory and Applications — 9787040569773

94 **Weinan E, Tiejun Li, Eric Vanden-Eijnden,** Applied Stochastic Analysis — 9787040556391

95 **David R. Mazur,** Combinatorics: A Guided Tour — 9787040556339

96 **David Bressoud,** A Radical Approach to Real Analysis — 9787040556322

97 **H. Iwaniec,** Lectures on the Riemann Zeta Function — 9787040556308

98 **Terence Tao,** Hilbert's Fifth Problem and Related Topics — 9787040556292

99 **David Lannes,** The Water Waves Problem: Mathematical Analysis and Asymptotics — 9787040556346

100 **Seán Dineen,** Probability Theory in Finance: A Mathematical Guide to the Black-Scholes Formula, Second Edition — 9787040556353

101 **Steven G. Krantz,** Complex Analysis: The Geometric Viewpoint, Second Edition — 9787040556360

102 **A. Borel, N. Wallach,** Continuous Cohomology, Discrete Subgroups, and Representations of Reductive Groups, Second Edition — 9787040556377

103 **H. S. M. Coxeter,** Non-Euclidean Geometry, Sixth Edition — 9787040556384

104 **Kurt O. Friedrichs**, Mathematical Methods of Electromagnetic Theory — 9787040569810

105 **Jeffrey Rauch,** Hyperbolic Partial Differential Equations and Geometric Optics — 9787040569803

106 **Alberto Bressan,** Lecture Notes on Functional Analysis: With Applications to Linear Partial Differential Equations — 9787040569773

107 **Wilhelm Schlag,** A Course in Complex Analysis and Riemann Surfaces — 9787040569797

108 **Emmanuel Kowalski,** An Introduction to the Representation Theory of Groups — 9787040569780

109 **Steven G. Krantz,** A Guide to Complex Variables — 9787040570229

110 **Rajendra Bhatia,** Fourier Series — 9787040570236

111 **Fernando Q. Gouvêa,** A Guide to Groups, Rings, and Fields — 9787040569995

112 **Steven G. Krantz,** A Guide to Functional Analysis — 9787040570243

113 **Alberto Torchinsky,** Problems in Real and Functional Analysis — 9787040570250

114 **Fritz Colonius,Wolfgang Kliemann,** Dynamical Systems and Linear Algebra — 9787040570212

115 **Jerome K. Percus, Stephen Childress,** Mathematical Models in Developmental Biology — 9787040570342

116 **Steven G. Krantz,** A Panorama of Harmonic Analysis — 9787040570274

117 **Ralph P. Boas, Revised by Harold P. Boas,** Invitation to Complex Analysis, Second Edition — 9787040570281

118 **William H. Meeks III, Joaquín Pérez,** A Survey on Classical Minimal Surface Theory — 9787040570298

119 **Peter Duren,** Invitation to Classical Analysis — 9787040570304

120 **Helene Shapiro,** Linear Algebra and Matrices: Topics for a Second Course — 9787040570311

121 **Ralph P. Boas, Revised and updated by Harold P. Boas,** A Primer of Real Functions, Fourth Edition — 9787040570328

122 **John P. D'Angelo,** Inequalities from Complex Analysis — 9787040570335

123 **Terence Tao,** Expansion in Finite Simple Groups of Lie Type — 9787040592979

124 **Goro Shimura,** Arithmetic and Analytic Theories of Quadratic Forms and Clifford Groups — 9787040592986

125 **Victor M. Buchstaber, Taras E. Panov,** Toric Topology — 9787040592993

126 **Benson Farb, Richard Hain, Eduard Looijenga,** Moduli Spaces of Riemann Surfaces — 9787040593099

127 **Charles Livingston,** Knot Theory — 9787040593075

128 **Mladen Bestvina, Michah Sageev, Karen Vogtmann,** Geometric Group Theory — 9787040593105

129 **Edward Packel,** The Mathematics of Games and Gambling, Second Edition — 9787040593082

130 **Kiran S. Kedlaya, Bjorn Poonen, Ravi Vakil,** The William Lowell Putnam Mathematical Competition 1985–2000: Problems, Solutions, and Commentary — 9787040593129

131 **Ida Kantor, Jiří Matoušek, Robert Šámal,** Mathematics++: Selected Topics Beyond the Basic Courses — 9787040593136

132 **John Roe,** Winding Around: The Winding Number in Topology, Geometry, and Analysis — 9787040593143

133 **Wolfgang Kühnel, Translated by Bruce Hunt,** Differential Geometry: Curves–Surfaces–Manifolds, Third Edition — 9787040593150

134 **Daniel W. Stroock,** Mathematics of Probability — 9787040593020

135 **Jennifer Schultens,** Introduction to 3-Manifolds — 9787040593037

136 **Gábor Székelyhidi,** An Introduction to Extremal Kähler Metrics — 9787040593044

137 **Diane Maclagan, Bernd Sturmfels,** Introduction to Tropical Geometry — 9787040593051

138 **Barry Simon,** Real Analysis : A Comprehensive Course in Analysis, Part 1 — 9787040593068

139 **Barry Simon,** Basic Complex Analysis: A Comprehensive Course in Analysis, Part 2A — 9787040593006

140 **Barry Simon,** Advanced Complex Analysis: A Comprehensive Course in Analysis, Part 2B — 9 787040 593013 >

141 **Barry Simon,** Harmonic Analysis: A Comprehensive Course in Analysis, Part 3 — 9 787040 593112 >

142 **Barry Simon,** Operator Theory: A Comprehensive Course in Analysis, Part 4 — 9 787040 593167 >

143 **Sergei Ovchinnikov,** Number Systems: An Introduction to Algebra and Analysis — 9 787040 611632 >

144 **Ian M. Musson,** Lie Superalgebras and Enveloping Algebras — 9 787040 611625 >

145 **John B. Conway,** A Course in Abstract Analysis — 9 787040 611618 >

146 **V. I. Arnold,** Lectures and Problems: A Gift to Young Mathematicians — 9 787040 611601 >

147 **Pascal Cherrier, Albert Milani,** Linear and Quasi-linear Evolution Equations in Hilbert Spaces — 9 787040 611595 >

148 **John M. Lee,** Axiomatic Geometry — 9 787040 612561 >

149 **Shiri Artstein-Avidan, Apostolos Giannopoulos, Vitali D. Milman,** Asymptotic Geometric Analysis, Part I — 9 787040 612578 >

150 **Gregory Berkolaiko, Peter Kuchment,** Introduction to Quantum Graphs — 9 787040 612608 >

151 **Libor Šnobl, Pavel Winternitz,** Classification and Identification of Lie Algebras — 9 787040 612585 >

152 **Maciej Zworski,** Semiclassical Analysis — 9 787040 612592 >

153 **Thomas Garrity, Richard Belshoff, Lynette Boos, Ryan Brown, Carl Lienert, David Murphy, Junalyn Navarra-Madsen, Pedro Poitevin, Shawn Robinson, Brian Snyder, Caryn Werner,** Algebraic Geometry: A Problem Solving Approach — 9 787040 612004 >

154 **Luis Barreira, Yakov Pesin,** Introduction to Smooth Ergodic Theory — 9 787040 611991 >

155 **László Lovász,** Large Networks and Graph Limits — 9 787040 612035 >

156 **Jean-Marie De Koninck, Florian Luca,** Analytic Number Theory: Exploring the Anatomy of Integers — 9 787040 612028 >

157 **María Cristina Pereyra, Lesley A. Ward,** Harmonic Analysis: From Fourier to Wavelets — 9 787040 612011 >

158 **Gail S. Nelson,** A User-Friendly Introduction to Lebesgue Measure and Integration — 9 787040 612783 >

159 **Mark Kot,** A First Course in the Calculus of Variations — 9 787040 612172 >

160 **Michael Aizenman, Simone Warzel,** Random Operators: Disorder Effects on Quantum Spectra and Dynamics — 9 787040 612165 >

161 **Charles A. Weibel,** The K-book: An Introduction to Algebraic K-theory — 9 787040 612158 >

162 **Aaron N. Siegel,** Combinatorial Game Theory — 9 787040 612141 >